Structural Engineer License Review Problems and Solutions

FOR CIVIL AND STRUCTURAL ENGINEERS

Second Edition

ALAN WILLIAMS, Ph.D., S.E., C.ENG.

Registered Structural Engineer, California; Chartered Engineer, United Kingdom;
Senior Engineer, Department of Transportation, State of California

Engineering Press   **Austin, Texas**

ISBN 1-57645-016-3

Engineering Press
P.O. Box 200129
Austin, TX 78720-0129
Phone (800) 800-1651
FAX   (800) 700-1651

CONTENTS

Structural Engineer License Review
Problems and Solutions

INTRODUCTION

REASON FOR THE NEW EDITION

The publication of this new edition is necessitated by the updating of the structural design codes[1,2,3,4,5,6,7]. Since the publication of the first edition of this text, several revised provisions have been introduced into the codes, reflecting the continuing changes occurring in structural design practice. This opportunity has been taken to update the text to conform to the requirements of the fourth printing of the 1994 edition of the Uniform Building Code[1]. The format of the new code has been completely revised and many new technical provisions have been introduced. For reinforced concrete design, the revised provisions include the treatment of bond and anchorage, two-way slab systems, torsion, and column design. In structural steel design, new provisions include the requirements for ordinary moment frames and braced frames. In the design of wood structures, allowable wood stresses have been modified, the determination of stability and volume factors has been introduced, and the design capacity of connectors has been changed. The Section on masonry design has been revised in its entirety. In addition, the Appendix previously containing calculator programs for the HP–28S calculator, which is no longer supported by Hewlett Packard, now provides design programs written for the new HP–48G calculator.

OBJECTIVE OF THE BOOK

The purpose of this textbook is to assist candidates in preparing for the qualifying examination for registration as a Structural Engineer. It is intended to serve as a comprehensive guide and reference for self study of the subject, with the emphasis placed on those analytical and design methods which lead to the quickest and simplest solution of any particular problem. The text is illustrated with the solution of more than seventy problems taken from recent examination papers, and all questions have been updated to conform to the requirements of the current 1994 edition of the Uniform Building Code[1]. In addition, an appendix contains eighteen calculator programs, written for the HP–48G calculator, to facilitate the solution of frequently occurring calculation procedures. These are easily applied to give a rapid solution to the problem while also presenting all the intermediate stages in the calculation. Comprehensive clarification and interpretation of the applicable Building Codes and Standard Specifications are provided and extensive reference publications are cited to reflect current design procedures.

NATIONAL COUNCIL STRUCTURAL ENGINEERING EXAMINATION

The special structural engineering examination of the National Council of Examiners for Engineering and Surveying (NCEES) is utilized by many state licensing boards to provide the qualifying examination for candidates seeking registration as a Structural Engineer. The examination questions are intended to assess the candidate's knowledge and understanding of basic structural engineering analysis and design principles and to determine the candidate's familiarity with design office procedures, standard references and current codes[1,2,3,4,5,6,7]. Removal of examination booklets from the examination room is prohibited and past examination papers are not publicly available. However, a handbook[8], published by NCEES, is available which illustrates representative problems which have appeared in past examinations. The content of the NCEES Special Structural examination is similar to the examination administered by the California state licensing board.

CALIFORNIA STATE BOARD STRUCTURAL ENGINEERING EXAMINATION

Registration as a Structural Engineer is required by all Civil Engineers in California who wish to design school and hospital structures and buildings exceeding 160 feet in height. Authority to use the title "Structural Engineer" is conferred in California by the California State Board of Registration for Professional Engineers and Land Surveyors. The title is awarded to those applicants who satisfy the following requirements:

(i)	Possession of a valid registration as a Civil Engineer in California.
(ii)	Completion of three years of qualifying experience in responsible change of structural engineering work.
(iii)	Provide satisfactory references from four existing Structural Engineers.
(iv)	Obtain a passing score in the 16-hour written structural engineer examination.

In addition to California, the examination is utilized by the States of Nevada, Washington, Idaho, and Hawaii.

The written examination consists of four separate four-hour papers, parts A, B, C and D. Part A examines general structural principles and seismic design. Part B examines structural steel design principles. Part C examines reinforced and prestressed concrete design principles. Part D examines timber design with one question on masonry design. Each paper is allotted a total score of twenty five points and generally contains four to six problems with the maximum number of points awarded to each problem indicated. Each paper contains one multiple-choice question with the remainder being in essay type format. All questions are compulsory in Part A and Part D. In Part B and Part C, the candidate may choose one alternative problem covering the design of bridges or bridge elements. In allocating the time to be spent on each problem, the candidate should allow 9.6 minutes for each point awarded to the problem.

All four parts of the examination are open-book, allowing the candidate to use textbooks, handbooks and bound reference materials. A battery-operated, nonprinting, silent calculator is allowed but loose materials and writing pads are prohibited. To complete all problems within the allotted time, the candidate must work quickly and decisively without spending time searching for solution procedures. Copies of past examination papers, prior to the 1996 examination, are obtainable from the California State Board of Registration[9]. The papers for the 1996 and subsequent examinations will not be made available and may not be removed from the examination room by the candidates.

CALIFORNIA STATE BOARD PROFESSIONAL ENGINEER REGISTRATION

In addition to passing the NCEES eight-hour Professional Engineer examination, the California State Board of Registration requires the candidate to obtain a passing score in two special supplemental two-hour examinations. These examinations cover the topics of seismic principles and engineering surveying principles. The national examination addresses the full scope of civil engineering practice while the Special Civil Engineer examinations assess the candidate's level of competence in the two areas of practice that are of particular concern to the California profession. The scope of the seismic examination is comparable to the seismic design requirements of Part A of the Structural Engineer examination.

ORGANIZATION OF THE TEXT

The text is organized into five separate sections with the sequence of the material following the same order as it does in the California Structural Engineer examination. Chapter 1 covers questions involving general structural principles, the determination of lateral forces and the design of the lateral-force-resisting system. In addition to problems from the Structural Engineer examination papers published by the California State Board of Registration, this chapter also contains appropriate questions from recent Special Seismic Principles examinations which form part of the California Professional Engineer registration process. Chapters 2, 3 and 4 deal with steel, concrete and timber problems, respectively. Chapter 5 covers masonry design problems. All problems in Chapters 2, 3, 4 and 5 are taken from recent California Structural Engineer examinations and cover a full range of the subject matter presented in the examinations.

A header on the question and solution pages indicates the year in which the question was set, the Section from which it is taken, and the number of the problem. Thus, 90A-2. means question two from Section A in the 1990 examination. All Figures associated with a problem are placed at the end of the solution to that problem. The Figure numbers are indicative of the questions to which they refer. References are located at the end of each Chapter.

In a book of this type, errors may be expected to occur and the author will appreciate being informed of any errors that may be found in the text. Any suggestions or comments may be sent to the author at the publisher's address.

The notation adopted in the text is that used in the Uniform Building Code[1] and conforms to current usage. Occasional duplication of the meaning of a particular symbol has been unavoidable, but the symbols are explained in the text as they occur and ambiguities should not arise.

The author wishes to express his gratitude to the California State Board of Registration for Professional Engineers and Land Surveyors for their kind permission to reproduce material from past examination papers.

The following abbreviations are used in the text to denote commonly occurring reference sources:

UBC[1]
AISC[2]
AASHTO[3]
ACI[4]
NDS[5]
NAVFAC[10]
SEAOC[11]

3

REFERENCES

1. International Conference of Building Officials. *Uniform Building Code – 1994.* Whittier, CA, fourth printing February 1996. (5360 South Workman Mill Road, Whittier, CA 90601. Tel: 310-699-0541)

2. American Institute of Steel Construction. *Manual of Steel Construction, Ninth Edition.* Chicago, IL, 1989 (1 East Wacker Drive, Chicago, IL 60601. Tel: 312-670-2400)

3. American Association of State Highway and Transportation Officials. *Standard Specifications for Highway Bridges, Fifteenth Edition.* Washington, DC, 1992. (AASHTO, 444 North Capitol Street, N.W., Washington, DC 20001. Tel: 202-624-5800)

4. American Concrete Institute. *Building Code Requirements and Commentary for Reinforced Concrete (ACI 318-95).* Farmington Hills, MI, 1995. (Box 9094, Farmington Hills, MI 48333. Tel: 810-848-3800)

5. American Forest and Paper Association. *National Design Specification for Wood Construction, Eleventh Edition (ANSI/NFoPA NDS-1991).* Washington, DC, 1991.

6. American Forest and Paper Association. *Commentary on the National Design Specification for Wood Construction, Eleventh Edition (ANSI/NFoPA NDS-1991).* Washington, DC, 1991.

7. American Concrete Institute. *Building Code Requirements for Masonry Structures (ACI 530-95).* Farmington Hills, MI, 1995. (Box 9094, Farmington Hills, MI 48333. Tel: 810-848-3800)

8. National Council of Examiners for Engineering and Surveying. *Handbook for Structural Engineers.* Clemson, SC, 1991. (Box 1686, Clemson, SC 29633-1686. Tel: 803-654-6824)

9. Board of Registration for Professional Engineers and Land Surveyors. *Examination Papers.* Sacramento, CA. (Box 349002, Sacramento, CA 95834-9002. Tel: 916-263-2222)

10. Corps of Engineers. *Seismic Design for Buildings.* NAVFAC, Technical Manual P-355, Washington, DC, 1982. (NTIS, U.S. Department of Commerce, Springfield, VA 22161. Tel: 703-487-4650)

11. Structural Engineers Association of California, *Recommended Lateral Force requirements and Commentary*, Sacramento, CA, 1990 (SEAOC, Box 19440, Sacramento, CA 95819-0440. Tel: 213-385-4424)

1

GENERAL STRUCTURAL PRINCIPLES AND SEISMIC DESIGN

SECTION 1.1

DETERMINATION OF LATERAL FORCES

PROBLEMS:

1990 A–1
1988 A–3
April 1989 Seismic Principles Examination 301A
April 1988 Seismic Principles Examination 301

STRUCTURAL ENGINEER EXAMINATION – 1990 ═══════════════════════

PROBLEM A–1 – WT. 5.0 POINTS $\times 9.6 = 48^{mix.}$

GIVEN: A typical interior frame of a steel, ordinary moment-resisting frame (OMRF) provides the lateral force resisting system for a one-story retail structure.

CRITERIA: Roof dead load = 80 psf
Curtain wall dead load = 20 psf on surface area
Basic ground snow load = 35 psf
Snow exposure coefficient C_e =0.9
Basic wind speed = 70 mph, exposure B
Seismic zone 3
Site coefficient S = 1.5

Assumptions:
- Frames are at 25 feet on center
- Columns are pinned at the base
- $I_{(col)} = I_{(beam)}$
- Do not consider torsion
- The curtain wall is vertically and horizontally supported at the top and bottom of the frame as shown
- The Building Official does not allow a reduction of roof snow load in combination with seismic loads

REQUIRED: Determine the following loads and the reactions on the frame. Show on a diagram for each case:

A. Dead load
B. Snow load on roof
C. Wind forces
D. Seismic forces

SOLUTION

A. DEAD LOAD

The distributed roof load is given by

$$w \qquad = 80 \times 25/1000 \qquad = 2 \text{ kips per foot}$$

and this produces a vertical force in each column of

$$V_2 = V_3 \qquad = 2 \times 34/2 \qquad = 34 \text{ kips}$$

6

The curtain wall applies vertical loads to each column of

$$V_1 = V_4 \quad = 20 \times 25 \times 20/1000 = 10 \text{ kips}$$

The total vertical reaction at each pin is, thus

$$R_1 = R_4 \quad = 34 + 10 \quad = 44 \text{ kips}$$

The horizontal reactions at each pin due to the roof load may be determined by the moment distribution procedure[1,2]. Due to the symmetry of the structure and the loading, no sway occurs and only joint rotations need to be considered. Advantage may be taken of the hinged supports and the symmetry to modify the stiffness factors, as shown in the Table, so that distribution is required in only half of the structure and there is no carry-over between the two halves. The initial fixed-end moment at joint 2 is

$$M_{23}^F \quad = -2 \times 34 \times 34/12 = -193 \text{ kip feet}$$

The convention adopted is that clockwise moments acting from the joint on a member are positive and the distribution, for the left half of the frame, is shown in the following Table.

Joint	1	2	2
Member	12	21	23
Relative EI/L	1/16	1/16	1/34
Modified stiffness	14	3/16	2/34
Distribution factor	1	0.76	0.24
Carry-over factor	1/2→		
		← 0	
Fixed-end moments	0	0	-193
Distribution and carry-over	0	147	46
Final moments, lb ft.	0	147	-147

The final moments in the right half of the frame are equal and of opposite sense to those obtained for the left half. The horizontal reaction at joint 1 is obtained by taking moments about joint 2 for member 12, then

$$H_1 \quad = 147/16 \quad = 9.2 \text{ kips acting inward}$$

Loads and reactions are shown in Figure 90A-1.

B. SNOW LOAD

From the UBC Appendix, Chapter 16, Division I, the roof snow load is given by UBC Formula (37-1A) as

$$P_f \quad = C_e I P_g$$

where C_e is the snow exposure coefficient, I is the importance factor given in UBC Table A-16-A, and P_g is the basic ground snow load.

7

Thus the roof snow load is

$$P_f = 0.9 \times 1 \times 35$$
$$= 31.5 \text{ pounds per square foot}$$

The snow density is calculated from UBC Formula (41–2) as

$$D = 0.13P_g + 14.0$$
$$= 0.13 \times 35 + 14$$
$$= 18.6 \text{ pounds per cubic foot}$$

which is less than 35 pounds per cubic foot as is required.

The height of the balanced snow load is

$$h_b = P_f/D$$
$$= 31.5/18.6$$
$$= 1.7 \text{ feet}$$

Then, applying Formula UBC (41–3) and assuming the parapet projects a distance $h_r = 4$ feet above the top of the roof

$$(h_r - h_b)/h_b = (4 - 1.7)/1.7$$
$$= 1.35$$

which exceeds 0.2, indicating that drift loads must be considered.

The height of the drift produced at the parapets is determined from UBC Section 1641.5 and from UBC Figure A–16–9, where the roof width W_b is taken as 50 feet minimum and the basic ground snow load P_g is 35 pounds per square foot. The height of the drift is

$$h_{dp} = 0.5h_d$$
$$= 0.5 \times 2.6$$
$$= 1.3 \text{ feet}$$
$$< h_r - h_b \dots \text{satisfactory}$$

Hence, the height of the drift at each parapet is 1.3 feet and the base width of each drift is given by Section 1641.2 as the lesser of

$$W_d = 4(h_r - h_b)$$
$$= 4 \times (4 - 1.7)$$
$$= 9.2 \text{ feet}$$

or
$$W_d = 4h_{dp}$$
$$= 4 \times 1.3$$
$$= 5.2 \text{ feet} \dots \text{governs}$$

The maximum intensity of the snow load at the parapet is given by UBC Formula (41–4) as

$$P_m = D(h_{dp} + h_b)$$
$$= 18.6(1.3 + 1.7)$$
$$= 55.8 \text{ pounds per square foot}$$

The effect of the snow load on the frame may be determined by combining the two separate loading conditions, a uniformly distributed load of $31.5 \times 25/1000 = 0.788$ kips per foot over the full width of the roof and an additional triangular load at each parapet with a maximum intensity of $25(55.8 - 31.5)/1000 = 0.608$ kips per foot.

The vertical reactions at each pin are given by

$$
\begin{aligned}
P_u &= \text{Distributed load} = 0.788 \times 34/2 &&= 13.4 \text{ kips} \\
P_d &= \text{Triangular load} = 0.608 \times 5.2/2 &&= \underline{1.6} \\
&\quad \text{Total} &&\ 15.0
\end{aligned}
$$

The horizontal reaction due to the distributed load is

$$
\begin{aligned}
H_1 &= 9.2 \times 0.788/2 \\
&= 3.6 \text{ kips acting inward}
\end{aligned}
$$

The horizontal reaction due to the triangular loads may be determined by moment distribution. The initial fixed end moment at joint 2 is[3,4]

$$
\begin{aligned}
M_{23}^{F} &= -P_d W_d (2\ell - W_d)/6\ell \\
&= -1.6 \times 5.2(2 \times 34 - 5.2)/(6 \times 34) \\
&= -2.56 \text{ kip feet}
\end{aligned}
$$

and the horizontal reaction at joint 1 is given by

$$
\begin{aligned}
H_1 &= 9.2 \times 2.56/193 \\
&= 0.1 \text{ kips acting inward}
\end{aligned}
$$

The total horizontal reaction is $3.6 + 0.1 = 3.7$ kips and the loads and reactions are shown in Figure 90A-1.

C. WIND FORCES[5,6,7,8]

Since the structure is less than 200 feet high and is not a gabled frame, the projected area method (Method 2) detailed in UBC Chapter 16 may be utilized to determine the wind forces acting on the structure. Assuming the structure may be classified as enclosed, UBC Table 16-H gives the values of the pressure coefficient C_q as 1.3 on the vertical projected area and 0.7 upward on the horizontal projected area. The wind stagnation pressure q_s for a basic wind speed of 70 miles per hour is given by UBC Table 16-F as 12.6 pounds per square foot. For exposure B and a structure height of 20 feet, UBC Table 16-G gives a value for the combined height, exposure and gust factor coefficient C_e of 0.62 for a height of 15 feet and 0.67 for a height of 20 feet. UBC Table 16-K gives a value of unity for the importance factor I_w. By applying UBC Formula (18-1), the design wind pressure on the right hand side curtain wall, for wind acting right to left, is obtained as

$$
\begin{aligned}
p &= C_e C_q q_s I_w \\
&= 0.62 \times 1.3 \times 12.6 \times 1 \ \ldots \text{height 0-15 feet} \\
&= 10.16 \text{ pounds per square foot on the vertical projected area}
\end{aligned}
$$

and

$$
\begin{aligned}
&= 0.67 \times 1.3 \times 12.6 \times 1 \ \ldots \text{height 15-20 feet} \\
&= 10.97 \text{ pounds per square foot on the vertical projected area}
\end{aligned}
$$

The step–function loading on the curtain wall produces horizontal forces on column 34 of

$$H_3 = 10.16 \times 25 \times 15 \times 7.5/16{,}000 + 10.97 \times 25 \times 5 \times 17.5/16{,}000$$
$$= 3.3 \text{ kips acting to the left}$$

and $\quad H_4 = 10.16 \times 25 \times 15/1000 + 10.97 \times 25 \times 5/1000 - 3.3$
$$= 1.9 \text{ kips acting to the left}$$

The vertical reactions at the joints due to the horizontal loading on the curtain wall are

$$R_1 = 3.3 \times 16/34$$
$$= 1.55 \text{ kips acting upward}$$
$$R_2 = -1.55 \text{ kips acting downward}$$

The design wind pressure on the roof is

$$p = 0.67 \times 0.7 \times 12.6 \times 1$$
$$= 5.91 \text{ pounds per square foot on the horizontal projected area}$$

The distributed load on the roof beam is

$$5.91 \times 25/1{,}000 = 0.148 \text{ kips per foot acting upward}$$

and this produces horizontal reactions at joints 1 and 4 of

$$H_1 = H_4 = -9.2 \times 0.148/2$$
$$= -0.68 \text{ kips acting outward}$$

and a vertical reaction at each pin of

$$R_1 = R_4 = -0.148 \times 34/2$$
$$= -2.52 \text{ kips acting downward}$$

The total horizontal reaction at joint 4, for both vertical and horizontal wind loads, is given by

$$H_4 = -0.68 - 3.3/2 - 1.9$$
$$= -4.23 \text{ kips acting outward}$$

Similarly, at joint 1, the horizontal reaction is

$$H_1 = -0.68 + 3.3/2$$
$$= 0.97 \text{ kips acting inward}$$

The total vertical reaction at joint 4, for both vertical and horizontal wind loads is given by

$$R_4 = -2.52 - 1.55$$
$$= -4.07 \text{ kips acting downward}$$

Similarly at joint 1, the vertical reaction is

$$R_1 = -2.52 + 1.55$$
$$= -0.97 \text{ kips acting downward}$$

$$V = \frac{ZICW}{R_w}$$

(handwritten annotations: Table 16-I, TABLE 16-K, FORM. (28.2) UBC 1628.2.1, TABLE 16-N)

D. SEISMIC FORCES

From the UBC Chapter 16, the structure period is given by UBC Formula (28–3) as

$$T = C_t(h_n)^{3/4}$$

where C_t = 0.035 for a steel moment–resisting frame with roof height h_n = 16 feet

Thus the structure period is

$$T = 0.035 \times (16)^{3/4}$$
$$= 0.28 \text{ seconds}$$

The force coefficient is given by UBC Formula (28– 2) as

$$C = 1.25S/T^{2/3}$$ *(handwritten: > 2.75)*

where S = 1.5 … site coefficient *(handwritten: for unknown soil)*

Thus the force coefficient is

$$C = 1.25 \times 1.5/(0.28)^{2/3}$$
$$= 4.38$$
$$> 2.75 \text{ … hence use the upper limit of}$$ *(handwritten: UBC 1628.2.1)*
$$C = \boxed{2.75}$$

The lateral force produced by the seismic effect is given by UBC Formula (28–1) as

$$V = (ZIC/R_w)W$$

where the zone factor Z for seismic zone 3 is obtained from UBC Table 16–I as 0.3, the importance factor I for a standard occupancy structure is obtained from UBC Table 16–K as 1.0 and the response factor R_w for an ordinary steel moment-resisting frame is obtained from UBC Table 16–N as 6. Thus the seismic force is given by

(handwritten: Z, J, Rw)

$$V = (0.3 \times 1 \times 2.75/6)W$$
$$= 0.1375W$$

The distributed roof load of 80 pounds per square foot produces a lateral force at beam level of

$$V = 0.1375 \times 80 \times 25 \times 34/1000$$
$$= 9.35 \text{ kips acting to the left}$$

(handwritten: .1375×80×25×34 = 9.4 k)

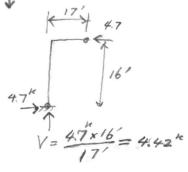

The horizontal reactions at joints 1 and 4, due to the roof seismic force acting right to left, are

$$H_1 = H_4 = 9.35/2$$
$$= 4.675 \text{ kips acting to the right}$$

The vertical reactions at points 1 and 4, due to the roof seismic force, are

$$R_1 = 9.35 \times 16/34$$
$$= 4.4 \text{ kips acting upward}$$

and R_4 = –4.4 kips acting downward

(handwritten: V = 4.7×16'/17' = 4.42 k)

The curtain wall dead loads of 20 pounds per square foot produce horizontal forces on the columns of

$.1375 \times 20^{p/s} \times 25'$

$=68.8 p/s$

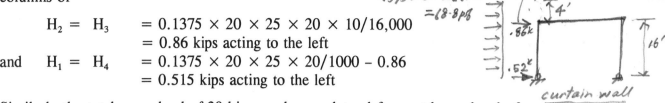

$$H_2 = H_3 = 0.1375 \times 20 \times 25 \times 20 \times 10/16,000$$
$$= 0.86 \text{ kips acting to the left}$$

and $\quad H_1 = H_4 = 0.1375 \times 20 \times 25 \times 20/1000 - 0.86$
$$= 0.515 \text{ kips acting to the left}$$

Similarly the total snow load of 30 kips produces a lateral force at beam level of

$$V = 0.1375 \times 30$$
$$= 4.13 \text{ kips}$$

$30 \times .1375$
$= 4.13$

Hence the snow load produces reactions at joints 1 and 4 of

$$H_1 = H_4 = 4.13/2$$
$$= 2.06 \text{ kips acting to the right}$$
$$R_1 = 4.13 \times 16/34$$
$$= 1.94 \text{ kips acting upward}$$
$$R_4 = -1.94 \text{ kips acting downward}$$

The total horizontal reactions, considering the roof, snow and curtain wall loads, are given by

$$H_1 = H_4 = 4.675 + 0.86 + 0.515 + 2.06$$
$$= 8.11 \text{ kips acting to the right}$$

The total vertical reaction at joint 4, considering the snow, roof and curtain wall loads, is

$$R_4 = -4.4 - 2 \times 0.86 \times 16/34 - 1.94$$
$$= -7.15 \text{ kips acting downward}$$

Similarly at joint 1, the vertical reaction is

$$R_1 = 4.4 + 2 \times 0.86 \times 16/34 + 1.94$$
$$= 7.15 \text{ kips acting upward}$$

The loads and reactions are shown in Figure 90A-1.

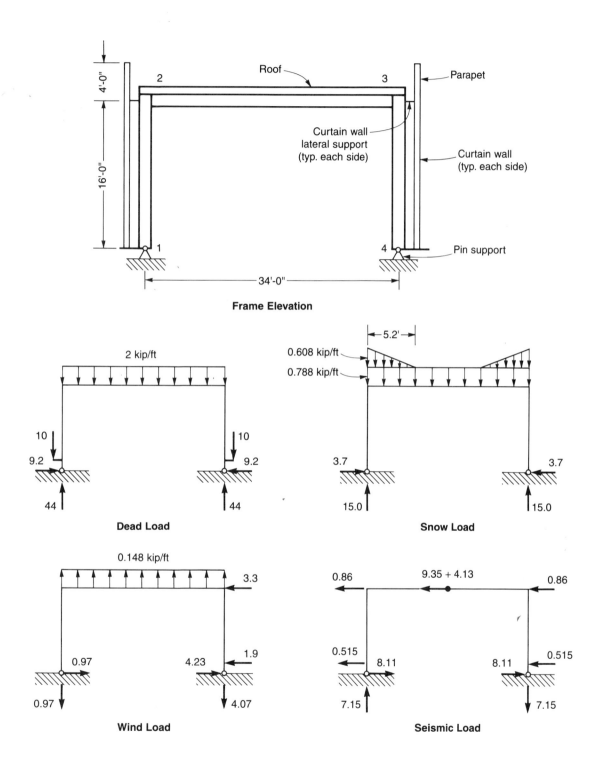

Frame Elevation

Dead Load

Snow Load

Wind Load

Seismic Load

FIGURE 90A–1

STRUCTURAL ENGINEER EXAMINATION – 1988 ═══════════════════

PROBLEM A-3 – WT. 4.0 POINTS

GIVEN: A four-story multi-use building has a structural steel special moment-resisting frame and a nonbearing curtain wall. The building plan "A" and elevation "B" are shown.

CRITERIA: Roof dead load = 60 psf
Roof live load = 20 psf (no snow load)
Floor dead load = 60 psf + 20 psf partition load on floors 3 and 4
Floor live load:
  Floor 2 (public assembly use) = 100 psf
  Floors 3 and 4 (office use) = 85 psf (not including partitions)
Average curtain wall weight = 20 psf
Basic wind speed = 90 mph, exposure B
Seismic zone 4

Assumptions:
- The primary occupancy of the building is for office use
- Column weights are included in the floor dead load

REQUIRED: 1. Determine the following design loadings in psf on point "X" at the parapet at column C-2:
  a. Design wind loading
  b. Design seismic loading (normal to the cladding)
2. Determine the following design loadings in psf on point "Y" at Floor 4 at column C-4:
  a. Design wind loading (both inward and outward)
3. Determine the following design loadings in psf on point "Z" at Floor 3 at column C-3:
  a. Design wind loading (both inward and outward)
  b. Design seismic loading (normal to the cladding)
4. Determine the following design loadings in psf on point "Z" at Floor 3 at column C-3 based on a 2 ft. × 2 ft. curtain wall panel:
  a. Design wind loading on a connection fastener (normal to the cladding)
  b. Design seismic loading on a connection fastener (normal to the cladding)
5. Calculate the design dead load, live load, and total load at each floor on column C-2. Use the live load reductions specified in the UBC.

SOLUTION

WIND FORCES[5,6,7,8]

Assume that the structure may be considered as enclosed as defined by UBC Section 1614 and that pressures are required for cladding elements with a tributary area not exceeding ten square feet. Hence, the pressure coefficients may not be reduced as specified in the footnotes to UBC Table 16–H and the design wind pressure is obtained from UBC Formula (18–1) as

$$p = C_e C_q q_s I_w$$

where
- C_e = height, exposure and gust factor given in UBC Table 16–G
- C_q = pressure coefficient given in UBC Table 16–H
- q_s = wind stagnation pressure given in UBC Table 16–F
  = 20.8 pounds per square foot for a wind speed of 90 miles per hour
- I_w = importance factor given in UBC Table 16–K
  = 1.0

Then $p = 20.8 C_e C_q$

SEISMIC FORCES

The design seismic force on elements and components is obtained from UBC Formula (30–1) as

$$F_p = Z I_p C_p W_p$$

where
- Z = 0.4 for zone 4 from UBC Table 16–I
- I_p = 1.0 from UBC Table 16–K
- C_p = force factor given in UBC Table 16–O

and
- W_p = weight of the curtain wall
  = 20 pounds per square foot

Then $F_p = 8 C_p$

1a. WIND LOAD AT POINT X

For exposure B and a roof height of 64 feet, the value of C_e is 0.97. For a parapet, the value of C_q is 1.3 inward and outward and the design wind load at point X is

$$p = 20.8 \times 0.97 \times 1.3$$
$$= 26.2 \text{ pounds per square foot inward and outward}$$

1b. SEISMIC LOAD AT POINT X

For a parapet, the value of C_p is 2.0 from UBC Table 16-O, item I.1a. and the design seismic load at point X is

$$F_p = 8 \times 2$$
$$= 16 \text{ pounds per square foot normal to the parapet.}$$

2. WIND LOAD AT POINT Y

For outward pressures, UBC Section 1620 indicates that the value of C_e shall be based on the mean roof height of 64 feet, giving a value for C_e of 0.97. For inward pressures, the value of C_e is based on the actual height of the element of 48 feet, giving a value for C_e of 0.88. The value of C_q for a wall corner element is 1.5 outward and 1.2 inward and is applied over a distance from the corner of the lesser of

	L	$= 10$ feet
or	L	$= 0.1 \times$ least width
		$= 6$ feet, which governs.

The design wind load at point Y is

	p	$= 20.8 \times 0.97 \times 1.5$
		$= 30.3$ pounds per square foot outward
and	p	$= 20.8 \times 0.88 \times 1.2$
		$= 22.0$ pounds per square foot inward

3a. WIND LOAD AT POINT Z

For outward pressures, the value of C_e based on a mean roof height of 64 feet, as indicated by UBC Section 1620, is 0.97. For inward pressures, the value of C_e based on the actual height of the element of 32 feet is 0.78. The value of C_q for a wall element is 1.2 outward and inward. The design wind load at point Z is

	p	$= 20.8 \times 0.97 \times 1.2$
		$= 24.2$ pounds per square foot outward
and	p	$= 20.8 \times 0.78 \times 1.2$
		$= 19.5$ pounds per square foot inward

3b. SEISMIC LOAD AT POINT Z

For a wall, the value of C_p is 0.75 from UBC Table 16–O, item 1.1b. The design seismic load at point Z is

$$F_p = 8 \times 0.75$$
$$= 6 \text{ pounds per square foot normal to the cladding}$$

4a. WIND LOAD ON CLADDING CONNECTION AT Z

The values of C_e and C_q for a cladding connection are the same as for the cladding and the design wind load of a connection at point Z is

$$p = 24.2 \text{ pounds per square foot outward}$$
and
$$p = 19.5 \text{ pounds per square foot inward}$$

4b. SEISMIC LOAD ON CONNECTION FASTENER AT Z

Connection fasteners, in accordance with UBC Section 1631.2.4.2, shall be designed for four time the seismic forces calculated by UBC Formula (30-1). The design seismic load on a fastener at point Z is

$$p = 6 \times 4$$
$$= 24 \text{ pounds per square foot normal to the cladding}$$

5. DESIGN LOAD ON COLUMN C-2

LIVE LOAD REDUCTIONS

The column tributary area is given by

$$A = 22 \times 15$$
$$= 330 \text{ square feet}$$

Since the tributary area exceeds 150 square feet, both roof and floor live loads may be reduced as indicated in UBC Section 1606. The reduced roof live load is obtained from UBC Table 16-C, Method 1, as

$$L_R = 16 \text{ pounds per square foot}$$

On floors three and four, the total dead load including partitions is given by

$$D = 60 + 20$$
$$= 80 \text{ pounds per square foot}$$

and the live load is given by

$$L = 85 \text{ pounds per square foot}$$

The allowable reduction in live load at floor four is the lesser of

$$R = 60\%, \text{ supporting load from two levels only}$$

or (6-2)
$$R = 23.1(1 + D/L)$$
$$= 23.1[1 + (60 + 80)/(20 + 85)]$$
$$= 53.9\%$$

or (6-1)
$$R = r(A - 150)$$
$$= 0.08(2 \times 330 - 150)$$
$$= 40.8\%, \text{ which governs.}$$

Hence the fourth floor live load is

$$L_4 = 330(20 + 85)(1 - 0.408) = 20{,}512 \text{ pounds}$$

The allowable reduction in live load at floor three is the lesser of

$$R = 60\%, \text{ supporting load from three levels}$$

or (6-2)
$$R = 23.1 [1 + (60 + 2 \times 80)/(20 + 2 \times 85)]$$
$$= 49.8\%, \text{ which governs}$$

or
$$R = r(A - 150)$$
$$= 0.08(3 \times 330 - 150)$$
$$= 67.2\%$$

Hence the third floor live load is

$$L_3 = 330(20 + 2 \times 85)(1 - 0.498) = 31{,}475$$

At floor two, no reduction is allowed in the public assembly loading of 100 pounds per square foot.

18

COLUMN LOADING

Fourth floor to roof:

Roof dead load	$= 60 \times 330$	$=$	19,800
Roof live load	$= 16 \times 330$	$=$	5,280
Wall at fourth story	$= 20 \times 22 \times 19$	$=$	8,360
Total design load above fourth floor			33,440 pounds

Third floor to fourth floor:

Fourth floor dead load	$= 80 \times 330$	$=$	26,400
Fourth floor live load addition	$= 20,512 - 5,280$	$=$	15,232
Wall at third story	$= 20 \times 22 \times 16$	$=$	7,040
Total design load above third floor			82,112 pounds

Second floor to third floor;

Third floor dead load	$= 80 \times 330$	$=$	26,400
Third floor live load addition	$= 31,475 - 20,512$	$=$	10,963
Wall at second story	$= 20 \times 22 \times 16$	$=$	7,040
Total design load above second floor			126,515 pounds

First floor to second floor:

Second floor dead load	$= 60 \times 330$	$=$	19,800
Second floor live load addition	$= 100 \times 330$	$=$	33,000
Wall at first story	$= 20 \times 22 \times 16$	$=$	7,040
Total design load above first floor			186,355 pounds

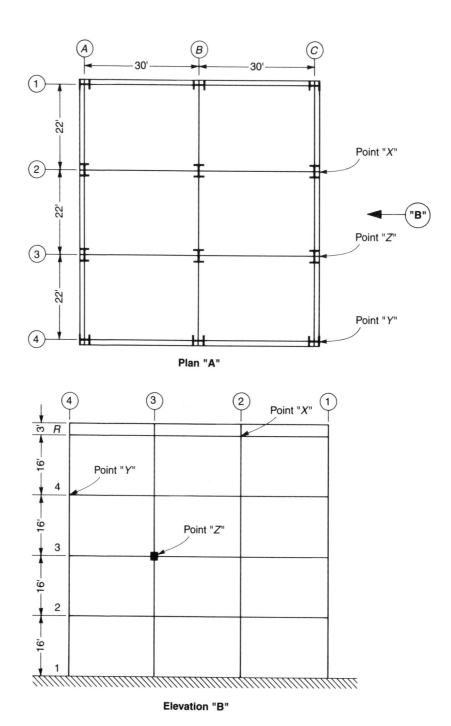

Plan "A"

Elevation "B"

FIGURE 88A-3

20

SEISMIC PRINCIPLES EXAMINATION – APRIL 1989 ════════

PROBLEM 301A – WT. 10 POINTS (one hour)

SITUATION: A building, as shown in the Figure, is to be used as a police station in San Francisco, California. It is to be a wood structure.

The following criteria apply:
Roof dead load $= 16$ psf
Window wall and shear wall weight $= 12$ psf
Force coefficient $C = 2.75$
Response modification factor $R_w = 6$

REQUIREMENTS: (a)   Wt. 4 pts
Calculate the maximum roof shear in each direction.
(b)   Wt. 2 pts
Calculate the maximum chord force.
(c)   Wt. 2 pts
Calculate the maximum shear in the shear walls at line A.
(d)   Wt. 2 pts
Determine the nailing requirement for a shear wall stress of 390 lbs. per ft. using ⅜" C–D plywood and 6d nails.

SOLUTION

LATERAL FORCE

The seismic force acting on the structure is obtained from UBC Section 1625 Formula (28–1) as

	V	$= (ZIC/R_w)W$
where	Z	$= 0.4$ for zone 4 from UBC Table 16–I
	I	$= 1.25$ from UBC Table 16–K for an essential facility
	C	$= 2.75$ as given
	R_w	$= 6$ as given
and	W	$=$ structure dead load
Then	V	$= (0.4 \times 1.25 \times 2.75/6)W$
		$= 0.229W$

DEAD LOAD

The structure dead loads at the roof level are given by

Roof	$= 16 \times 80 \times 40$	$=$	51200 pounds
North wall	$= 12 \times 80 \times 10/2$	$=$	4800
South wall		$=$	4800
East wall	$= 12 \times 40 \times 10/2$	$=$	2400
West wall		$=$	2400

a. DIAPHRAGM SHEAR

Diaphragm shear may be determined by employing the techniques of flexible roof diaphragm design[9].

N–S SEISMIC

The dead load tributary to the roof diaphragm in the North–South direction is due to the North wall, South wall and roof and is given by

$$W = 51,200 + 4800 + 4800 = 60,800 \text{ pounds}$$

The seismic force on the roof diaphragm in the North–South direction is given by

$$
\begin{aligned}
V &= 0.229W \\
&= 0.229 \times 60,800 \\
&= 13,923 \text{ pounds}
\end{aligned}
$$

Assuming that the roof diaphragm is flexible, the diaphragm shear along the East and West walls is given by

$$
\begin{aligned}
q &= V/(2 \times 40) \\
&= 13,923/80 \\
&= 174 \text{ pounds per linear foot}
\end{aligned}
$$

E–W SEISMIC

The dead load tributary to the roof diaphragm in the East–West direction is due to the East wall, West wall and roof and is given by

$$W = 51{,}200 + 2400 + 2400$$
$$= 56{,}000 \text{ pounds}$$

The seismic force on the roof diaphragm in the East–West direction is given by

$$V = 0.229W$$
$$= 0.229 \times 56{,}000$$
$$= 12{,}824 \text{ pounds}$$

The diaphragm shear along the North and South walls is given by

$$q = V/(2 \times 80)$$
$$= 12{,}824/160$$
$$= 80.2 \text{ pounds per linear foot}$$

b. CHORD FORCE

By inspection, the maximum chord force occurs simultaneously in the North wall and the South wall due to the seismic force in the North–South direction. From Figure 89SA–1, the bending moment in the roof diaphragm is given by

$$M = VL/8$$
$$= 13{,}923 \times 80/8$$
$$= 139{,}230 \text{ pounds feet}$$

The maximum chord force is given by

$$F = M/B$$
$$= 139{,}230/40$$
$$= 3481 \text{ pounds}$$

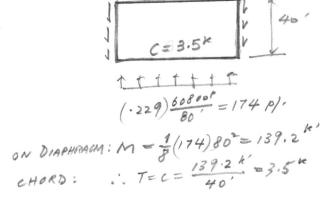

$$(.229)\frac{60800^{\#}}{80'} = 174 \, P/'$$

$$\text{ON DIAPHRAGM}: \quad M = \frac{1}{8}(174)80^{2} = 139.2^{k'}$$

$$\text{CHORD}: \quad \therefore T = C = \frac{139.2^{k'}}{40'} = 3.5^{k}$$

c. WALL SHEAR

The total length of shear walls on line A is twenty feet and their self weight is given by

$$W = 12 \times 10 \times 20$$
$$= 2400 \text{ pounds}$$

The seismic force produced by the self weight of the shear walls is given by

$$V = 0.229W$$
$$= 0.229 \times 2400$$
$$= 550 \text{ pounds}$$

The total shear force on the shear walls is due to the self weight of the shear walls plus the seismic force on the roof diaphragm acting in the East–West direction and is given by

$$V' = 550 + 12{,}824/2 = 6962 \text{ pounds}$$

and the shear per unit length in the shear walls is

$$q' = V'/20 = 348 \text{ pounds per linear foot.}$$

d. NAILING REQUIREMENTS

The nail spacing required, to provide a shear capacity of 390 pounds per linear foot using ⅜ inch C–D plywood and six penny nails with 1.25 inch penetration, may be obtained from UBC Chapter 23, Table 23–I–K–1. Assuming two inch nominal framing is provided and all edges are blocked, the required nail spacing is

at all plywood edges	=	3 inches
at intermediate framing members	=	12 inches

and this provides a shear capacity of 390 pounds per linear foot.

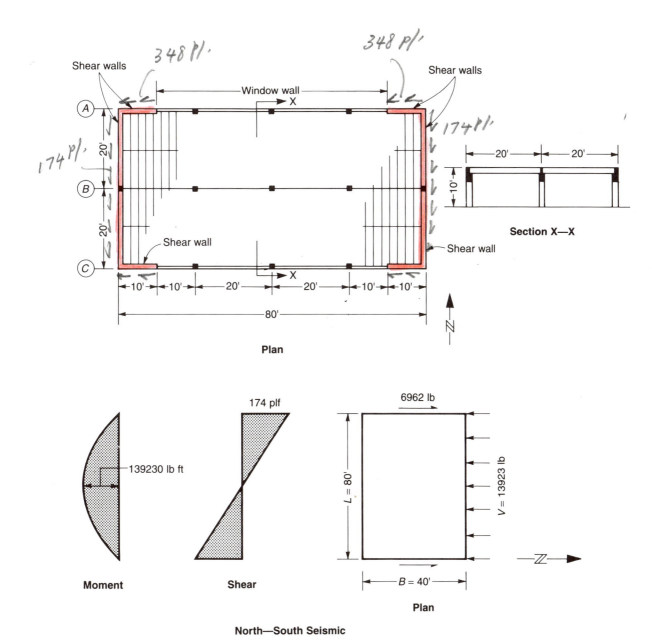

348 P/· 348 P/·

174 P/· 174 P/·

Shear walls

Window wall

Shear walls

A

20'

B

20'

C

Shear wall

Shear wall

←10'→←10'→←—20'—→←—20'—→←10'→←10'→

←————————————80'————————————→

Plan

N

Section X—X

←—20'—→←—20'—→

10'

139230 lb ft

174 plf

6962 lb

L = 80'

V = 13923 lb

B = 40'

N

Moment　　**Shear**　　**Plan**

North—South Seismic

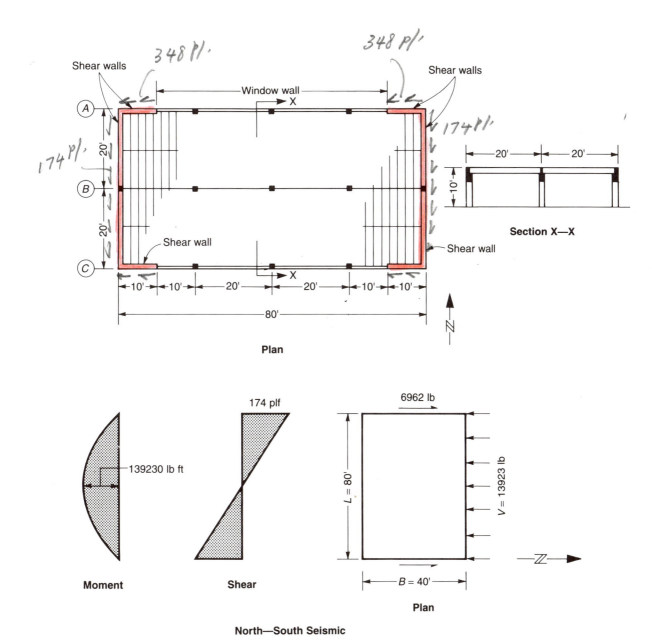

FIGURE 89SA–1

25

89SA-1.5

SEISMIC PRINCIPLES EXAMINATION – APRIL 1988 ══════════

PROBLEM 301 (One Hour)

SITUATION: Consider the plywood diaphragm on masonry (CMU) walls shown. The following data apply:

Roof dead load	=	15 psf
Wall weight	=	50 psf
Seismic force V	=	0.186W

REQUIREMENTS: (a)   Determine the maximum diaphragm shear at Lines 1 and 5.
               (b)   Determine the maximum diaphragm shear at Lines A and C.
               (c)   Determine the maximum force, chord or drag at the intersection of Grids C and 3.
               (d)   Determine the maximum force, chord or drag at the intersection of Grids C and 2.
               (e)   Determine the maximum force, chord or drag at the intersection of Grids B and 5.
               (f)   If the masonry wall is added at Line 4 as shown, determine the maximum drag force at the intersection of Grids B and 4.

SOLUTION

a. EAST–WEST SEISMIC

The lateral seismic forces acting at the roof diaphragm level are given by

Roof	$= 0.186 \times 15 \times 24$	$= 67$ pounds per linear foot
East wall	$= 0.186 \times 50 \times 10/2$	$= 47$
West wall		$= 47$

The forces acting on the roof diaphragm are shown in Figure 88SA–1(ii) and the total applied force is

$$
\begin{aligned}
V &= 47 \times 20 + (67 + 47)80 \\
&= 940 + 9120 \\
&= 10{,}060 \text{ pounds}
\end{aligned}
$$

The lateral forces acting on the diaphragm at lines 1 and 5 are

$$
\begin{aligned}
R_5 &= 9120/2 + 940 \times 10/80 \\
&= 4678 \text{ pounds} \\
R_1 &= V - R_5 \\
&= 10{,}060 - 4678 \\
&= 5383 \text{ pounds}
\end{aligned}
$$

26

The diaphragm shears along lines 1 and 5 are

$$q_1 \quad = R_1/24$$
$$= 5383/24$$
$$= 224 \text{ pounds per linear foot}$$
$$q_5 \quad = R_5/24$$
$$= 4678/24$$
$$= 195 \text{ pounds per linear foot}$$

b. NORTH–SOUTH SEISMIC

The lateral seismic forces acting at the roof diaphragm level are given by

Roof	$= 0.186 \times 15 \times 80$	$= 223$ pounds per linear foot
North wall	$= 0.186 \times 50 \times 10/2$	$= 47$
South wall		$= 47$

The forces acting on the roof diaphragm are shown in Figure 88SA–1(ii) and the total applied force is

$$V \quad = 47 \times 12 + (223 + 47)24$$
$$= 564 + 6480$$
$$= 7044 \text{ pounds}$$

The lateral forces acting on the diaphragm at lines A and C are

$$R_C \quad = 6480/2 + 564 \times 6/24$$
$$= 3381 \text{ pounds}$$
$$R_A \quad = V - R_C$$
$$= 7044 - 3381$$
$$= 3663 \text{ pounds}$$

The diaphragm shears along lines A and C are

$$q_A \quad = R_A/80$$
$$= 3663/80$$
$$= 46 \text{ pounds per linear foot}$$
$$q_C \quad = R_C/80$$
$$= 3381/80$$
$$= 42 \text{ pounds per linear foot}$$

c. CHORD AND DRAG FORCE AT C3

The shear distribution and net shear at line C, due to the North–South seismic force, are shown in Figure 88SA-1(iii) and the drag force at C3 is given by

$$
\begin{aligned}
F_{C3} &= 40q_C \\
&= 40 \times 42 \\
&= 1680 \text{ pounds}
\end{aligned}
$$

As shown in Figure 88SA-1(ii), the bending moment in the roof diaphragm at C3 due to the East–West seismic force is

$$
M_3 = 95{,}920 \text{ pounds feet}
$$

and the corresponding chord force is

$$
\begin{aligned}
F'_{C3} &= M_3/24 \\
&= 95{,}920/24 \\
&= 3997 \text{ pounds, which governs.}
\end{aligned}
$$

d. CHORD AND DRAG FORCE AT C2

The drag force at C2, due to the North–South seismic force, is obtained from Figure 88SA-(iii) as

$$
\begin{aligned}
F_{C2} &= 60q_C \\
&= 60 \times 42 \\
&= 2520 \text{ pounds}
\end{aligned}
$$

As shown in Figure 88SA-1(ii), the bending moment in the roof diaphragm at C2 due to the East–West seismic force is

$$
M_2 = 75{,}460 \text{ pounds feet}
$$

and the corresponding chord force is

$$
\begin{aligned}
F'_{C2} &= M_2/24 \\
&= 75{,}460/24 \\
&= 3144 \text{ pounds, which governs.}
\end{aligned}
$$

e. CHORD AND DRAG FORCE AT B5

As shown in Figure 88SA-1(ii), the bending moment in the roof diaphragm at B5 due to the North–South seismic force is

$$
M_B = 21{,}132 \text{ pounds feet}
$$

The corresponding chord force is

$$F'_{B5} = M_B/80$$
$$= 21,132/80$$
$$= 264 \text{ pounds}$$

The drag force at B5, due to the East-West seismic force, is given by

$$F_{B5} = 12q_5$$
$$= 12 \times 195$$
$$= 2340 \text{ pounds, which governs}$$

f. DRAG FORCE AT B4

Adding an additional shear wall at line 4 effectively subdivides the roof diaphragm into two simply supported segments. These are span 14 and span 45. The seismic forces acting at the roof diaphragm level, in the East-West direction, are shown in Figure 88SA-1(iii) and the lateral forces acting on the diaphragm are

$$R_{45} = 114 \times 10$$
$$= 1140 \text{ pounds}$$

and

$$R_{41} = 114 \times 30 + 47 \times 20 \times 10/60$$
$$= 3577 \text{ pounds}$$

The diaphragm shears on either side of the shear wall at line 4 are

$$q_{45} = R_{45}/24$$
$$= 1140/24$$
$$= 47.5 \text{ pounds per linear foot}$$

and

$$q_{41} = R_{41}/24$$
$$= 3577/24$$
$$= 149 \text{ pounds per linear foot}$$

The shear in the shear wall at roof level is given by

$$q_W = (R_{45} + R_{41})/12$$
$$= 4717/12$$
$$= 393 \text{ pounds per linear foot}$$

The shear distribution and net shear at line 4 are shown in Figure 88SA-1(iii) and the drag force at B4 is given by

$$F_{B4} = 12(q_{41} + q_{45})$$
$$= 12(149 + 47.5)$$
$$= 12 \times 196.5$$
$$= 2358 \text{ pounds}$$

29

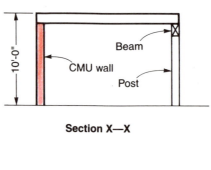

Section X—X

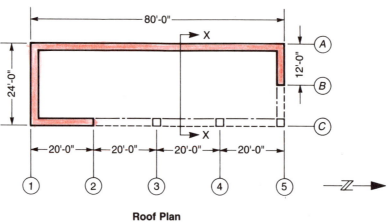

Roof Plan

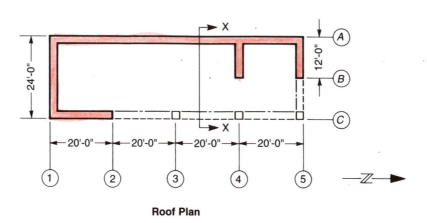

Roof Plan

FIGURE 88SA–1(i)

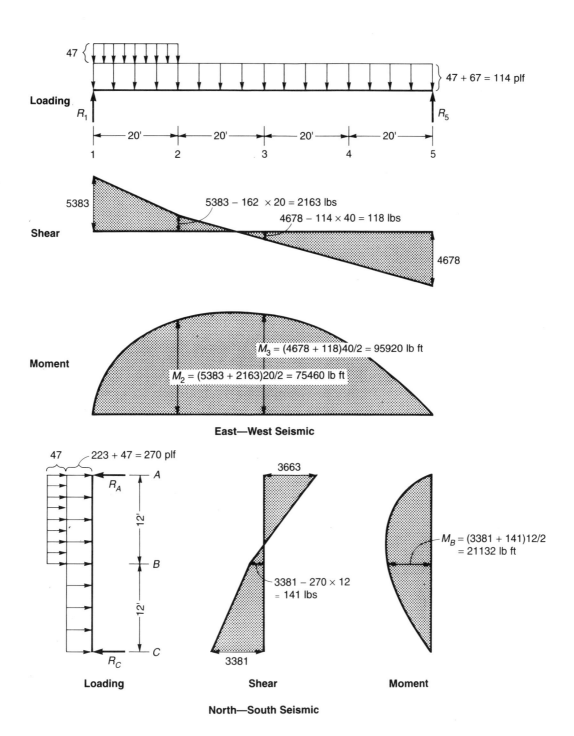

FIGURE 88SA-1(ii)

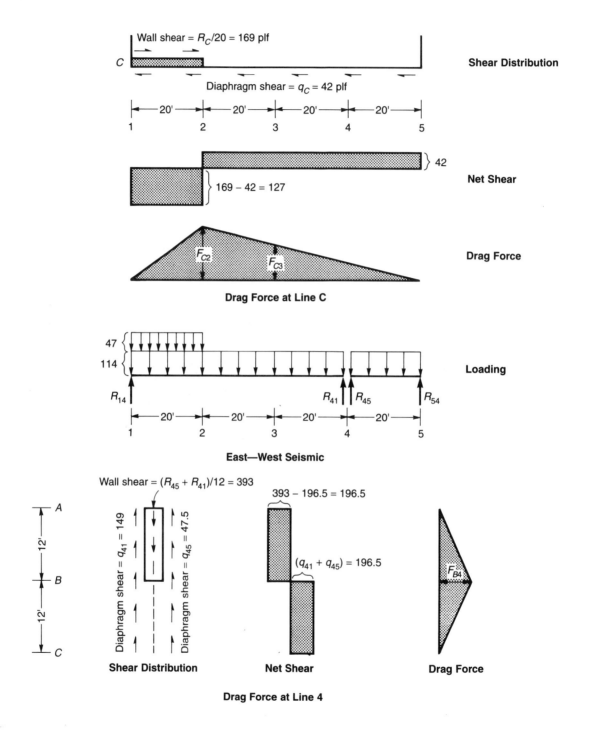

Shear Distribution

Wall shear = $R_C/20$ = 169 plf

C

Diaphragm shear = q_C = 42 plf

| 20' | 20' | 20' | 20' |

1   2   3   4   5

Net Shear

} 42

} 169 – 42 = 127

Drag Force

F_{C2} F_{C3}

Drag Force at Line C

Loading

47 {

114 {

R_{14} R_{41} | R_{45} R_{54}

| 20' | 20' | 20' | 20' |

1   2   3   4   5

East—West Seismic

Wall shear = $(R_{45} + R_{41})/12$ = 393

A

12'

B

12'

C

Diaphragm shear = q_{41} = 149

Diaphragm shear = q_{45} = 47.5

393 – 196.5 = 196.5

$(q_{41} + q_{45})$ = 196.5

F_{B4}

Shear Distribution   **Net Shear**   **Drag Force**

Drag Force at Line 4

FIGURE 88SA–1(iii)

32

STRUCTURAL ENGINEER REGISTRATION

CHAPTER 1

GENERAL STRUCTURAL PRINCIPLES AND SEISMIC DESIGN

SECTION 1.2

STRUCTURAL DEFORMATIONS

PROBLEMS:

1990 A–3
1988 A–1
1988 A–5
1986 A–1

STRUCTURAL ENGINEER EXAMINATION – 1990 ══════════════════

PROBLEM A-3 – WT. 6.0 POINTS

GIVEN: A one story wood structure is provided with a steel gable rigid frame at the east end bay.

CRITERIA: Basic wind speed 70 mph, exposure B, wind pressure to be determined by UBC Method 1.

Assumptions:
- Structure is classified as enclosed
- Ignore axial deformation
- Ignore dead and live load

REQUIRED:
1. Determine the wind pressures for the primary structure and show your results with a pressure diagram on Figure A.
2. Determine the horizontal deflection at point "a," using the moment diagram shown on Figure B. Provide all calculations.
3. Sketch the deflected shape of the gable frame for the loading given in Item 2. Show all joint rotations and points of inflection. No calculations are required.

SOLUTION

1. WIND PRESSURE[5,6,7,8]

The design wind pressure is obtained from UBC Formula (18–1) as

$$p = C_e C_q q_s I_w$$

where C_e = height, exposure and gust factor given in UBC Table 16–G

C_q = pressure coefficient given in UBC Table 16–H

q_s = wind stagnation pressure given in UBC Table 16–F
= 12.6 psf for a wind speed of 70 mph

and I_w = importance factor given in UBC Table 16–K
= 1.0

Then $p = 12.6 C_e C_q$

Applying UBC Method 1, the normal force method detailed in UBC Table 16–H, the wind pressure on the primary structure may be obtained at the various locations on the frame.

For the windward wall, which is fifteen feet in height, the value of C_e is 0.62 for exposure B, the value of C_q is 0.8 inward, and the design wind pressure is

$$p \quad = 12.6 \times 0.62 \times 0.8$$
$$= 6.25 \text{ pounds per square foot, inward}$$

For the leeward wall, UBC Section 1619.2 indicates that the value of C_e shall be based on the mean roof height of 23.75 feet, giving a value of C_e of 0.71. The value for C_q is 0.5 outward and the design wind pressure is

$$p \quad = 12.6 \times 0.71 \times 0.5$$
$$= 4.47 \text{ pounds per square foot, outward}$$

For the leeward roof, the value of C_e is also based on the mean roof height of 23.75 feet, giving a value for C_e of 0.71. The value for C_q is 0.7 outward and the design wind pressure is

$$p \quad = 12.6 \times 0.71 \times 0.7$$
$$= 6.26 \text{ pounds per square foot, outward}$$

For the windward roof, the value of C_e is based on the mean roof height of 21.25 feet, giving a value for C_e of 0.68. For a roof slope exceeding 2:12 and less than 9:12, the value for C_q is 0.9 outward or 0.3 inward and the design wind pressure is

$$p \quad = 12.6 \times 0.68 \times 0.9$$
$$= 7.71 \text{ pounds per square foot, outward}$$
$$p \quad = 12.6 \times 0.68 \times 0.3$$
$$= 2.57 \text{ pounds per square foot, inward}$$

The wind pressure diagram is shown in Figure 90A–3(ii)

2. HORIZONTAL DEFLECTION

The horizontal deflection at point "a" is most readily determined by the virtual work method[2,4]. The cut-back structure may be obtained by introducing a horizontal release at the right hand support to produce a roller support as shown in Figure 90A-3(ii). The horizontal deflection at point "a" in the original structure is obtained by considering unit virtual load applied horizontally outward to the cut-back structure at point "a." Then, the horizontal deflection due to the applied loads may be determined by evaluating the expression.

$$\delta = \int Mm\,ds/EI$$

where M is the bending moment at any section in the real structure due to the applied loads, m is the bending moment in the cut-back structure due to the unit virtual load, the effects of axial and shear forces are neglected, and the integral extends over all members of the structure. The moments, m, are shown on the cut-back structure in Figure 90A-3(ii) and the integral is most readily evaluated by applying the volume integration technique. Then,

$$
\begin{aligned}
EI\delta &= 15' \times 15' \times 46/3 \text{ (for the left hand column)} \\
&+ 32.5' \times 15(46 + 39.3)/2 \text{ (for the left hand rafter)} \\
&+ 19.5' \times 15(2 \times 39.3 - 123.8)/6 \text{ (for the right hand rafter)} \\
&= 22,038
\end{aligned}
$$

and,

$$
\begin{aligned}
\delta &= 22,038 \times 12^3/(882 \times 29,000) \\
&= 1.49 \text{ inches.}
\end{aligned}
$$

3. DEFLECTED SHAPE

The deflected shape of the frame is shown in Figure 90A-3(ii) and, from inspection, the joint rotations are as indicated. A point of inflection occurs in the right hand rafter at a point 4.7 feet from the ridge where the bending moment changes from positive to negative. ======

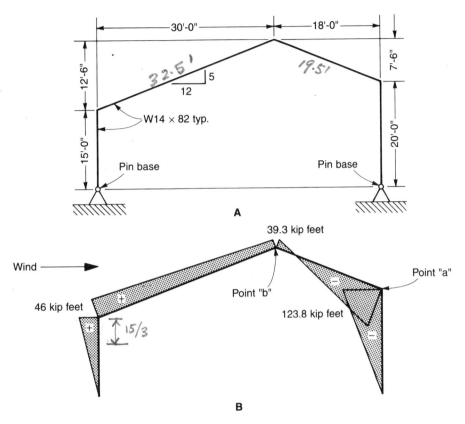

A

B

Note: Moment diagram is drawn on compression side of member

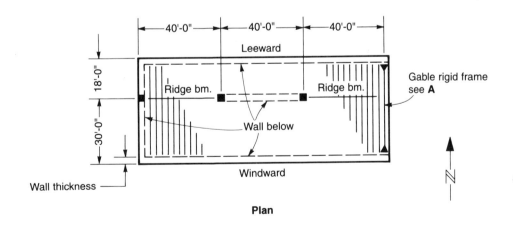

Plan

FIGURE 90A-3(i)

37

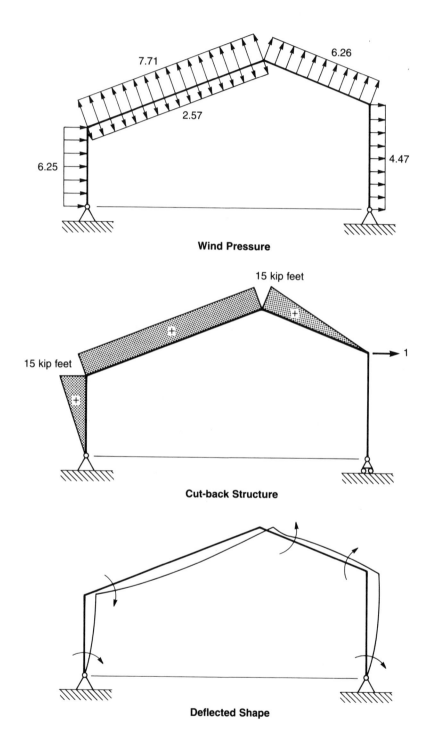

Wind Pressure

Cut-back Structure

Deflected Shape

FIGURE 90-3(ii)

STRUCTURAL ENGINEER EXAMINATION – 1988

PROBLEM A–1 – WT. 3.0 POINTS

GIVEN: A steel, special moment–resisting frame as shown.

CRITERIA: Modulus of elasticity E = 29,000 kips per square inch.
           All columns are fixed at the top. For bottom condition, see sketch.
           The continuous beam is infinitely stiff.

REQUIRED: 1.   Calculate the story drift due to the seismic load of 90 kips.
          2.   Determine whether your calculated drift is in compliance with the
               UBC requirements.
          3.   If non–structural masonry walls are required to fill in each bay, what are
               the minimum spacings you would recommend between wall and column,
               wall and beams, in order to comply with the UBC requirements.

SOLUTION

FRAME STIFFNESS

The frame stiffness may be obtained by summing the stiffnesses of the individual columns. The stiffness of column A, which is fixed at both ends, is given by

$$K_A = 12EI/L^3$$
$$= 12E \times 1500/(14^3 \times 1728)$$
$$= 6.56E/1728$$

The stiffness of column B, which is hinged at one end is given by

$$K_B = 3EI/L^3$$
$$= 3E \times 1800/(17^3 \times 1728)$$
$$= 1.10E/1728$$

Column C, which contains two hinges, constitutes a mechanism and its stiffness is

$$K_C = 0$$

The stiffness of column D, which is hinged at one end is given by

$$K_D = 3EI/L^3$$
$$= 3E \times 2100/(22^3 \times 1728)$$
$$= 0.59E/1728$$

The stiffness of column E, which is fixed at both ends, is given by

$$K_E = 12EI/L^3$$
$$= 12E \times 3000/(25^3 \times 1728)$$
$$= 2.30E/1728$$

The total frame stiffness, for lateral loading, is

$$K_T = K_A + K_B + K_C + K_D + K_E$$
$$= (6.56 + 1.10 + 0 + 0.59 + 2.30)E/1728$$
$$= 10.55E/1728$$

1. STORY DRIFT

The lateral displacement of the frame, or story drift, due to the seismic load V of ninety kips is given by

$$\Delta_H = V/K_T$$
$$= 90/K_T$$
$$= 90 \times 1728/(10.55 \times 29,000)$$
$$= 0.51 \text{ inches}$$

2. ALLOWABLE DRIFT

For a steel, special moment-resisting frame, the response factor R_W is obtained from UBC Table 16-N as 12. Assuming the building has a fundamental period of less than 0.7 seconds, UBC Section 1628.8.2 limits the allowable drift to the lesser of

$$\Delta_A = 0.005 \text{ (story height)}$$

or

$$\Delta_A = 0.04 \text{ (story height)}/R_W$$
$$= 0.04 \text{ (story height)}/12$$
$$= 0.0033 \text{ (story height), which governs.}$$

Hence, the allowable drift is

$$\Delta_A = 0.0033 \text{ (story height)}$$
$$= 0.0033 \times 14 \times 12$$
$$= 0.56 \text{ inches}$$
$$> 0.51 \text{ inches}$$

Hence, the calculated drift is in compliance with UBC requirements.

3. BUILDING SEPARATION

It is required that all parts of a building be separated a sufficient distance to permit independent seismic motion without impact between adjacent parts. UBC Section 1631.2.4.2 specifies that the separation shall allow for $3(R_w/8)$ times the displacement due to the prescribed seismic forces.

Assuming that the displacements of the masonry walls are negligible, only the lateral and vertical displacements of the steel frame need be considered. The calculated lateral displacement of the frame is

$$\Delta_H \quad = 0.51 \text{ inches}$$

and the required horizontal separation between columns and the masonry walls is

$$\begin{aligned}
\Delta_{H(req)} &= \Delta_H \times 3R_w/8 \\
&= 0.51 \times 3 \times 12/8 \\
&= 2.3 \text{ inches}
\end{aligned}$$

The calculated vertical displacement of the frame is

$$\begin{aligned}
\Delta_V &= L_A - (L_A^2 - \Delta_H^2)^{1/2} \\
&= 14 \times 12 - [(14 \times 12)^2 - (0.51)^2]^{1/2} \\
&= 0.0007 \text{ inches}
\end{aligned}$$

and the required vertical separation between beams and the masonry walls is

$$\begin{aligned}
\Delta_{V(req)} &= \Delta_V \times 3R_w/8 \\
&= 0.00077 \times 3 \times 12/8 \\
&= 0.0035 \text{ inches}
\end{aligned}$$

Use 0.5 inch minimum as specified in UBC Section 1631.2.4.2.

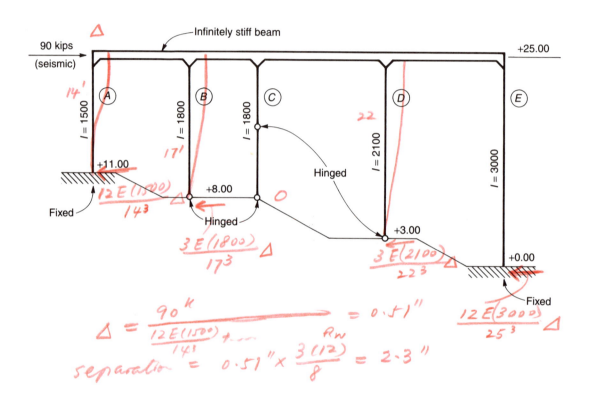

FIGURE 88A-1

STRUCTURAL ENGINEER EXAMINATION – 1988 ══════════════

PROBLEM A-5 – WT. 4.0 POINTS

GIVEN: W18 × 35 and TS 3 × 3 × ¼ brace as shown.

CRITERIA: All steel members have a yield strength of $F_y = 36$ ksi.
The steel beam is adequately braced at both flanges.
Ignore the beam shortening due to axial load.
$E = 29,000$ ksi; $I_{(beam)} = 510 in^4$; $A_{(diagonal)} = 2.59 in^2$
$M_{r(beam)} = 115 k\text{-}ft$ allowable; $P_{(tube)} = 56$ kips allowable.
The tributary uniform dead load is 600 plf (beam weight included).

REQUIRED: 1. Determined the maximum allowable uniform live load in plf for the beam.
2. Calculate the deflection at the center under the maximum loading condition.

SOLUTION

1. ALLOWABLE LIVE LOAD

The structure is one degree redundant, and it is convenient to consider the tensile force R in the tubular steel brace as the redundant. The cut-back structure is produced by removing this redundant and the applied loads and a unit virtual load replacing R are applied, in turn, to the cut-back structure. The redundant force is obtained from the virtual work method[2,4] as

$$R = -(\Sigma PuL/AE + \int Mm\,ds/EI)/(\Sigma u^2 L/AE + \int m^2 ds/EI)$$

where P and u are the member forces in the cut-back structure due to the applied loads and unit virtual load and M and m are the bending moments at any point in a member of the cut-back structure due to the applied loads and the unit virtual load, respectively. The summations are considered to extend over all the members of the actual structure with $P_{24} = 0$ and $u_{24} = 1.0$.

Figure 88A-5 shows the allowable dead plus live load w applied to the cut-back structure and the unit virtual load applied to the cut-back structure.

The term $\Sigma PuL/AE = 0$ since P is zero for all members.

The term $\int Mm\,ds/EI$ is applicable to members 12 and 23 with

$$
\begin{aligned}
M_2 &= w \times 30^2/8 \\
&= 112.5w \text{ pounds feet}
\end{aligned}
$$

and
$$
\begin{aligned}
m_2 &= -0.707 \times 30/4 \\
&= -5.303 \text{ pounds feet}
\end{aligned}
$$

Using the volume integration technique

$$\int Mm\,ds/EI = (5/12)M_2m_2L_{13}/EI$$
$$= -(5/12) \times 112.5w \times 5.303 \times 30 \times 1728/510E$$
$$= -25{,}265w/E \text{ pounds inch}$$

The term $\Sigma u^2L/AE$ is applicable only to member 24 with $u_{24} = 1.0$, since the shortening of the beam, member 12, is ignored.

$$\Sigma u^2L/AE = 1.0^2 \times 15 \times 1.414 \times 12/2.59E$$
$$= 98/E \text{ pounds inch}$$

The term $\int m^2\,ds/EI$ is applicable to members 12 and 23 and using the volume integration technique

$$\int m^2\,ds/EI = (1/3)m_2^2L_{13}/EI$$
$$= (1/3) \times (5.303)^2 \times 30 \times 1728/510E$$
$$= 953/E \text{ pounds inch}$$

Then, the force in member 24 is given by

$$R = 25{,}265w/(98 + 953)$$
$$= 24.04w \text{ pounds}$$

The vertical component of R is

$$V_2 = 0.707 \times 24.04w$$
$$= 17w \text{ pounds}$$

The vertical reaction at support 3 is

$$V_3 = (30w - 17w)/2$$
$$= 6.5w \text{ pounds}$$

The shear in the beam is zero at a distance from support 3 of

$$a = V_3/w$$
$$= 6.5w/w$$
$$= 6.5 \text{ feet}$$

The maximum sagging moment occurs in the beam at this point and has the magnitude

$$M_s = 6.5V_3 - w \times (6.5)^2/2$$
$$= 6.5 \times 6.5w - w \times (6.5)^2/2$$
$$= w \times (6.5)^2/2$$
$$= 21.125w \text{ pounds feet.}$$

The maximum hogging moment occurs at the beam center and has the magnitude

$$M_H = Rm_2 + M_2$$
$$= -24.04w \times 5.303 + 112.5w$$
$$= -15w \text{ pounds feet.}$$

Hence, sagging moment governs the load which may be applied to the beam and the maximum allowable total load is

$$w = 115,000/21.125$$
$$= 5444 \text{ pounds per foot}$$

The maximum capacity of the tubular steel brace is 56 kips and this limits the maximum allowable total load on the beam to

$$w = 56,000/24.04$$
$$= 2329 \text{ which governs}$$

Hence, the maximum allowable uniform live load on the beam is given by

$$(w - 600) = 1729 \text{ pounds per linear foot.}$$

2. CENTRAL DEFLECTION

The central deflection of the beam is caused by the extension of the tubular steel brace which is given by

$$\delta_{42} = (RL/AE)_{42}$$
$$= 24.04 \times 2329 \times 15 \times 1.414 \times 12/2.59 \times 29,000$$
$$= 0.19 \text{ inches}$$

The central deflection produced is given by the vertical component of this extension which is

$$\delta_2 = 1.414 \times \delta_{42}$$
$$= 1.414 \times 0.19$$
$$= 0.27 \text{ inch}$$

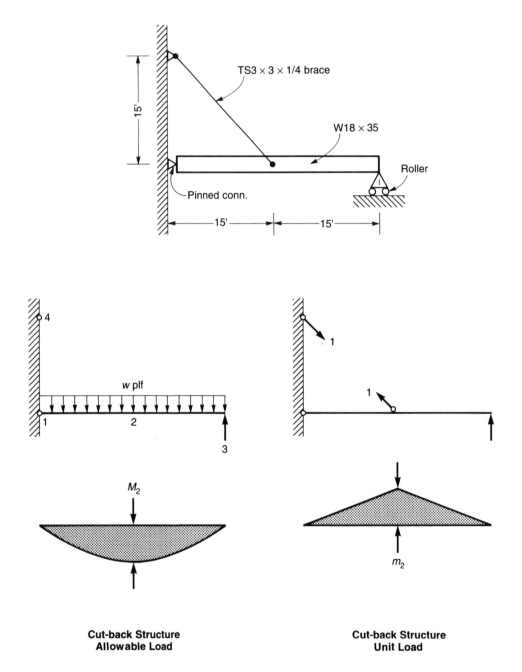

FIGURE 88A–5

STRUCTURAL ENGINEER EXAMINATION – 1986 ===============

PROBLEM A–1 – WT. 3.0 POINTS

GIVEN: A single–story steel frame as shown.

CRITERIA: All members are W18 × 50 sections.

        E   = 29,000 ksi, I = 800 in^4.

REQUIRED: 1) Determine the frame moments and forces for a 20kips seismic force at roof level. Show all calculations.
           2) Draw the resulting moment diagram.
           3) Draw the deflected shape of frame. No calculations are required.

SOLUTION

1. FRAME FORCES

The frame forces may be determined by the moment distribution procedure[1,2] by assuming that the twenty kips seismic force produces a lateral sway displacement of δ. The initial fixed–end moments due to this sway displacement are

$$M_{BA}^F = M_{FE}^F = M_{CD}^F = -3EI\delta/L^3$$
$$= -60x \ldots \text{say}$$

The moments M_{BA}^F and M_{FE}^F are the final sway moments in the frame whilst M_{CD}^F must be distributed to the adjacent members CA and CE. Advantage may be taken of the hinged supports to modify the stiffness factors, as shown in the table, so that there is no carry–over to the hinged ends.

Joint		C	
Member	CA	CD	CE
Relative EI/L	1/10	1/15	1/30
Modified stiffness	3/10	3/15	3/30
Distribution factor	1/2	1/3	1/6
Fixed–end moments	0	– 60x	0
Distribution	30x	20x	10x
Final sway moments, M^S	30x	–40x	10x

The sway equation is obtained by applying the equation of virtual work to the small displacements involved and is

$$-M_{BA} - M_{FE} - M_{CD} = 20 \times 15$$
$$= 300 \text{ kip feet}$$

Substituting the final sway moments into the sway equation gives

$$60x + 60x + 40x = 300$$
and $$x = 300/160$$
$$= 1.875$$

The actual moments are obtained by multiplying the final sway moments by x. Then

$$M_{BA} = M_{FE} = -60 \times 1.875$$
$$= -112.5 \text{ kip feet}$$
$$M_{CD} = -40 \times 1.875$$
$$= -75 \text{ kip feet}$$
$$M_{CA} = 30 \times 1.875$$
$$= 56.25 \text{ kip feet}$$
$$M_{CE} = 10 \times 1.875$$
$$= 18.75 \text{ kip feet}$$

The horizontal reactions at the supports are

$$H_B = M_{BA}/15$$
$$= -112.5/15$$
$$= -7.5 \text{ kips to the left}$$
$$H_D = M_{CD}/15$$
$$= -75/15$$
$$= -5.0 \text{ kips to the left}$$
$$H_F = -20 - H_B - H_D$$
$$= -20 + 7.5 + 5.0$$
$$= -7.5 \text{ kips to the left}$$

The vertical reactions at the supports are

$$V_B = -M_{CA}/L_{AC}$$
$$= -56.25/10$$
$$= -5.625 \text{ kips downward}$$
$$V_F = M_{CE}/L_{CE}$$
$$= 18.75/30$$
$$= 0.625 \text{ kips upward}$$
$$V_D = -V_B - V_F$$
$$= 5.625 - 0.625$$
$$= 5.0 \text{ kips upward}$$

2. MOMENT DIAGRAM

The moment diagram, drawn on the compression side of the members, is shown in Figure 86A-1.

3. DEFLECTED SHAPE

The deflected shape of the frame is shown in Figure 86A-1

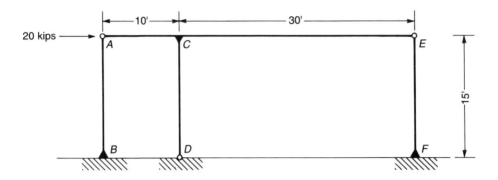

Frame Elevation

Legend

▲ "Fixed" connection

○ "Pinned" connection capable of resisting uplift

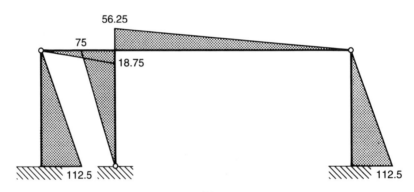

Moment Diagram

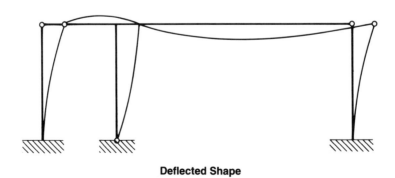

Deflected Shape

FIGURE 86A-1

50

STRUCTURAL ENGINEER REGISTRATION

CHAPTER 1

GENERAL STRUCTURAL PRINCIPLES AND SEISMIC DESIGN

SECTION 1.3

RIGIDITY AND TORSION

PROBLEMS

STRUCTURAL ENGINEER EXAMINATION – 1990 ══════════

PROBLEM A-4 – WT. 7.0 POINTS

GIVEN: A 24,000 pound (includes members and platform) grain bin is to be supported as shown.

CRITERIA: Seismic zone 4
Importance factor $I = 1.5$
Force coefficient $C = 2.75$

Assumptions:
- All joints are pinned
- Neglect wind loads
- Horizontal plane at top of platform acts as a rigid diaphragm
- Neglect accidental torsion

REQUIRED: 1.   Determine the lateral seismic force acting at the center of gravity of the grain bin.

                2.   Assuming a horizontal seismic force of 14.4 kips acting at point "F":

                     A. Determine forces in braces C and D for the transverse direction. Show all calculations for the rotational forces.

                     B. Determine the forces in braces A and B for the longitudinal direction

                     C. Determine the maximum net uplift force in column "E".

SOLUTION

1. LATERAL SEISMIC FORCE

The structure consists of a nonbuilding structure as defined by UBC Section 1632.5 and the lateral seismic force is given by UBC Formula (28-1) as

           V $= (ZIC/R_w)W$

where Z $= 0.4$ for zone 4 from UBC Table 16-I

             I $=$ importance factor of 1.5, as given

             C $=$ force coefficient of 2.75, as given

             R_w $= 4$ from UBC Table 16-P, item 7, for bins on braced or unbraced legs

and W $=$ total load of 24 kips, as given

The ratio of C/R_w $= 2.75/4$

                            $= 0.69$

                            > 0.5 as required by UBC Section 1632.5

Hence the value selected for R_w is satisfactory.

Then $\quad$ V $\quad = (0.4 \times 1.5 \times 2.75/4)24$
$\quad\quad\quad\quad = 0.4125 \times 24$
$\quad\quad\quad\quad = 9.9$ kips

STIFFNESS OF FRAME 1

Assuming that the platform beam is of infinite area, its elastic shortening may be neglected, and only the columns and braces contribute to the stiffness of the frames.

The stiffness of frame 1 is obtained by applying a unit virtual load to the frame, as shown in Figure 90A–4. The horizontal displacement at the point of application of the load is determined by the virtual work method[2,4] by evaluating the expression

$\delta \quad = \Sigma u^2 L / AE$

where u is the force in a member due to the unit virtual load, L is the length of the member, A is the sectional area of the member, and E is the modulus of elasticity of the member. The summation is obtained by tabulating the relevant values as shown

Member	L	A	u	u^2L/A
Brace 13	11.31	2.59	1.414	8.731
Column 34	8.00	11.70	−1.000	0.684
Total	–	–	–	9.415

The horizontal displacement of point 3 is given by

$\delta \quad = \Sigma u^2 L / AE$
$\quad\quad = 9.415 \times 12/29,000$
$\quad\quad = 0.00390$ inches

The lateral stiffness of frame 1, which is defined as the force required to produce unit horizontal displacement, is given by

$K_1 \quad = 1/\delta$
$\quad\quad = 1/0.00390$
$\quad\quad = 256.4$ kips per inch

STIFFNESS OF FRAME 2

The stiffness of frame 3 is obtained in a similar manner to frame 1and the relevant values are tabulated below.

Member	L	A	u	u^2L/A
Brace 24	12.81	4.59	−1.601	7.151
Column 12	10.00	11.70	1.250	1.336
Total	–	–	–	8.487

The horizontal displacement of point 3 is given by

$$\delta = \Sigma u^2L/AE$$
$$= 8.487 \times 12/29,000$$
$$= 0.00351 \text{ inches}$$

The lateral stiffness of frame 3 is given by

$$K_3 = 1/\delta$$
$$= 1/0.00351$$
$$= 284.8 \text{ kips per inch}$$

STIFFNESS OF FRAME 2

Unit horizontal load applied to joint 4 of frame 2 produces no forces in the columns. Since the elastic shortening of the platform beam may be neglected, only the two braces contribute to the stiffness of the frame.

Resolving vertically at joint 3 provides the equation

$$f_A \sin(53.13°) - f_B \sin(59.04°) = 0$$

where f_A and f_B are the forces in members A and B due to the unit load.

Then $\quad 0.8f_A - 0.8576f_B = 0$

Resolving horizontally at joint 3 provides the equation

$$f_A \cos(53.13°) + f_B \cos(59.04°) = 1.0$$
$$0.6f_A + 0.5146f_B = 1.0$$

Solving simultaneously gives the member forces

$$f_A = 0.9259$$
$$f_B = 0.8637$$

The relevant values are tabulated below.

Member	L	A	u	u^2L/A
Brace 13 Brace 35	10.00 11.66	2.59 2.59	0.9259 −0.8637	3.310 3.358
Total	–	–	–	6.668

The horizontal displacement of point 4 is given by

$$\begin{aligned}
\delta \quad &= \Sigma u^2L/AE \\
&= 6.668 \times 12/29{,}000 \\
&= 0.00276 \text{ inches}
\end{aligned}$$

The lateral stiffness of frame 2 is given by

$$\begin{aligned}
K_2 \quad &= 1/\delta \\
&= 1/0.00276 \\
&= 362.4 \text{ kips per inch}
\end{aligned}$$

The relative stiffnesses of frames 1, 2 and 3 are

$$256.4 : 362.4 : 284.8 = 1.0 : 1.41 : 1.11$$

2A. TRANSVERSE SEISMIC FORCE

Since the horizontal platform acts as a rigid diaphragm and the transverse seismic force is applied eccentrically, in determining the distribution of lateral force to the frames, torsional effects must be considered.

CENTER OF RIGIDITY

By inspection, the center of rigidity lies on the longitudinal axis of the structure and is located a distance from the horizontal axis through column E of

$$y_r \quad = 4 \text{ feet}$$

The center of rigidity, for transverse forces, is obtained by taking moments of the stiffnesses of frames 1 and 3 about the vertical axis through column E and is given by

$$\begin{aligned}
x_r \quad &= (K_1 \times 0 + K_3 \times 12)/(K_1 + K_3) \\
&= (1.0 \times 0 + 1.11 \times 12)/(1.0 + 1.11) \\
&= 6.3 \text{ feet}
\end{aligned}$$

The distance of each frame from the center of rigidity is

$$
\begin{aligned}
r_1 &= 6.3 \text{ feet} \\
r_3 &= 12 - 6.3 \\
&= 5.7 \text{ feet} \\
r_2 &= 4.0 \text{ feet}
\end{aligned}
$$

The polar moment of inertia of the frames $\Sigma r^2 K$ is

$$
\begin{aligned}
\Sigma r^2 K &= K_1 r_1^2 + K_3 r_3^2 + 2K_2 r_2^2 \\
&= 1.0 \times 6.3^2 + 1.11 \times 5.7^2 + 2.0 \times 1.41 \times 4.0^2 \\
&= 121 \text{ square feet}
\end{aligned}
$$

The sum of the frame stiffnesses for transverse seismic force is

$$
\begin{aligned}
\Sigma K_y &= K_1 + K_3 \\
&= 1.0 + 1.11 \\
&= 2.11
\end{aligned}
$$

APPLIED TORSION

From the problem statement, the transverse force is located eight feet from the vertical axis through column E and the eccentricity is

$$
\begin{aligned}
e_x &= 8 - x_r \\
&= 8 - 6.3 \\
&= 1.7 \text{ feet}
\end{aligned}
$$

In accordance with the problem statement, accidental eccentricity does not have to be considered. From the problem statement, the lateral force acting has a magnitude of 14.4 kips and the torsional moment acting about the center of rigidity is

$$
\begin{aligned}
T &= 14.4 e_x \\
&= 14.4 \times 1.7 \\
&= 24.5 \text{ kip feet}
\end{aligned}
$$

FRAME FORCES

The forces produced in a frame are the sum of the in-plane shear forces and the torsional forces. The in-plane shear force is

$$
\begin{aligned}
F_s &= 14.4 K_y / \Sigma K_y \\
&= 14.4 K_y / 2.11 \\
&= 6.82 K_y
\end{aligned}
$$

The torsional shear force is

$$
\begin{aligned}
F_T &= T r K / \Sigma r^2 K \\
&= 24.5 r K / 121 \\
&= 0.20 r K
\end{aligned}
$$

The total force in a frame is

$$F_F = F_S + F_T$$
$$= 6.82k_y + 0.20rK$$

and the torsional shear force is, in accordance with UBC Section 1603.3.3, neglected when of opposite sense to the in-plane shear force.

The total force acting on frame 1 is given by

$$F_{F1} = F_{S1} + F_{T1}$$
$$= 6.82K_1 + 0.20r_1K_1$$
$$= 6.82 \times 1.0 - 0.20 \times 6.3 \times 1.0$$
$$= 6.82 \times 1.0 \dots \text{ neglecting negative torsional forces}$$
$$= 6.82 \text{ kips}$$

The force in brace C is given by

$$F_C = 6.82 \times 1.414$$
$$= 9.64 \text{ kips}$$

The total force acting on frame 3 is given by

$$F_{F3} = F_{S3} + F_{T3}$$
$$= 6.82K_3 + 0.20r_3K_3$$
$$= 6.82 \times 1.11 + 0.20 \times 5.7 \times 1.11$$
$$= 7.57 + 1.27$$
$$= 8.84 \text{ kips}$$

The force in brace D is given by

$$F_D = 8.84 \times 1.601$$
$$= 14.14 \text{ kips}$$

2B. LONGITUDINAL SEISMIC FORCE

Since the longitudinal seismic force acts through the center of rigidity and accidental eccentricity does not have to be considered, no torsional effects are produced under this loading condition.

The in-plane shear force in frame 2 is given by

$$F_S = 14.4K_x/\Sigma K_x$$
$$= 14.4K_2/(2 \times K_2)$$
$$= 14.4 \times 1.41/(2 \times 1.41)$$
$$= 7.20 \text{ kips}$$

The force acting in brace A is given by

$$F_A = 7.20 \times 0.9259$$
$$= 6.67 \text{ kips}$$

The force acting in brace B is given by

$$F_B = 7.20 \times 0.8637$$
$$= 6.22 \text{ kips}$$

2C. UPLIFT FORCE

The platform beam may be assumed infinitely rigid and without support from the bracing members A and B of frame 2. Then the vertical reaction produced at column E by the self weight of the grain bin is given by

$$R_D = 0.5(24 \times 4/12)$$
$$= 4.0 \text{ kips}$$

The uplift on column E, due to the longitudinal seismic force, is produced by the vertical component of the force in bracing member A of frame 2. This uplift force is given by

$$R_L = F_A \sin(53.13°)$$
$$= 6.67 \times 0.80$$
$$= 5.34 \text{ kips}$$

The uplift on column E, due to the transverse seismic force, is produced by the tensile force in the column of frame 1 plus the vertical component of the force in the bracing member A of frame 2. The force in bracing member A is due to the torsional effects of the transverse force and is given by

$$F_A = 0.9259F_T$$
$$= 0.9259Tr_2K_2/\Sigma r^2 K$$
$$= 0.9259 \times 0.20 \times r_2 K_2$$
$$= 0.9259 \times 0.20 \times 4.0 \times 1.41$$
$$= 1.04 \text{ kips}$$

The uplift on column E due to the transverse seismic force is given by

$$R_T = F_C \sin(45°) + F_A \sin(53.13°)$$
$$= 9.64 \times 0.707 + 1.04 \times 0.80$$
$$= 7.65 \text{ kips}$$
$$> 5.34$$

and the transverse seismic force governs.

The maximum net uplift on column E is given by

$$R = R_T - R_D$$
$$= 7.65 - 4.0$$
$$= 3.65 \text{ kips}$$

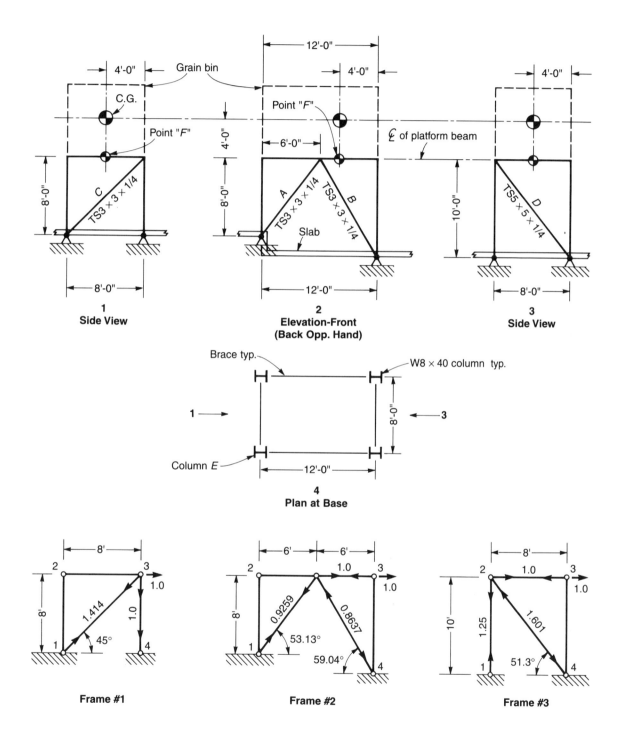

FIGURE 90A–4

59

STRUCTURAL ENGINEER EXAMINATION – 1989

PROBLEM A–4 – WT. 6.0 POINTS

GIVEN: The roof plan, cross section and wall elevation of an auditorium for a maximum occupancy of 400 people. Roof and walls are of cast–in–place concrete.

CRITERIA: • Seismic zone 4

Assumptions:
- Roof diaphragm is rigid.
- Show all calculations and any additional assumptions.
- Do not add accidental torsion as required by the code, UBC Section 1628.5.
- All walls are 10" thick.
- For calculating weights, neglect wall openings.

REQUIRED;   1. Compute the seismic base shear.
2. Calculate the relative rigidities of each wall, assuming the wall is fixed at the base only.
3. Distribute a 100 kip base shear in the "Y" direction using the following assumed rigidities:

R_A =   1.25
R_B =   1.25
R_C =   1.0

SOLUTION

DEAD LOAD

Assuming that the roof diaphragm is constructed in concrete, its thickness is given by

$$h \quad = 100/150$$
$$= 0.67 \text{ feet}$$

The structure dead loads at base level are given by

Roof	= 0.1 × 50 × 50	= 250.00 kips
Wall A + Wall B	= 0.125 × 20 × 14.33 × 2	= 71.65
Wall C	= 0.125 × 20 × 19.33	= 48.33
Total		369.98

1. BASE SHEAR

From UBC Section 1628.2.2, the structure period is given by UBC Formula (28–3) as

$$T = C_t(h_n)^{3/4}$$

where C_t = 0.02 for a shear wall type structure
and h_n = the roof height of fifteen feet
Then T = 0.02 $(15)^{3/4}$
= 0.152 seconds

The force coefficient is given by UBC Formula (28–2) as

$$C = 1.25 \, S/T^{2/3}$$

where S = 1.5 from UBC Table 16–J for a site with unknown soil properties
Then C = 1.25 × 1.5/$(0.152)^{2/3}$
= 6.58
> 2.75 ... hence use the upper limit of
C = 2.75

The base shear is given by UBC Formula (28–1) as

$$V = (ZIC/R_W)W$$

where Z = 0.4 for zone 4 from UBC Table 16–I.
I = 1.0 from UBC Table 16–K for a special occupancy structure.
C = 2.75 as calculated.
R_W = 6 from Table 16–N for a bearing wall system
and W = 369.98 kips as calculated for the total dead load.
Then V = (0.4 × 2.75/6) 369.98
= 0.183 × 369.98
= 67.83 kips

2. WALL RIGIDITIES

The deflection of a pier due to a unit applied load is given by

$$\delta = \delta_F + \delta_S$$

where δ_F = deflection due to flexure
= $4(H/L)^3/Et$ for a cantilever pier
= $(H/L)^3/Et$ for a pier fixed at top and bottom
with H = height of pier
L = length of pier
E = modulus of elasticity of pier
t = thickness of pier
and δ_S = deflection due to shear
= $3(H/L)/Et$

61

The rigidity, or stiffness, of a pier is given by

$$R = 1/\delta$$

Using calculator program A.5.2 in Appendix A, the rigidity of wall C is obtained as follows:

H	L	Type	$Et\delta_F$	$Et\delta_S$	$Et\delta$	R/Et
19.33	20	Cant	3.61	2.90	6.51	0.154

Several techniques are available for determining the rigidities of walls with openings. The technique given in NAVFAC Appendix C is considered the most accurate of the concise methods and is adopted here. The deflection of the wall is first obtained as though it is a solid wall. From this is subtracted the deflection of that section of the wall which contains the openings. The deflection of each pier, formed by the openings, is now added back.

For wall A and wall B, the procedure is as follows

Pier	H	L	Type	$Et\delta_F$	$Et\delta_S$	$Et\delta$	R/Et
Wall	14.33	20	Cant	1.471	2.150	3.621	–
Mid Strip	6	20	Cant	–0.108	–0.900	–1.008	–
1	6	3	Fixed	8.000	6.000	–	0.0714
2	6	2.5	Fixed	13.824	7.200	–	0.0476
3	6	2.5	Fixed	13.824	7.200	–	0.0476
4	6	3	Fixed	8.000	6.000	–	0.0714
1+2+3+4	–	–	–	–	–	4.202	←0.238
Total	–	–	–	–	–	6.815→	0.147

The relative rigidities of walls A, B and C are

$$R_A = R_B = 0.147/0.147$$
$$= 1.0$$

and $$R_C = 0.154/0.147$$
$$= 1.048$$

3. BASE SHEAR DISTRIBUTION

Since the roof diaphragm is rigid, in determining the distribution of the base shear to the shear walls, torsional effects must be considered.

CENTER OF MASS

Assume that the given base shear of 100 kips allows for the mass of the roof and all walls, including wall B and wall C. Hence, in locating the center of mass for seismic loads in the Y-direction, it is appropriate to include the dead load of the roof and all walls.

Taking moments about the bottom left hand corner of the roof, the distance of the center of mass from the left hand edge is

$$\bar{x} = (250 \times 25 + 35.83 \times 21 + 35.83 \times 10 + 48.33 \times 49.58)/369.98$$
$$= 26.37 \text{ feet}$$

CENTER OF RIGIDITY

In locating the center of rigidity for seismic loads in the Y-direction, wall A, which has no stiffness in the Y-direction, is omitted. Taking moments about the bottom left hand corner of the roof, the distance of the center of rigidity from the left hand edge is

$$x_r = (1.25 \times 10 + 1.0 \times 49.58)/(1.0 + 1.25)$$
$$= 27.60 \text{ feet}$$

For seismic loads in the X-direction, walls B and C are omitted and by inspection, the center of rigidity lies on the longitudinal axis of wall A. Then, the distance of the center of rigidity from the bottom edge of the roof is

$$y_r = 40 \text{ feet}$$

The distance r of each wall from the center of rigidity is

$$r_A = 40 - 40$$
$$= 0 \text{ feet}$$
$$r_B = 27.60 - 10$$
$$= 17.60$$
$$r_C = 49.58 - 27.60$$
$$= 21.98$$

The polar moment of inertia of the walls $\Sigma r^2 R$ is

$$\Sigma r^2 R = R_A \times r_A^2 + R_B \times r_B^2 + R_C \times r_C^2$$
$$= 1.25 \times 0 + 1.25(17.60)^2 + 1.0(21.98)^2$$
$$= 870 \text{ square feet}$$

The sum of the shear wall rigidities for seismic loads in the Y-direction ΣR_y is

$$\Sigma R_y = R_B + R_C$$
$$= 1.25 + 1.0$$
$$= 2.25$$

APPLIED TORSION

For seismic loads in the Y-direction, the eccentricity is

$$e_x = x_r - \bar{x}$$
$$= 27.60 - 26.37$$
$$= 1.23 \text{ feet}$$

The torsional moment acting about the center of rigidity is

$$T = Ve_x$$
$$= 100 \times 1.23$$
$$= 123 \text{ kip feet}$$

From the problem statement, accidental torsion does not have to be considered.

WALL FORCES

The force produced in a wall by the base shear V acting in the Y-direction is the sum of the in-plane shear force and the torsional shear force. The in-plane shear force is

$$F_S = VR_y/\Sigma R_y$$
$$= 100R_y/2.25$$
$$= 44.44R_y \text{ kips}$$

The torsional shear force is

$$F_T = TrR/\Sigma r^2 R$$
$$= 123rR/870$$
$$= 0.14rR \text{ kips}$$

The total force in a wall is

$$F_F = F_S + F_T \text{ kips}$$

and the torsional shear force F_T in accordance with UBC Section 1603.3.3 is neglected when of opposite sense to the in-plane shear force F_S.

The wall forces are given by

Wall	R_y	F_S	rR	F_T	F_F
A	0	0	0	0	0
B	1.25	55.56	21.98	3.08	58.64
C	1.00	44.44	21.98	–3.08	44.44

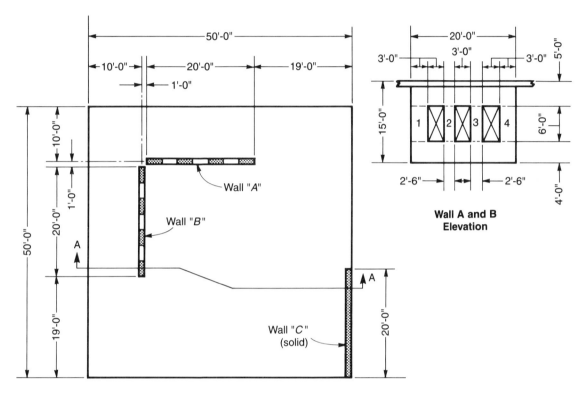

Roof Plan

**Wall A and B
Elevation**

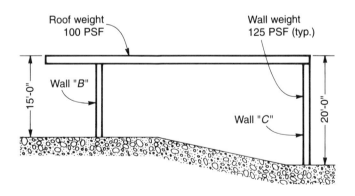

Roof weight
100 PSF

Wall weight
125 PSF (typ.)

Wall "B"

Wall "C"

Section A—A

FIGURE 89A–4

65

STRUCTURAL ENGINEER EXAMINATION – 1988

PROBLEM A–4 – WT. 5.0 POINTS

GIVEN: The concrete roof slab is supported by 3 steel tube columns, each of TS $10 \times 10 \times \frac{1}{2}$ as shown.

CRITERIA: Seismic zone 4, $R_w = 4$, $C = 2.75$
Steel tube column $F_y = 46$ kips per square inch
Wt. of concrete = 150 pounds per cubic foot

Assumptions:
The concrete slab is adequate.
Weight of steel columns may be neglected.

REQUIRED: 1. Determine the maximum reactions at the base of each column due to the seismic forces in East – West direction only.
2. Determine the adequacy of all steel columns.

SOLUTION

DEAD LOAD

The dead load of the roof slab is

$$W \quad = 0.15 \times 1.0 \times 24^2$$
$$= 86.4 \text{ kips}$$

BASE SHEAR

From UBC Section 1628.2 the base shear is given by Formula (28-1) as

	V	$= (ZIC/R_w)W$
where	Z	= 0.4 for zone 4 from Table 16–I
	I	= 1.0 from Table 16–K for a non-essential facility
	C	= 2.75 as given
	R_w	= 4 as given
and	W	= 86.4 as calculated
Then	V	$= (0.4 \times 1.0 \times 2.75/4)86.4$
		$= 0.275 \times 86.4$
		= 23.76 kips

66

CENTER OF MASS

By inspection, the center of mass lies at the center of the concrete roof slab and its distance from column 2, as indicated in Figure 88A-4(ii), is

$$\bar{y} = 9 \text{ feet}$$

COLUMN RIGIDITIES

The rigidity of each column is identical in the North-South direction and in the East-West direction and the rigidities are given by

$$
\begin{aligned}
R_1 = R_3 &= 3EI/L^3 \\
&= 3EI/13^3 \\
&= 3EI/2197
\end{aligned}
$$

and
$$
\begin{aligned}
R_2 &= 3EI/L^3 \\
&= 3EI/10^3 \\
&= 3EI/1000
\end{aligned}
$$

The relative rigidities of the columns are

$$
\begin{aligned}
R_1 = R_3 &= 2197/2197 \\
&= 1.0
\end{aligned}
$$

and
$$
\begin{aligned}
R_2 &= 2197/1000 \\
&= 2.197
\end{aligned}
$$

CENTER OF RIGIDITY

For seismic loads in the North-South direction, the distance of the center of rigidity from column 2 is obtained by taking moments about 2 and, as indicated in Figure 88A-4(ii), is

$$
\begin{aligned}
x_r &= (R_1 + R_3) \times 18/\Sigma R \\
&= (1.0 + 1.0) \times 18/(1.0 + 1.0 + 2.197) \\
&= 36/4.197 \\
&= 8.58 \text{ feet}
\end{aligned}
$$

For seismic loads in the East-West direction, the distance of the center of rigidity from column 2 is

$$
\begin{aligned}
y_r &= R_3 \times 18/\Sigma R \\
&= 1.0 \times 18/4.197 \\
&= 4.29 \text{ feet}
\end{aligned}
$$

The distances of column 3 from the center of rigidity are shown in Figure 88A-4(ii).

APPLIED TORSION

For seismic loads in the East–West direction, the eccentricity is

$$e_y = \bar{y} - y_r$$
$$= 9 - 4.29$$
$$= 4.71 \text{ feet}$$

To comply with UBC, Section 1628.5, accidental eccentricity must be considered and this amounts to five percent of the roof slab dimension perpendicular to the direction of the seismic force. Neglecting torsional irregularity effects, the accidental eccentricity is given by

$$e_a = \pm 0.05 \times 24$$
$$= \pm 1.2 \text{ feet}$$

The net eccentricity is then

$$e = 4.71 \pm 1.2$$
$$= 5.91 \text{ or } 3.51 \text{ and the critical value is}$$
$$e = 5.91 \text{ feet}$$

The torsional moment acting about the center of rigidity is

$$T = Ve$$
$$= 23.76 \times 5.91$$
$$= 140.42 \text{ kip feet}$$

FORCES DUE TO UNIT ROTATION

Imposing a unit clockwise rotation on the roof slab produces forces in the columns which are given by the expression

$$F = rR$$

and these forces are shown in Figure 88A–4(ii).

MOMENT TO PRODUCE UNIT ROTATION

In order to produce a unit rotation of the roof slab, a moment must be applied to the center of rigidity and this is given by

$$M = \Sigma rF$$
$$= 18.84 \times 18 + 13.71 \times 18$$
$$= 586 \text{ kip feet}$$

FORCES DUE TO APPLIED TORSION

The torsion T, applied to the roof slab, produces forces in the columns which are given by the expression

$$F_T = FT/M$$
$$= F \times 140.42/586$$
$$= 0.24F \text{ kips, and these forces are shown in Figure 88A–4(ii).}$$

FORCES DUE TO LATERAL SEISMIC FORCE

For seismic loads in the East–West direction, the in–plane shear forces produced in the columns are given by

$$F_S = VR_x/\Sigma R_x$$

where R_x is the rigidity of a column in the East–West direction. Then,

$$
\begin{aligned}
F_S &= 23.76R_x/(1 + 2.197 + 1) \\
&= 23.76R_x/4.197 \\
&= 5.66R_x \text{ kips}
\end{aligned}
$$

and the forces are shown in Figure 88A–4(ii).

1. COLUMN REACTIONS DUE TO SEISMIC LOADS

The total shear forces produced in the columns are

$$F_F = F_S + F_T$$

and, to comply with UBC Section 1603.3, the torsional shear force, F_T, is neglected when of opposite sense to the in–plane shear force, F_S. The total shear forces are shown in Figure 88A–4(ii) and the X and Y components may be resolved as indicated.

Because of the hinge at the top of each column, no moments are produced in the roof slab by seismic loads, and hence, no axial forces are produced in the columns. Moments are produced, by seismic loads, at the bottom of the columns. For column 1, the moments produced about the x-axis and the y-axis are

$$
\begin{aligned}
M_{1x} &= 2.26 \times 13 \times 12 \\
&= 353 \text{ kip inches} \\
M_{1y} &= 5.66 \times 13 \times 12 \\
&= 883 \text{ kip inches}
\end{aligned}
$$

For column 2, the corresponding moments are

$$
\begin{aligned}
M_{2x} &= 4.52 \times 10 \times 12 \\
&= 542 \text{ kip inches} \\
M_{2y} &= 12.44 \times 10 \times 12 \\
&= 1493 \text{ kip inches}
\end{aligned}
$$

For column 3, the corresponding moments are

$$
\begin{aligned}
M_{3x} &= 2.26 \times 13 \times 12 \\
&= 353 \text{ kip inches} \\
M_{3y} &= 8.95 \times 13 \times 12 \\
&= 1396 \text{ kip inches}
\end{aligned}
$$

2. ADEQUACY OF THE COLUMNS

AXIAL LOADS

The column tributary areas are shown in Figure 88A-4(ii) and the column axial loads are given by

$$
\begin{aligned}
P_1 \quad &= W/4 \\
&= 86.4/4 \\
&= 21.6 \text{ kips} \\
P_2 = P_3 &= 3W/8 \\
&= 32.4 \text{ kips}
\end{aligned}
$$

COLUMN 3

The properties of the 10" x 10" x ½" structural tubing are obtained from AISC Table 1–94 as

$$
\begin{aligned}
A \quad &= 18.4 \text{ inches}^2 \\
S \quad &= 54.2 \text{ inches}^3 \\
r \quad &= 3.84 \text{ inches}
\end{aligned}
$$

To establish the allowable flexural stress in the tubing, compliance with UBC Section 2251 B5.1 must be determined. The limiting width-thickness ratio for a compact section is

$$
\begin{aligned}
b/t \quad &\leq 190/(F_y)^{1/2} \\
&\leq 190/(46)^{1/2} \\
&\leq 28
\end{aligned}
$$

The actual width-thickness ratio of the section is

$$
\begin{aligned}
b/t \quad &= 10/0.5 \\
&= 20
\end{aligned}
$$

which is satisfactory. In addition, the laterally unsupported length is limited by UBC Section 2251 F3.1 to

$$
\begin{aligned}
\ell_b \quad &\leq 1200 \, b/F_y \\
&\leq 1200 \times 10/46 \\
&\leq 260 \text{ inches}
\end{aligned}
$$

The actual unbraced length of the column is

$$
\begin{aligned}
\ell_b \quad &= 13 \times 12 \\
&= 156 \text{ inches}
\end{aligned}
$$

which is satisfactory. Hence, the section is compact and the allowable bending stress is

$$
\begin{aligned}
F_b \quad &= 0.66F \\
&= 0.66 \times 46 \\
&= 30.4 \text{ kips per square inch}
\end{aligned}
$$

The actual bending stresses in the column are

$$f_{bx} = M_{3x}/S$$
$$= 353/54.2$$
$$= 6.50 \text{ kips per square inch}$$

and

$$f_{by} = M_{3y}/S$$
$$= 1396/54.2$$
$$= 25.76 \text{ kips per square inch}$$

The effective length factor for a column fixed at its base and free to rotate and translate at its top is obtained from AISC Table C-C2.1 item (e) as

$$K = 2.1$$

and the effective slenderness ratio is

$$K\ell/r = 2.1 \times 13 \times 12/3.84$$
$$= 85.3$$

The allowable compressive stress is then obtained from AISC Table C-50 as

$$F_a = 17.93 \text{ kips per square inch}$$

The actual compressive stress in the column is given by

$$f_a = P_3/A$$
$$= 32.4/18.4$$
$$= 1.76 \text{ kips per square inch}$$
$$< 0.15 \times F_a$$

Hence, in checking the adequacy of the column for combined axial compression and bending, in accordance with UBC Section 2251 H1, Formula (H1-3) is applicable and

$$f_a/F_a + f_{bx}/F_b + f_{by}/F_b = 1.76/17.93 + 6.5/30.4 + 25.76/30.4$$
$$= 0.098 + 0.214 + 0.847$$
$$= 1.16$$
$$< 1.0 \times 1.33$$

Hence, column 3 is adequate as the combined stresses do not exceed the one-third increase, allowed by UBC Section 1603.5, for seismic forces.

COLUMN 1

By inspection, column 1 is adequate

COLUMN 2

The actual bending stresses in column 2 are

$$f_{bx} = M_{2x}/S$$
$$= 542/54.2$$
$$= 10.00 \text{ kips per square inch}$$

and $f_{by} = M_{2y}/S$
$$= 1493/54.2$$
$$= 27.55 \text{ kips per square inch}$$

The effective slenderness ratio for column 2 is

$$K\ell/r = 2.1 \times 10 \times 12/3.84$$
$$= 65.6$$

and the allowable compressive stress is obtained from AISC Table C–50 as

$$F_a = 21.74 \text{ kips per square inch}$$

The actual compressive stress in the column is given by

$$f_a = P_2/A$$
$$= 32.4/18.4$$
$$= 1.76 \text{ kips per square inch}$$
$$< 0.15 \times F_a$$

Hence Formula (H1–3) is again applicable and

$$f_a/F_a + f_{bx}/F_b + f_{by}/F_b = 1.76/21.74 + 10/30.4 + 27.55/30.4$$
$$= 0.081 + 0.329 + 0.906$$
$$= 1.32$$
$$< 1.0 \times 1.33$$

Hence column 2 is adequate ═══════════════════════════

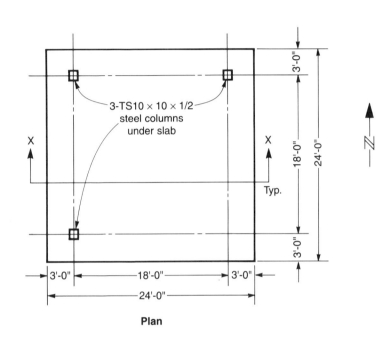

Plan

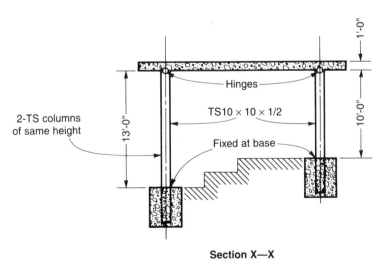

Section X—X

FIGURE 88A-4(i)

73

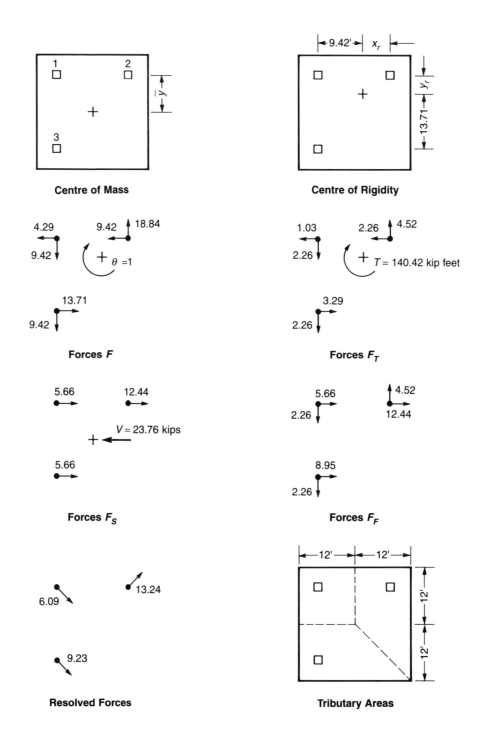

FIGURE 88A-4(ii)

STRUCTURAL ENGINEER EXAMINATION – 1987 ══════════

PROBLEM A–4 – WT. 4.0 POINTS

GIVEN: The roof plan and front elevation for a one–story building located in seismic zone 3. All shear walls are load bearing and the relative rigidity for each wall is noted on the plan. The roof deck is a rigid diaphragm. The design for vertical loads is adequate.

CRITERIA: Roof weight is 50 psf.
Shear wall weight is 70 psf.
Weight of the window wall is zero.
Do not add accidental torsion as required by the UBC Section 1628.5.

REQUIRED: 1. Compute the lateral seismic forces.
2. Distribute the lateral forces to the four shear walls.
3. Using the results from 2
   A. Compute the roof diaphragm chord force at point "A".
   B. Compute the roof diaphragm drag force at point "B".

SOLUTION

DEAD LOAD

The lateral forces developed at diaphragm level are required and the dead loads tributary to the diaphragm are given by

Roof	$= W_R = 0.05 \times 40 \times 20$	$= 40$ kips
North wall	$= W_3 = 0.07 \times 8 \times 14.29/2$	$= 4$
South wall	$= W_1 = 0.07 \times 12 \times 14.29/2$	$= 6$
East wall	$= W_2 = 0.07 \times 10 \times 14.29/2$	$= 5$
West wall	$= W_4$	$= 5$

DIAPHRAGM SHEAR

From UBC Section 1268.2.2, the structure period is given by UBC Formula (28-3) as

$$T = C_t(h_n)^{3/4}$$

where C_t = 0.02 for a shear wall structure
and h_n = the roof height of 14.29 feet
Then T = $0.02(14.29)^{3/4}$
= 0.147 seconds

The force coefficient is given by UBC Formula (2812-2) as

$$C = 1.25S/T^{2/3}$$

where S = 1.5 from UBC Table 16-J for a site with unknown soil properties
Then, C = $1.25 \times 1.5/(0.147)^{2/3}$
= 6.74
> 2.75 hence use the upper limit of
C = 2.75

The diaphragm shear is given by UBC Formula (28-1) as

$$V = (ZIC/R_w)W$$

Z = 0.3 for zone 3 from UBC Table 16-I
I = 1.0 from UBC Table 16-K for a standard occupancy structure
C = 2.75 as calculated
R_w = 6 from UBC Table 16-N for a bearing wall system
W = structure dead load
Then V = $(0.3 \times 1.0 \times 2.75/6)W$
= 0.1375W

1. LATERAL SEISMIC FORCES

NORTH-SOUTH SEISMIC

The dead load tributary to the roof diaphragm in the North-South direction is due to the North wall, South wall and roof and is given by

$$W_{NS} = W_3 + W_1 + W_R$$
$$= 4 + 6 + 40$$
$$= 50 \text{ kips}$$

The seismic force on the roof diaphragm in the North-South direction is given by

$$V_{NS} = 0.1375W_{NS}$$
$$= 0.1375 \times 50$$
$$= 6.88 \text{ kips}$$

EAST-WEST SEISMIC

The dead load tributary to the roof diaphragm in the East-West direction is due to the East wall, West wall, and roof and is given by

$$
\begin{aligned}
W_{EW} &= W_2 + W_4 + W_R \\
&= 5 + 5 + 40 \\
&= 50 \text{ kips}
\end{aligned}
$$

The seismic force on the roof diaphragm in the East-West direction is given by

$$
\begin{aligned}
V_{EW} &= 0.1375 W_{EW} \\
&= 0.1375 \times 50 \\
&= 6.88 \text{ kips}
\end{aligned}
$$

2. LATERAL FORCE DISTRIBUTION

Since the roof diaphragm is rigid, in determining the distribution of lateral seismic forces to the shear walls, torsional effects must be considered.

CENTER OF MASS

NORTH-SOUTH SEISMIC

The calculated seismic force in the North-South direction does not include the mass of the East and West shear walls and, hence, in locating the center of mass for North-South forces the East and West walls are omitted.

Taking moments about the bottom right hand corner of the roof, the distance of the center of mass from the right hand edge is

$$
\begin{aligned}
\bar{x} &= (W_R \times 20 + W_3 \times 16 + W_1 \times 18)/W_{NS} \\
&= (40 \times 20 + 4 \times 16 + 6 \times 18)/W_{NS} \\
&= 972/50 \\
&= 19.44 \text{ feet}
\end{aligned}
$$

EAST-WEST SEISMIC

The calculated seismic force in the East-West direction does not include the mass of the North and South shear walls and, hence, in locating the center of mass for East-West forces the North and South walls are omitted.

By inspection, the distance of the center of mass from the bottom edge of the roof is

$$
\bar{y} = 10 \text{ feet}
$$

CENTER OF RIGIDITY

NORTH–SOUTH SEISMIC

In locating the center of rigidity for seismic loads in the North–South direction, the North and South walls, which have no stiffness in this direction, are omitted. By inspection, the distance of the center of rigidity from the right hand edge of the roof is

$$x_r \quad = 14 \text{ feet}$$

EAST–WEST SEISMIC

For seismic loads in the East–West direction, the East and West walls, which have no stiffness in this direction, are omitted. Taking moments about the bottom right hand corner of the roof, the distance of the center of rigidity from the bottom edge of the roof is

$$
\begin{aligned}
y_r \quad &= (R_3 \times 20 + R_1 \times 0)/(R_3 + R_1) \\
&= (0.37 \times 20 + 1.00 \times 0)/(0.37 + 1.00) \\
&= 7.40/1.37 \\
&= 5.4 \text{ feet}
\end{aligned}
$$

STRUCTURE PROPERTIES

The distance of each wall from the center of rigidity is

$$
\begin{aligned}
r_3 \quad &= 20 - y_r \quad &= 14.6 \text{ feet} \\
r_1 \quad &= y_r \quad &= 5.4 \\
r_2 \quad &= x_r \quad &= 14.0 \\
r_4 \quad &= 28 - x_r \quad &= 14.0
\end{aligned}
$$

The polar moment of inertia of the wall is

$$
\begin{aligned}
\Sigma r^2 R \quad &= r_3^2 R_3 + r_1^2 R_1 + r_2^2 R_2 + r_4^2 R_4 \\
&= 14.6^2 \times 0.37 + 5.4^2 \times 1.00 + 14^2 \times 0.65 + 14^2 \times 0.65 \\
&= 363 \text{ square feet}
\end{aligned}
$$

The sum of the shear wall rigidities for seismic loads in the North–South direction is

$$
\begin{aligned}
\Sigma R_{NS} \quad &= R_2 + R_4 \\
&= 0.65 + 0.65 \\
&= 1.30
\end{aligned}
$$

The sum of the shear wall rigidities for seismic loads in the East–West direction is

$$
\begin{aligned}
\Sigma R_{EW} \quad &= R_3 + R_1 \\
&= 0.37 + 1.00 \\
&= 1.37
\end{aligned}
$$

APPLIED TORSION

NORTH–SOUTH SEISMIC

For seismic loads in the North–South direction, the eccentricity is

$$
\begin{aligned}
e_x \quad &= \bar{x} - x_r \\
&= 19.44 - 14 \\
&= 5.44 \text{ feet}
\end{aligned}
$$

The torsional moment acting about the center of rigidity is

$$
\begin{aligned}
T_{NS} \quad &= V_{NS}e_x \\
&= 6.88 \times 5.44 \\
&= 37.43 \text{ kip feet}
\end{aligned}
$$

Accidental torsion is neglected in accordance with the problem statement.

EAST–WEST SEISMIC

For seismic loads in the East–West direction, the eccentricity is

$$
\begin{aligned}
e_y \quad &= \bar{y} - y_r \\
&= 10 - 5.4 \\
&= 4.6 \text{ feet}
\end{aligned}
$$

The torsional moment acting about the center of rigidity is

$$
\begin{aligned}
T_{EW} \quad &= V_{EW}e_y \\
&= 6.88 \times 4.6 \\
&= 31.65 \text{ kip feet}
\end{aligned}
$$

WALLS FORCES

The force produced in a wall by the seismic force on the roof diaphragm is the sum of the in-plane shear force, the torsional shear force and the shear force generated by the self weight of the wall.

NORTH–SOUTH SEISMIC

The in-plane shear force is given by

$$
\begin{aligned}
F_s \quad &= V_{NS}R_{NS}/\Sigma R_{NS} \\
&= 6.88R_{NS}/1.3 \\
&= 5.29R_{NS} \text{ kips}
\end{aligned}
$$

where R_{NS} is the rigidity of a wall in the North–South direction.

The torsional shear force is given by

$$F_T = T_{NS}rR/\Sigma r^2 R$$
$$= 37.43rR/363$$
$$= 0.103rR \text{ kips}$$

and the torsional shear force is neglected when of opposite sense to the in-plane shear force in accordance with UBC Section 1603.3.3.

The shear force due to the self-weight of the wall is given by

$$F_W = (ZIC/R_W)W_N$$
$$= 0.1375W_N \text{ kips}$$

where W_N is the total weight of a wall with its longitudinal axis in the North-South direction.

The total force in a wall is

$$F_F = F_S + F_T + F_W$$

and the total and individual forces are shown in the following Table.

Wall	R_{NS}	F_S	rR	F_T	W_N	F_W	F_F
3	0.00	0.00	5.40	0.56	0.00	0.00	0.56
1	0.00	0.00	5.40	0.56	0.00	0.00	0.56
2	0.65	3.44	9.10	−0.94	10.00	1.38	4.82
4	0.65	3.44	9.10	0.94	10.00	1.38	5.76

EAST-WEST SEISMIC

The in-plane shear force is given by

$$F_S = V_{EW}R_{EW}/\Sigma R_{EW}$$
$$= 6.88R_{EW}/1.37$$
$$= 5.02R_{EW} \text{ kips}$$

where R_{EW} is the rigidity of a wall in the East-West direction.

The torsional shear force is given by

$$F_T = T_{EW}rR/\Sigma r^2 R$$
$$= 31.65rR/363$$
$$= 0.087rR \text{ kips}$$

and the torsional shear force is neglected when of opposite sense to the in-plane shear force in accordance with UBC Section 1603.3.3.

The shear force due to the self weight of the wall is given by

$$F_W = (ZIC/R_W)W_E$$
$$= 0.1375W_E \text{ kips}$$

where W_E is the total weight of a wall with its longitudinal axis in the East-West direction.

The total force in a wall is

$$F_F = F_S + F_T + F_W$$

and the total and individual forces are shown in the following Table.

Wall	R_{EW}	F_S	rR	F_T	W_E	F_W	F_F
3	0.37	1.86	5.40	0.47	8.00	1.10	3.43
1	1.00	5.02	5.40	−0.47	12.00	1.65	6.67
2	0.00	0.00	9.10	0.79	0.00	0.00	0.79
4	0.00	0.00	9.10	0.79	0.00	0.00	0.79

3A. CHORD FORCE AT A

The lateral force on the roof diaphragm to the West of point A, due to the North–South seismic force, is produced by the self weight of the roof and is given by

$$w = 0.1375 \times 0.05 \times 20$$
$$= 0.1375 \text{ kips per linear foot}$$

The bending moment at point A due to the lateral force is

$$M_A = w \times 12^2/2$$
$$= 9.90 \text{ kip feet}$$

The resulting chord force at point A is given by

$$F_A = M_A/20$$
$$= 9.9/20$$
$$= 0.495 \text{ kips}$$

3B. DRAG FORCE AT B

The diaphragm shear along the South edge of the roof diaphragm, due to the East–West seismic force, is produced by the in–plane shear force F_S and the torsional shear force F_T. However, the torsional shear force is neglected as it is of opposite sense to the in–plane shear force. Hence, the diaphragm shear is given by

$$q_S = F_S/40$$
$$= 5.02/40$$
$$= 0.126 \text{ kips per linear foot}$$

The resultant drag force at point B is

$$F_B = 16q_S$$
$$= 16 \times 0.126$$
$$= 2.01 \text{ kips}$$

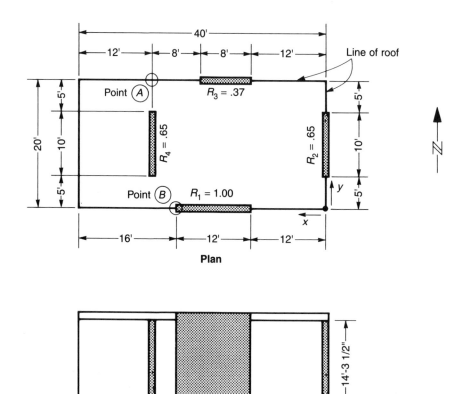

Plan

Elevation

FIGURE 87-4

STRUCTURAL ENGINEER EXAMINATION – 1984 ══════════

PROBLEM A–6 – WT. 4.0 POINTS

GIVEN: The layout of three masonry shear walls is shown in sketches "A", "B", and "C".

CRITERIA: Modulus of elasticity of masonry $E_m = 1,500,000$ psi.
All walls are 8 inches thick.
Assume walls (piers) "B" and "C" are fixed at the top and the base. Piers at wall "A" are fixed base only.
Assume the diaphragm is rigid.

REQUIRED: (Use "by hand" calculations)
A. Calculate the relative rigidity of all three shear walls "A", "B", and "C".
B. Calculate the relative rigidity of shear wall "C" in the direction shown in sketch "D". Neglect the weak axis rigidity of the wall.

SOLUTION

A. WALL RIGIDITIES

The deflection of a pier due to a unit load applied at the top edge is given by

$$\delta = \delta_F + \delta_S$$

where δ_F = deflection due to flexure
= $4(H/L)^3/Et$ for a cantilever pier
= $(H/L)^3/Et$ for a pier fixed at top and bottom

with H = height of pier
L = length of pier
E = modulus of elasticity of pier
t = thickness of pier

and δ_S = deflection due to shear
= $1.2\,H/GA$
= $3(H/L)/Et$

with G = rigidity modulus of pier
= $0.4E$

and A = cross sectional area of pier
= tL

The rigidity, or stiffness, of a pier is given by

$$R = 1/\delta$$

The wall rigidities may be obtained by using the calculator program A.5.2 in Appendix A. The rigidity of a wall with openings is most accurately determined by the technique proposed by NAVFAC. In this method, the deflection of the wall is first obtained as though it is a solid wall. From this is subtracted the deflection of that portion of the wall which contains the openings. The deflection of each pier, formed by the openings, is now added back.

WALL A

Pier	H	L	Type	$Et\delta_F$	$Et\delta_S$	$Et\delta$	R/Et
1	20	10	Cant	32	6	38	0.0263
2	20	15	Cant	9.48	4	13.48	0.0742
Total	–	–	–	–	–	–	0.1005

The actual rigidity of wall A is

$$
\begin{aligned}
R \quad &= 0.1005\ Et \\
&= 0.1005 \times 1500 \times 8 \\
&= 1206 \text{ kips per inch}
\end{aligned}
$$

WALL B

Pier	H	L	Type	$Et\delta_F$	$Et\delta_S$	$Et\delta$	R/Et
Wall	20	20	Fixed	1	3	4.000	–
1+2+4	10	20	Fixed	–0.125	–1.500	–1.625	–
1	10	10	Fixed	1	3	–	0.250
2	10	5	Fixed	8	6	–	0.071
1+2	–	–	–	–	–	3.115	←0.321
Total	–	–	–	–	–	5.490→	0.182

The actual rigidity of wall B is

$$
\begin{aligned}
R \quad &= 0.182 Et \\
&= 0.182 \times 1500 \times 8 \\
&= 2185 \text{ kips per inch}
\end{aligned}
$$

WALL C

Pier	H	L	$Et\delta_F$	$Et\delta_S$	$Et\delta$	R/Et	$Et\delta$
Wall	20	35	0.1866	0.7143	1.9009	–	–
1+2+3+4+6+7	10	35	–0.0233	–0.8572	–0.8805	–	–
2+3+4+6	10	20	0.1250	1.5000	–	–	1.6250
2+3+6	5	20	–0.0156	–0.7500	–	–	–0.7656
2	5	5	1.0	3.0	–	0.2500	–
3	5	5	1.0	3.0	–	0.2500	–
2+3	–	–	–	–	–	0.5000→	2.0000
2+3+4	–	–	–	–	–	0.3497	←2.8594
1	10	10	1.0	3.0	–	0.2500	–
1+2+3+4	–	–	–	–	1.6674	←0.5997	–
Total	–	–	–	–	2.6878→	0.3721	–

The actual rigidity of wall C is

$$
\begin{aligned}
R &= 0.3721Et \\
  &= 0.3721 \times 1500 \times 8 \\
  &= 4465 \text{ kips per inch}
\end{aligned}
$$

B. RIGIDITY AT AN ANGLE

Introducing a unit displacement to a wall along its longitudinal axis requires the application of a force of magnitude R where R is the rigidity of the wall along its longitudinal axis. As shown in Figure 84A–6, this produces a displacement and a force along the x axis, inclined at an angle $\theta°$ to the wall axis, of

$$\delta_x = 1/\cos\theta$$

and $\quad F_x = R\cos\theta$

The rigidity of the wall along the x–axis is given by

$$
\begin{aligned}
R_x &= F_x/\delta_x \\
    &= R\cos^2\theta
\end{aligned}
$$

Hence, the rigidity of wall C at an angle of 30° to its longitudinal axis is

$$
\begin{aligned}
R_x &= R\cos^2\theta \\
    &= 4465 \cos^2 30° \\
    &= 3349 \text{ kips per inch}
\end{aligned}
$$

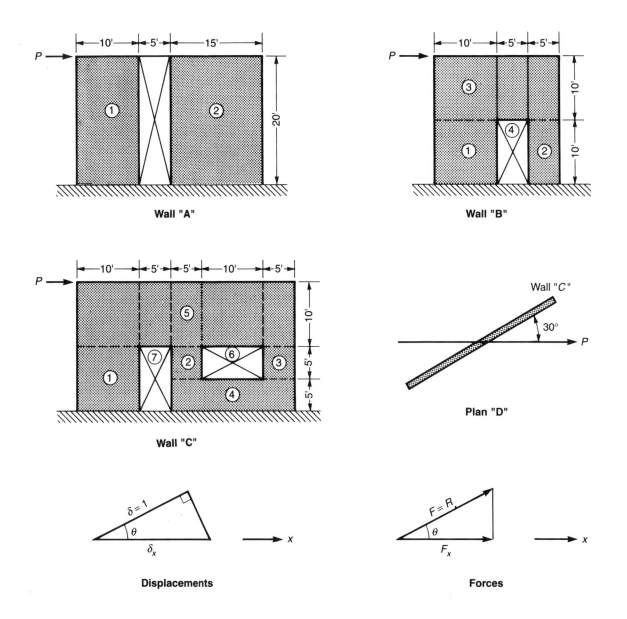

FIGURE 84A-6

STRUCTURAL ENGINEER REGISTRATION

CHAPTER 1

GENERAL STRUCTURAL PRINCIPLES AND SEISMIC DESIGN

SECTION 1.4

RESPONSE SPECTRA AND MODE SHAPES

PROBLEMS:

STRUCTURAL ENGINEER EXAMINATION – 1991 ========

PROBLEM A–2 – WT. 7.0 POINTS

GIVEN: A five-story special moment-resisting steel frame building. The building
             mass distribution and story heights are shown. The building fundamental
             mode shape has been determined by computer analysis and is shown. The
             fundamental period has been calculated by rational analysis to be T = 0.90
             seconds.
             ● Table 1 – Site specific response spectra data

CRITERIA: Seismic zone 4
             Soil type 2
             Importance factor 1.0

             Assumptions:
             ● Only the fundamental (first) mode response need be considered for
               dynamic analysis.

REQUIRED: 1. Using the given mass distribution, mode shape, building period, and
                site specific response spectrum data; determine the total shear at the
                base for the building, using dynamic lateral force procedures.
             2. Using the given data, determine the design lateral force (shear at the
                base) using the static force procedure.
             3. Using the given site specific response spectrum, determine the
                minimum design story shear distribution in accordance with UBC
                Section 1629.
             4. Using the response spectrum data, determine the expected
                displacement at the top of the building and the story drift ratios.

SOLUTION

1. DYNAMIC PROCEDURES

For a special moment-resisting steel frame, the structural response factor is obtained from UBC
Table 16-N as

$$R_w = 12$$

For the given fundamental period of 0.90 seconds, the site specific spectral acceleration is
obtained from the given data as

$$S_a = 231.6 \text{ inches per sec}^2$$

The value of the design acceleration is

$$
\begin{aligned}
a &= S_a/R_w \\
&= 231.6/12 \\
&= 19.30 \text{ inches per sec}^2
\end{aligned}
$$

The effective modal weight[10,11] is given by

$$W^E = (\Sigma W_i \phi_i)^2 / \Sigma W_i \phi_i^2$$

where W_i = weight at floor level i

ϕ_i = mode shape component at floor level i

From the Table below, the effective modal weight is obtained as

$$
\begin{aligned}
W^E &= 1411^2/1024 \\
&= 1944
\end{aligned}
$$

The total base shear is, then

$$
\begin{aligned}
V_D &= W^E a/g \\
&= 1944 \times 19.3/386.4 \\
&= 97.11 \text{ kips}
\end{aligned}
$$

Level	W_i	ϕ_i	$W_i\phi_i$	$W_i\phi_i^2$	F_i	V_i
Roof	475	1.00	475	475	39.25	39.25
Floor 5	475	0.78	371	289	30.66	69.91
Floor 4	475	0.59	280	165	23.14	93.05
Floor 3	475	0.40	190	76	15.70	108.75
Floor 2	475	0.20	95	19	7.85	116.60
Total	2375	–	1411	1024	116.60	–

2. STATIC PROCEDURE

In the static force procedure, the UBC Section 1628 allows two procedures for calculating the value of the fundamental period and each of the values may be used to determine the base shear.

METHOD A

The fundamental period is given by UBC Formula (28-3) as

$$T = C_t(h_n)^{3/4}$$

where C_t = 0.035 for a steel moment resisting frame

and, h_n = the roof height of fifty eight feet

Then $T = 0.035(58)^{3/4}$

$$= 0.7356 \text{ seconds}$$

The force coefficient is given by UBC Formula (28-2) as

$$C = 1.25S/T^{2/3}$$

where $\quad S \quad = 1.2$ from UBC Table 16-J, for soil type 2

Hence $\quad C \quad = 1.25 \times 1.2/(0.7356)^{2/3}$

$$= 1.84$$
$$< 2.75$$

and $\quad C/R_w = 1.84/12$

$$= 0.153$$
$$> 0.075$$

Hence $\quad C \quad = 1.84$ is an acceptable value for the force coefficient.

The base shear is given by UBC Formula (28-1) as

$$V_A = (ZIC/R_w)W$$

where $\quad Z \quad = 0.4$ for zone 4 from UBC Table 16-I

$I \quad = 1.0$ as given

$W \quad =$ total dead load of 2375 kips

Then, $\quad V_A \quad = (0.4 \times 1.0 \times 1.84/12)\,2375$

$$= 0.0613 \times 2375$$
$$= 145.7 \text{ kips}$$

METHOD B

Using the given value for the fundamental period of 0.9 seconds, the lateral force coefficient is given by UBC Formula (28-2) as

$$C = 1.25S/T^{2/3}$$
$$= 1.25 \times 1.2/(0.9)^{2/3}$$

The base shear is given by UBC Formula (28-1) as

$$V_B = (ZIC/R_w)W$$
$$= (0.4 \times 1.0 \times 1.61/12)\,2375$$
$$= 0.0537 \times 2375$$
$$= 127.5 \text{ kips}$$

3. STORY SHEAR DISTRIBUTION

For a regular structure, UBC Section 1629.5.3 requires that the design base shear, based on a dynamic analysis procedure, be scaled up to the greater value given by

$$V = 0.90V_B$$
$$= 0.90 \times 127.5$$
$$= 114.7 \text{ kips}$$

or $\quad V \quad = 0.80V_A$

$$= 0.80 \times 145.7$$
$$= 116.6 \text{ kips, which governs.}$$

Then, the design lateral force at each level is given by

$$F_i \quad = VW_i\phi_i/\Sigma W_i\phi_i$$
$$= 116.6 W_i\phi_i/1411$$
$$= 0.0826 W_i\phi_i$$

The design force at each level and the shear V_i at each story is shown in the Table above.

4. EXPECTED DISPLACEMENTS

For the given fundamental period of 0.90 seconds, the site specific spectral displacement is

$$S_d \quad = 4.75 \text{ inches}$$

Then, the displacement at each level is given by

$$\delta_i \quad = P\phi_i S_d$$
where $\quad P \quad = \text{participation factor}$
$$= W^E/\Sigma W_i\phi_i$$
$$= 1944/1411$$
$$= 1.38$$
Then $\quad \delta_i \quad = 1.38 \times 4.75\phi_i$
$$= 6.56\phi_i$$

The drift at a given story is defined as the relative displacement of the upper and lower floor levels at that story and is given by

$$D_i \quad = \delta_i - \delta_{(i-1)}$$

The drift ratio at a given story is defined as the ratio of the story drift to the height of that story and is given by

$$R_i \quad = D_i/h_i$$

These values are shown in the Table below.

Level	Story	h_i	ϕ_i	δ_i	D_i	$R_i\%$
Roof	–	–	1.00	6.56	–	–
Floor 5	5	10	0.78	5.12	1.44	1.20
Floor 4	4	10	0.59	3.87	1.25	1.04
Floor 3	3	10	0.40	2.62	1.25	1.04
Floor 2	2	10	0.20	1.31	1.31	1.09
Floor 1	1	18	–	–	1.31	0.61

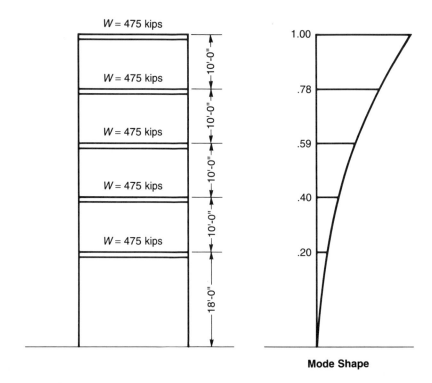

Mode Shape

PERIOD	S_a	S_v	S_d	FREQ.
SECONDS	INCH/SEC/SEC	INCHES/SEC	INCHES	RAD/SEC
0.10	308.80	4.91	0.08	62.83
0.20	386.00	12.29	0.39	31.42
0.30	386.00	18.43	0.88	20.94
0.40	386.00	24.57	1.56	15.71
0.50	386.00	30.72	2.44	12.57
0.60	347.40	33.18	3.17	10.47
0.70	301.10	33.55	3.74	8.98
0.80	258.60	32.93	4.19	7.85
0.90	231.60	33.18	4.75	6.98
1.00	208.40	33.17	5.28	6.28
1.50	154.40	36.86	8.80	4.19
2.00	115.80	36.86	11.73	3.14
2.50	96.50	38.40	15.28	2.51

Table 1 — Site Specific Response Spectra

FIGURE 91A–2

STRUCTURAL ENGINEER EXAMINATION – 1989 ====

PROBLEM A–3 – WT. 7.0 POINTS

GIVEN: A three–story office building, using special moment–resisting frames, symmetrical in both directions.

CRITERIA: Seismic zone 4.

REQUIRED: 1. (STATIC METHOD) Calculate the following using the relevant sections of the UBC:
       a. Find the seismic base shear using S = 1.5.
       b. Distribute the seismic base shear over the height of the building.
       c. Calculate the diaphragm force for each level.

          2. (DYNAMIC METHOD) Calculate the following using the two mode shapes given.
       a. The effective weight in each mode.
       b. The story force and shear in each mode.
       c. The combined story force using the square–root–of–the–sum–of–the–squares method.

         Useful equations:
            For $T \leq 0.65$ seconds : S_a = Spectral Acceleration = $0.3g$
            For $T > 0.65$ seconds : S_v = Spectral velocity = 12 inches per second

W^E	= Effective Weight	$= (\Sigma W_i \, \phi_i)^2 / \Sigma W_i \phi_i^2$
V	= Total shear	$= S_a W^E$
V_i	= Story Shears	$= (W_i \phi_i / \Sigma W_i \phi_i) V$
S_a	$= S_v / 61.5T$	

SOLUTION

1a. SEISMIC BASE SHEAR

From UBC Section 1628.1, the structure period is given by UBC Formula (28–3) as

	T	$= C_t (h_n)^{3/4}$
where	C_t	= 0.035 for a steel moment–resisting frame
and	h_n	= the roof height of thirty nine feet
Then	T	$= 0.035(39)^{3/4}$
		$= 0.546$ seconds

93

The force coefficient is given by UBC Formula (28–2) as

	C	$= 1.25S/T^{2/3}$
where	S	= 1.5 as given
Then	C	$= 1.25 \times 1.5/(0.546)^{2/3}$
		= 2.81
		> 2.75 hence use the upper limit of
	C	= 2.75

The base shear is given by UBC Formula (28–1) as

	V	$= (ZIC/R_w)W$
where	Z	= 0.4 for zone 4 from UBC Table 16–I
	I	= 1.0 from UBC Table 16–K for a standard occupancy structure
	C	= 2.75 as calculated
	R_w	= 12 from UBC Table 16–N for a steel moment–resisting space frame
	W	= total dead load
		= 700 + 850 + 850
		= 2400 kips
Then	V	$= (0.4 \times 1 \times 2.75/12)2400$
		$= 0.092 \times 2400$
		= 220 kips

1b. BASE SHEAR DISTRIBUTION

The base shear is distributed over the height of the structure in accordance with UBC Formula (28–8)

	F_x	$= (V - F_t)w_xh_x/\Sigma w_ih_i$
where	F_x	= the lateral force at level x
	V	= the base shear of 220 kips, as calculated
	F_t	= the additional concentrated force at roof level
		= 0, as T is less than 0.7 seconds
	w_x	= the structure dead load at level x
	h_x	= the height in feet above the base to level x
	Σw_ih_i	= the sum of the product $w \times h$ at all levels
		= 63,000 kips feet, from the Table below
Then	F_x	$= Vw_xh_x/\Sigma w_ih_i$
		$= 220w_xh_x/63,000$
		$= 0.00349w_xh_x$

and the force at each level, and the shear at each story V_x is shown in the following Table.

Level	w_x	h_x	$w_x h_x$	F_x	V_x
Roof	700	39	27,300	95.33	–
3rd Floor	850	27	22,950	80.14	95.33
2nd Floor	850	15	12,750	44.53	175.47
1st Floor	–	–	–	–	220.00
Total	2400	–	63,000	220.00	–

1c. DIAPHRAGM FORCE

The diaphragm force[6] at level x is given by UBC Formula (31-1) as

$$F_{px} = w_{px} \Sigma F_i / \Sigma w_i$$

where F_{px} = the diaphragm force at level x

w_{px} = weight of the diaphragm and tributary elements at level x

ΣF_i = the sum of the lateral forces at level x and above

Σw_i = the sum of the structure dead load w_x at level x and above

and $F_{px} \leq 0.75 ZIw_{px} = 0.30 w_{px}$

$F_{px} \geq 0.35 ZIw_{px} = 0.14 w_{px}$

Assuming that the weights of the walls are negligible compared with the diaphragm weight at each level, the weight tributary to each diaphragm is identical with the structure dead load at that level and, hence

$$w_{px} = w_x$$

The diaphragm force at each level is shown in the Table below.

Level	F_x	ΣF_i	w_x	Σw_i	$\Sigma F_i / \Sigma w_i$	w_{px}	F_{px}
Roof	95.33	95.33	700	700	0.14 Minimum	700	98
3rd Floor	80.14	175.47	850	1550	0.14 Minimum	850	119
2nd Floor	44.53	220.00	850	2400	0.14 Minimum	850	119

2a and 2b. DYNAMIC METHOD

The base shear and story forces may be obtained by the method of response spectrum analysis given in SEAOC Appendix 1F or in other reference texts[10,11]. For a two-dimensional structure, the total number of node points equals the number of stories. Each node is located at a floor level and has one degree of freedom in the horizontal direction.

For a given mode of vibration, the participation factor is defined by

$$P = \Sigma M_i \phi_i / M$$

where
M_i = mass at floor level i
ϕ_i = mode shape component for node point i for the given mode
M = modal mass
$= \Sigma M_i \phi_i^2$

and the summation extends over all the nodes in the structure.

The effective mass is defined by

$$M^E = \Sigma M_i \phi_i)^2 / \Sigma M_i \phi_i^2$$
$$= P \Sigma M_i \phi_i$$
$$= (\Sigma M_i \phi_i)^2 / M$$
$$= P^2 M$$

Similarly the effective weight is defined by

$$W^E = (\Sigma W_i \phi_i)^2 / \Sigma W_i \phi_i^2$$

where
W_i = weight at floor level i

The peak acceleration at a node is defined by

$$\ddot{x}_i = \phi_i P S_a$$

where
S_a = spectral acceleration corresponding to the natural period T for the given mode
$= 2\pi S_v / T$ inches per second^2

and
S_v = spectral velocity

The lateral force at a node is given by

$$F_i = M_i \ddot{x}_i$$
$$= M_i \phi_i P S_a$$

The total base shear is given by

$$V = \Sigma F_i$$
$$= P S_a \Sigma M_i \phi_i$$
$$= P^2 M S_a$$
$$= M^E S_a$$
$$= W^E S_a / g$$

where
g = acceleration due to gravity
$= 386$ inches per sec^2

The lateral force at a node may also be determined by distributing the base shear over the node points as

$$F_i = M_i \phi_i P S_a$$
$$= (M_i \phi_i / PM) V$$
$$= (M_i \phi_i / \Sigma M_i \phi_i) V$$
$$= (W_i \phi_i / \Sigma W_i \phi_i) V$$

FIRST MODE

The natural period of the first mode is given as

$$T \quad = 1.5 \text{ seconds}$$
$$> 0.65 \text{ seconds}$$

Hence, from the problem statement, the spectral velocity is

$$S_v \quad = 12 \text{ inches per second}$$

and the corresponding spectral acceleration is

$$S_a \quad = 2\pi S_v/T$$
$$= 2\pi \times 12/1.5$$
$$= 50.27 \text{ inches per second}^2$$
$$= 0.13g$$

From the Table below, where the weight at each floor level is used for convenience, the effective weight is given by

$$W^E \quad = (\Sigma W_i \phi_i)^2 / \Sigma W_i \phi_i^2$$
$$= 1488^2/1{,}141$$
$$= 1940 \text{ kips}$$

The base shear is given by

$$V \quad = W^E S_a/g$$
$$= 1940 \times 0.13g/g$$
$$= 252 \text{ kips}$$

The lateral force at each node is given by

$$F_i \quad = (W_i \phi_i / \Sigma W_i \phi_i) V$$
$$= (W_i \phi_i / 1488) 252$$
$$= 0.169 W_i \phi_i$$

and the force at each level and the shear at each story V_i is shown in the following Table.

Level	W_i	ϕ_i	$W_i \phi_i$	$W_i \phi_i^2$	F_i	V_i
Roof	700	1.000	700	700	118.6	–
3rd Floor	850	0.675	574	387	97.2	118.6
2nd Floor	850	0.252	214	54	36.2	215.8
1st Floor	–	–	–	–	–	252.0
Total	–	–	1488	1141	252.0	–

SECOND MODE

The natural period of the second mode is given as

$$T = 0.64 \text{ seconds}$$
$$< 0.65 \text{ seconds}$$

Hence, from the problem statement, the spectral acceleration is

$$S_a = 0.3g$$

From the Table below the effective weight is given by

$$W^E = 619^2/854$$
$$= 449 \text{ kips}$$

The base shear is given by

$$V = 449 \times 0.3g/g$$
$$= 134.7 \text{ kips}$$

The lateral force at each node is given by

$$F_i = (W_i\phi_i/619)134.7$$
$$= 0.2176W_i\phi_i$$

and the force at each level and the shear at each story V_i is shown in the Table below.

Level	W_i	ϕ_i	$W_i\phi_i$	$W_i\phi_i^2$	F_i	V_i
Roof	700	1.000	700	700	152.3	–
3rd Floor	850	0.250	212	53	46.1	152.3
2nd Floor	850	–0.345	–293	101	–63.7	198.4
1st Floor	–	–	–	–	–	134.7
Total	–	–	619	854	134.7	–

2c. COMBINED STORY FORCE

As a percentage of the total structural weight, the sum of the effective weights for the first two modes is given by

$$100(W_1^E + W_2^E)/W = 100(1940 + 449)/2400$$
$$= 99.5 \text{ percent}$$
$$> 90 \text{ percent}$$

Hence, combining the first two modes ensures that a minimum of 90 percent of the structural mass participates in the determination of the response parameters and UBC Section 1629.5.1 is satisfied. The combined force at each level for the two modes may be obtained by using the square-root-of-the-sum-of-the-squares method as detailed in SEAOC commentary Section 1F.5.b. This is acceptable for two-dimensional structures when the ratio of the periods of any higher mode to any lower mode is 0.75 or less. The combined force at level i is given by

$$F_{ci} = (F_{1i}^2 + F_{2i}^2)^{1/2}$$

where
F_{1i} = lateral force at level i for the first mode
F_{2i} = lateral force at level i for the second mode

The combined force at each level is shown in the Table below.

Level	F_{1i}	F_{2i}	F_{ci}
Roof	118.6	152.3	193
3rd Floor	97.2	46.1	108
2nd Floor	36.2	-63.7	73
Base	252.0	134.7	286

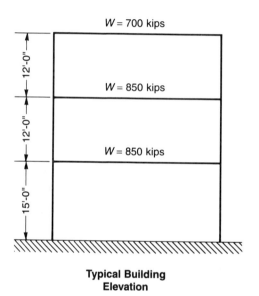

**Typical Building
Elevation**

LEVEL	MODE NUMBER	
	1(ϕ_1)	2(ϕ_2)
	MODE SHAPES	
Roof	1	1
3rd Floor	0.675	0.25
2nd Floor	0.252	−0.345
T = PERIOD (sec.)	1.5	0.64

**Building Mode
Shapes and Periods**

FIGURE 89A–3

STRUCTURAL ENGINEER EXAMINATION 1988 ═══════════════════════════

PROBLEM A-6 – WT. 4.0 POINTS

GIVEN: A portion of a canopy structure is to be constructed over the ambulance entrance to a hospital. Its mathematical model is shown in Figure "A". The beam can be considered infinitely rigid and the columns axially inextensible implying a deformation pattern as shown by the dashed lines.

CRITERIA: Response Spectrum : Design pseudo-velocity (S_v) response spectrum as shown in Figure "B".

Material Properties:
- A36 steel

Frame Properties:
- Weight   = W = 926 kips
- Lateral stiffness   = k = 148 kips per inch
- Viscous damping coefficient   = c = 1.88 kips-seconds per inch

Seismic Parameters:
- Seismic zone 3
- Response factor   = R_w = 4
- Site coefficient   = S = 1.5
- Gravitational acceleration   = g = 386 inches per sec^2

Assumptions:
- Frame behaves as a single-degree-of-freedom system.

REQUIRED: 1. Calculate the moment of inertia I for the columns that corresponds to the given lateral stiffness.
2. Calculate the frame's natural period of vibration using UBC Formula (28-5).
3. Calculate the UBC seismic base shear.
4. Calculate the exact (theoretical) undamped natural circular frequency and period of vibration.
5. Calculate the damping ratio.
6. From the response spectrum, determine the pseudo-velocity.
7. Calculate the corresponding pseudo-acceleration and the maximum relative displacement.
8. Calculate the resulting maximum base shear.

SOLUTION

1. COLUMN INERTIA

Since the frame stiffness is 148 kips per inch, the stiffness of each column is

$$k \quad = 148/2$$
$$= 74 \text{ kips per inch.}$$

The moment of inertia of each column, which is considered fixed at both ends, is given by

$$I \quad = kL^3/12E$$
$$= 74 \times 12^3 \times 1,728/12 \times 29,000$$
$$= 635 \text{ in}^4$$

2. PERIOD OF VIBRATION

The canopy may be considered a nonbuilding structure and, in accordance with UBC Section 1632.1.4, the fundamental period may be obtained by means of the Rayleigh method. From UBC Formula (28–5), the fundamental period for a multiple-degree-of-freedom system is

$$T \quad = 2\pi(\Sigma w_i\delta_i^2/g\Sigma f_i\delta_i)^{1/2}$$

For a single-degree-of-freedom system of weight W with a lateral force F producing a lateral displacement δ this formula reduces to

$$T \quad = 2\pi(W\delta^2/gF\delta)^{1/2}$$
$$= 2\pi(W\delta^2/gk\delta^2)^{1/2}$$
$$= 0.32(W/k)^{1/2}$$
$$= 0.32 \times (926/148)^{1/2}$$
$$= 0.80 \text{ seconds}$$

3. BASE SHEAR

For a nonbuilding structure, as defined in UBC Section 1632.5, the force coefficient is given by UBC Formula (28–2) as

$$C = 1.25S/T^{2/3}$$

where the site coefficient S is 1.5. Thus, the force coefficient is

$$C = 1.25 \times 1.5/(0.8)^{2/3}$$
$$= 2.18$$

which is less than 2.75 as required by UBC Section 1628.2.1 and is greater than $0.4R_w$ as required by UBC Section 1632.5.1. From UBC Formula (28–1) the base shear is

	V	$= (ZIC/R_w)W$
where	Z	$= 0.3$ for zone 3 from UBC Table 16–I
	I	$= 1.25$ for an essential facility as defined in UBC Table 16–K
	C	$= 2.18$ as calculated
	R_w	$= 4$ as given
and	W	$= 926$ kips as given
Then	V	$= (0.3 \times 1.25 \times 2.18/4)926$
		$= 0.204 \times 926$
		$= 189.25$ kips

4. UNDAMPED CIRCULAR FREQUENCY

The undamped circular frequency is given by[10,11]

	ω	$= (k/m)^{1/2}$
where	m	$= 926/386$ is the mass of the structure.
Then	ω	$= (148 \times 386/926)^{1/2}$
		$= 7.85$ radians per second

The period of vibration is

T	$= 2\pi/\omega$
	$= 2 \times 3.142/7.85$
	$= 0.80$ seconds

5. DAMPING RATIO

The damping ratio is given by

$$\xi = c/2m\omega$$

where   c = damping coefficient = 1.88 kips–seconds per inch, as given.

Then ξ = 1.88 × 386/(2 × 926 × 7.85)

         = 0.05

         = 5%

6. PSEUDO-VELOCITY

From the response spectra, for a period of 0.8 seconds and a damping ratio of 5%, the pseudo–velocity is obtained as

$$S_v = 0.5 \text{ feet per second}$$

7. PSEUDO-ACCELERATION

The pseudo–acceleration is given by

$$S_a = \omega S_v$$

         = 7.85 × 0.5

         = 3.93 feet per second per second

Assuming elastic behavior, the maximum relative displacement is given by

$$S_d = S_v/\omega$$

         = 0.5/7.85

         = 0.0637 feet

         = 0.764 inches

8. BASE SHEAR

Assuming elastic behavior, the maximum base shear is given by

$$V = kS_d$$

         = 148 × 0.764

         = 113 kips

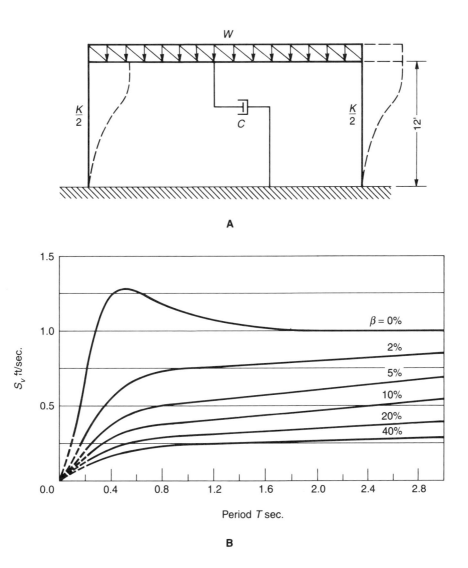

A

B

FIGURE 88A-6

STRUCTURAL ENGINEER EXAMINATION – 1982

PROBLEM A–6 – WT. 4.0 POINTS

GIVEN: The figure represents a three–story building. The effective dead loads are shown on each floor.

CRITERIA: The following dynamic properties of the plane frame are given:

Eigenvectors

$$[\Phi] = \begin{bmatrix} 1.680 & -1.208 & -0.714 \\ 1.220 & 0.704 & 1.697 \\ 0.572 & 1.385 & -0.984 \end{bmatrix}$$

$$[\Phi]^T[M][\Phi] = [I]$$

Eigenvalues

$$\{\omega\} = \begin{bmatrix} 8.77 \\ 25.18 \\ 48.13 \end{bmatrix} \text{ radians per second}$$

REQUIRED: a) Compute the participation factors
b) How can you check that your participation factors are correct?
c) Calculate the displacements of each floor based on the spectra given.
d) Calculate the interstory drift for each floor.

SOLUTION

a. and b. PARTICIPATION FACTORS

The participation factors for a multiple-degree-of-freedom system, are defined in matrix notation by[10]

$$\{P\} = [\Phi]^T[M]\{1\}/[\Phi]^T[M][\Phi]$$

where
- $\{P\}$ = column vector of participation factors for all modes considered
- $[\Phi]$ = mode shape matrix or eigenvectors
- $\{1\}$ = column vector of ones
- $[M]$ = diagonal matrix of lumped masses or mass matrix

and
- $[\Phi]^T[M][\Phi]$ = modal mass matrix
  - = $[I]$ as given in the problem statement
  - = identity matrix

Hence the matrix of eigenvectors is normalized and the column vector of participation factors is given by

$$\{P\} = [\Phi]^T[M]\{1\}$$

The mass matrix is

$$[M] = \begin{bmatrix} 80 & 0 & 0 \\ 0 & 80 & 0 \\ 0 & 0 & 120 \end{bmatrix} \; 1/g$$

Executing the necessary matrix operations gives the column vector of participation factors for the three modes as

$$\{P\} = \begin{bmatrix} 301.44 \\ 125.88 \\ -39.44 \end{bmatrix} \; 1/g$$

Summing the product of the participation factors and the eigenvectors for the first node gives

$$\Sigma P_j \phi_{1j} = (301.44 \times 1.608 - 125.88 \times 1.208 + 39.44 \times 0.714)/386$$
$$= 382.52/386$$
$$\approx 1.0$$

Hence the participation factors are correct[12]

c. FLOOR DISPLACEMENTS

The natural periods for each of the three modes is obtained from the given eigenvalues using the expression

and
- $T_n = 2\pi/\omega_n$
- $T_1 = 2\pi/8.77 = 0.717$ seconds
- $T_2 = 2\pi/25.18 = 0.250$
- $T_3 = 2\pi/48.13 = 0.131$

The spectral accelerations for each of the three modes are obtained from the response curve given and are

$$S_{a1} = 0.17g$$
$$S_{a2} = 0.80g$$
$$S_{a3} = 0.40g$$

The matrix of actual node displacements is defined in matrix notation by

$$[x] = [\Phi][P][S_d]$$
$$= [\Phi][P][S_a][\omega^2]^{-1}$$

where
$[P]$ = diagonal matrix of participation factors
$[S_d]$ = diagonal matrix of spectral displacements
$[S_a]$ = diagonal matrix of spectral accelerations
$[\omega^2]$ = diagonal matrix of squared modal frequencies

Executing the necessary matrix operations gives the matrix of node displacements for the three modes as

$$[x] = \begin{bmatrix} 1.126 & -0.192 & 0.0049 \\ 0.813 & 0.112 & -0.0115 \\ 0.381 & 0.220 & 0.0067 \end{bmatrix}$$

The total displacements at each node are obtained as the absolute value of the square-root-of-the-sum-of-the-squares for each row vector, as recommended in SEAOC commentary Section 1F.5.b, and are given by the column vector

$$[x_c] = \begin{bmatrix} 1.142 \\ 0.821 \\ 0.440 \end{bmatrix}$$

d. INTERSTORY DRIFT

The interstory drifts for the three modes are obtained by subtracting the row vector of node displacements for each node from the row vector of node displacements for the node above. The matrix of story drifts is then

$$[\Delta] = \begin{bmatrix} 0.313 & -0.304 & 0.0164 \\ 0.432 & -0.108 & -0.0182 \\ 0.381 & 0.220 & 0.0067 \end{bmatrix}$$

The interstory drifts for the combined displacements are similarly obtained as

$$[\Delta_c] = \begin{bmatrix} 0.321 \\ 0.381 \\ 0.440 \end{bmatrix}$$

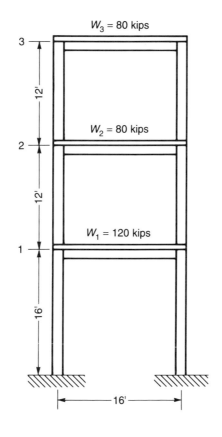

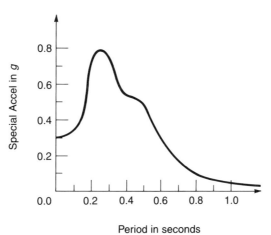

FIGURE 82A–6

STRUCTURAL ENGINEER EXAMINATION – 1974 ═══════════

PROBLEM A–6 – WT. 5.0 POINTS

GIVEN;

A two-story steel frame structure is shown. Consider the structure to have 7% damping. Each story is known to deflect 0.1 inch under a 10 kip story shear.

REQUIRED:

1. Determine the mathematical model for dynamic analysis, and summarize this in a sketch indicating story masses and stiffnesses.
2. Sketch the approximate 1st mode shape, and calculate the fundamental period of vibration of the mathematical model. Do not use UBC equations to calculate the period.
3. For the first mode assuming $T_1 = 0.5$ second, and that the mode shape is $\phi_2 = 1.0$ and $\phi_1 = .67$, calculate the 1st mode story forces.
4. What is the peak ground acceleration?

SOLUTION

DEAD LOAD

The dead load tributary to the roof level is obtained by summing the contributions from the roof self weight, the columns and the walls. Thus,

Roof	$= 0.2 \times 20 \times 20$	=	80.0 kips
Four walls	$= 0.01 \times 20 \times (12/2) \times 4$	=	4.8
Four columns	$= 0.05 \times (12/2) \times 4$	=	1.2
Total	$= W_2$	=	86.0

The dead load tributary to the second floor is

Roof	$= 0.2 \times 20 \times 20$	=	80.0 kips
Four walls	$= 0.01 \times 20 \times 12 \times 4$	=	9.6
Four columns	$= 0.05 \times 12 \times 4$	=	2.4
Total	$= W_1$	=	92.0

STORY STIFFNESS

The stiffness of each story is the shear force required to produce a unit displacement of that story and is given by

$$K_1 = K_2 \quad = \text{story shear/displacement}$$
$$= 10/0.1$$
$$= 100 \text{ kips per inch}$$

1. DYNAMIC MODEL

The two-story frame may be considered a shear structure with all mass concentrated at the rigid floor slabs and having one degree of freedom, a horizontal translation, at each floor slab. The dynamic model is shown in Figure 74A-6.

2. MODE SHAPE AND PERIOD

The dynamic behavior of multi-story frame structures may be analyzed by the application of Newton's second law of motion to the dynamic model described above. The shear at any story equals the product of the story stiffness and the drift of that story. The inertia force at any floor level equals the increment of shear force at that level. This may be expressed mathematically for a particular mode of vibration by

$$M_i \omega^2 x_i \quad = K_i \Delta_i - K_{i+1} \Delta_{i+1}$$

where
$$M_i \quad = \text{mass at floor level i}$$
$$\omega \quad = \text{natural frequency of the particular mode}$$
$$x_i \quad = \text{horizontal displacement of floor level i}$$
$$K_i \quad = \text{stiffness of story i}$$
$$\Delta_i \quad = \text{drift of story i}$$
$$\quad = x_i - x_{i-1}$$
$$K_i \Delta_i \quad = \text{shear force at story i}$$

An adaptation[13] of the Holzer iteration technique[14] may be used to obtain a solution. This consists of assuming an initial mode shape, with unit displacement at the top. From this is calculated the inertia, or incremental shear force, and the corresponding story shear, drift and floor displacement in terms of ω^2. Dividing these displacement values by the displacement at the top of the structure gives a revised mode shape. This revised mode shape may be used to provide a new initial mode shape and the iteration procedure repeated until agreement is obtained between the revised and initial mode shapes.

The procedure is illustrated in the following Table with the initial mode shape defined by

$$x_2 = 1.0$$
$$x_1 = 0.67$$

Floor Level i	Story	Floor Mass M_i	Story Stiff K_i	Initial Mode x_i	Incrm Shear $M_i\omega^2 x_i$	Story Shear $K_i\Delta_i$	Story Drift Δ_i	Floor Disp $\Sigma\Delta_i$	Revised Mode x_i
2		86/g		1.00	$86\omega^2/g$			$2.34\omega^2/g$	1.00
	2		100			$86\omega^2/g$	$0.86\omega^2/g$		
1		92/g		0.67	$62\omega^2/g$			$1.48\omega^2/g$	0.63
	1		100			$148\omega^2/g$	$1.48\omega^2/g$		

The iterative procedure is now repeated in the Table below with the initial mode shape defined by

$$x_2 = 1.00$$
$$x_1 = 0.63$$

Floor Level i	Story	Floor Mass M_i	Story Stiff K_i	Initial Mode x_i	Incrm Shear $M_i\omega^2 x_i$	Story Shear $K_i\Delta_i$	Story Drift Δ_i	Floor Disp $\Sigma\Delta_i$	Revised Mode x_i
2		86/g		1.00	$86\omega^2/g$			$2.30\omega^2/g$	1.00
	2		100			$86\omega^2/g$	$0.86\omega^2/g$		
1		92/g		0.63	$58\omega^2/g$			$1.44\omega^2/g$	0.63
	1		100			$144\omega^2/g$	$1.44\omega^2/g$		

The revised mode shape is identical to the initial shape and another iteration is unnecessary. The final mode shape is shown in Figure 74A-6.

The natural circular frequency of the first mode is obtained by equating the final value of the largest displacement component to its initial value. Thus,

$$2.30\omega^2/g = 1.00$$
$$\omega = (386/2.30)^{1/2}$$
$$= 12.96 \text{ radians per second.}$$

The fundamental period is given by

$$T = 2\pi/\omega$$
$$= 0.49 \text{ seconds}$$

3. LATERAL FORCES

The natural period for the first mode is given as

$$T = 0.5 \text{ seconds}$$

and, for a damping coefficient of seven percent, the corresponding spectral acceleration is obtained from the response spectra as

$$S_a = 0.2g$$

From the Table below, the participation factor is obtained as

$$P = \Sigma W_i \phi_i / \Sigma W_i \phi_i^2$$

where $\quad W_i$ = weight at floor level i

$\quad\quad\quad\quad \phi_i$ = mode shape component for node point i for the first mode

Then, $\quad\quad P = 144/123$

$$= 1.17$$

The lateral force at a node is given by

$$F_i = W_i \phi_i P S_a / g$$
$$= W_i \phi_i \times 1.17 \times 0.2g/g$$
$$= 0.234 W_i \phi_i$$

and the force at each level and the shear at each story, V_i, is shown in the Table below

Level	W_i	ϕ_i	$W_i\phi_i$	$W_i\phi_i^2$	F_i	V_i
Roof	86	1.00	86	86	20.1	–
2nd Floor	92	0.63	58	37	13.6	20.1
1st Floor	–	–	–	–	–	33.7
Total	–	–	144	123	33.7	–

4. PEAK GROUND ACCELERATION

The peak ground acceleration occurs at time $T = 0$ and is obtained from the response spectra as

$$S_a = 0.2g$$

113

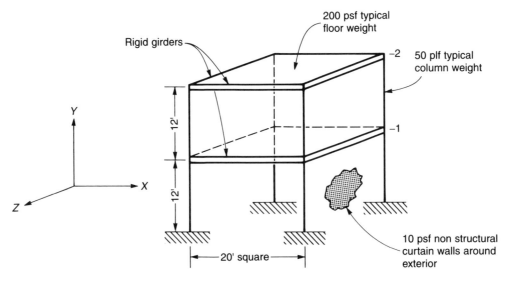

Symmetrical Structure

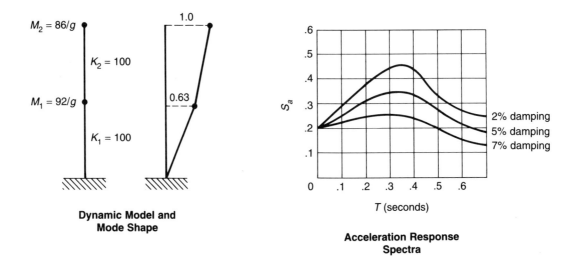

**Dynamic Model and
Mode Shape**

**Acceleration Response
Spectra**

FIGURE 74A-6

114

STRUCTURAL ENGINEER REGISTRATION

CHAPTER 1

GENERAL STRUCTURAL PRINCIPLES AND SEISMIC DESIGN

SECTION 1.5

MULTI-STORY STRUCTURES

PROBLEMS:

1990 A-2
1988 A-2
April 1989 Seismic Principles Examination 301B

STRUCTURAL ENGINEER EXAMINATION – 1990 ══════════════════

PROBLEM A–2 – Wt. 7.0 POINTS

GIVEN: A five story frame building with special moment-resisting frames in the north-south direction and concentric braced frames in the east-west direction.

CRITERIA: Seismic zone 4
             Importance factor $I = 1.0$
             Soil type S3

             Assumptions:
             • Rigid Diaphragms
             • Ignore torsion for the distribution of shear to the frames
             • Ignore orthogonal effects in calculating seismic axial forces
             • At every level, member sizes are identical in each direction
             • See Figures 1 through 6 for assumed floor dead load, dimensions, frame shear strength and frame stiffness

REQUIRED: 1. Determine the seismic base shear in each direction.
           2. Point out four Vertical and one Plan Irregularities for this building. For each type of irregularity, briefly discuss additional code requirements and procedures that shall be met for this building.
           3. For the column B2 between the ground and first level, determine the total design axial load combinations for dead, live and east-west seismic load per UBC Section 1603.6 and UBC Section 1628.7.2. The unfactored gravity load for this column is:
           Dead load   = 220 kips,   reduced live load   = 100 kips.

SOLUTION

1. SEISMIC BASE SHEAR

EAST-WEST DIRECTION:

Assume the structure acts as a building frame system in the east-west direction. From the UBC Section 1625, the structure period is given by UBC Formula (28-3) as

	T	$= C_t(h_n)^{3/4}$
where	C_t	$= 0.02$ for all other buildings
Thus	T	$= 0.02(60)^{3/4}$
		$= 0.43$ seconds

The force coefficient is given by UBC Formula (28-2) as

	C	$= 1.25S/T^{2/3}$
		$= 1.25 \times 1.5/(0.43)^{2/3}$
		$= 3.29$
		> 2.75 hence use the upper limit of
	C	$= 2.75$

The seismic dead load assuming no storage occupancies, partition loads, snow loads or permanent equipment loads is given by

	W	$= 1000 + 4 \times 500$
		$= 3000$ kips

The design base shear is given by UBC Formula (28-1) as

	V	$= (ZIC/R_w)W$
where	R_w	$= 8$ from UBC Table 16-N for a building frame system using concentric braced steel frames and assuming that the bracing carries only seismic loads
and	C/R_w	> 0.075 as is required by UBC Section 1628.2.1
Thus	V	$= (0.4 \times 1 \times 2.75/8)3000$
		$= 0.138 \times 3000$
		$= 414$ kips

NORTH–SOUTH DIRECTION:

From UBC Formula (28–3), the structure period is given by

$$T = C_t(h_n)^{3/4}$$

where C_t = 0.035 for a steel moment – resisting frame

h_n = 60 feet roof height

Thus T = $0.035(60)^{3/4}$

= 0.755 seconds

The force coefficient is given by UBC Formula (28–2) as

$$C = 1.25S/T^{2/3}$$

where S = 1.5 for soil type S3 from UBC Table 16–J

Thus C = $1.25 \times 1.5/(0.755)^{2/3}$

= 2.26

< 2.75 as is required by UBC Section 1628.2.1

The design base shear is given by UBC Formula (28–1) as

$$V = (ZIC/R_W)W$$

where Z = 0.4 for zone 4 from UBC Table 16–I

I = 1.0 for standard occupancy from UBC Table 16–K

R_W = 12 for a special moment-resisting frame from UBC Table 16–N

and C/R_W = 2.26/12

= 0.188

> 0.075 as is required by UBC Section 1628.2.1

Thus V = $(0.4 \times 1 \times 2.26/12) 3000$

= 0.075×3000

= 225 kips

2. VERTICAL IRREGULARITIES

(i)   The stiffness of the second story in the north–south direction is

K_2 = 2×350

= 700 kips per inch

The stiffness of the first story in the north–south direction is

K_1 = $2 \times 142 + 2 \times 95$

= 474

< 70% $\times K_2$ kips per inch

Hence the first story constitutes a soft story and is considered a vertical irregularity Type 1 in UBC Table 16–L Since the structure does not exceed five stories and is less than 65 feet in height, UBC Section 1627.8.2.3 permits the structure to be designed using the static lateral force procedure. However, if the dynamic force procedure is used, UBC Section 1629.5.3 specifies that the design base shear used shall not be less than that calculated by the static force procedure. The ten percent reduction allowed for regular structures is not permitted for irregular structures.

In addition, UBC Section 1809.4 specifies that, for regular structures, the force F_t at the top of the structure determined from UBC Formula (28-7) may be omitted when determining the overturning moment at the base. For irregular structures, the force F_t shall be included when determining the overturning moment.

(ii)   The effective weight of the second story is

$$W_2 = 500 \text{ kips}$$

The effective weight of the first story is

$$W_1 = 1000 \text{ kips}$$
$$> 150\% \times W_1$$

Hence this constitutes a weight irregularity Type 2 in UBC Table 16-L. The additional code requirements are identical with those given for the vertical irregularity Type 1.

(iii)   The horizontal dimension of the moment-resisting frame in the second story is

$$L_2 = 48 \text{ feet}$$

The horizontal dimension of the moment-resisting frame in the first story is

$$L_1 = 96 \text{ feet}$$
$$> 130\% \times L_2$$

Hence this constitutes a vertical geometric irregularity Type C in Table 23-M. The additional code requirements are identical with those given for the vertical irregularity Type 1.

(iv)   The shear strength of the second story in the north-south direction is

$$S_2 = 2 \times 650$$
$$= 1300 \text{ kips}$$

The shear strength of the first story in the north-south direction is

$$S_1 = 2 \times 250 + 2 \times 150$$
$$= 800 \text{ kips}$$
$$< 80\% \times S_2$$

Hence the first story constitutes a weak story and is considered a vertical irregularity Type 5 in UBC Table 16-L. Since the structure exceeds two stories and is more than 30 feet in height, the first story is required by UBC Section 1627.9.1 to be designed for a lateral force of $3(R_w/8)$ $= 36/8 = 4.5$ times the normal design force in the north-south direction.

119

PLAN IRREGULARITY

The lateral force resisting system in the east–west direction above the first story consists of braced frames on grid lines 2 and 4. In the first story, the brace lines are on grid lines 1 and 5. This out–of–plane offset of the vertical elements constitutes a plan irregularity Type 4 in UBC Table 16–M. The UBC Section 1631.2.9.6 specifies that, for this irregularity, connection of diaphragms and collectors to the vertical elements shall be designed without considering the usual one–third increase in allowable stresses. In addition UBC Section 1628.7.2 requires that the first story columns B2, B4, D2 and D4, which support the discontinuous braced frames in story two and above, be specially designed and detailed.

3. COLUMN B2 DESIGN LOAD

For the east–west direction, the structure period is

$$T = 0.43 \text{ seconds}$$
$$< 0.7 \text{ seconds}$$

and the concentrated force at the top of the structure is

$$F_t = 0$$

in accordance with UBC Section 1628.4.

Then, the lateral force[9] at level x is given by UBC Formula (28–8) as

$$F_x = V w_x h_x / \Sigma w_i h_i$$

where
$$V = 414 \text{ kips}$$
$$w_x = \text{seismic dead load at level x}$$
$$h_x = \text{height of level x}$$

and the summation $\Sigma w_i h_i$ extends over all levels. The lateral forces, shears and moments at each level are obtained in the following Table

Level	w_x	h_x	$w_x h_x / 1000$	F_x	V_x	M_x
Roof	500	60	30	129.4	–	–
4th floor	500	48	24	103.5	129.4	1,553
3rd floor	500	36	18	77.6	232.9	4,348
2nd floor	500	24	12	51.75	310.5	8,074
1st floor	1000	12	12	51.75	362.25	12,421
Ground	–	–	–	–	414.0	17,389
Total	3000	–	96	414.0	–	–

Then
$$F_x = 414 w_x h_x / 96{,}000$$
$$= 0.00431 w_x h_x$$

The axial force in column B2 due to the seismic forces is obtained by taking moments at the first level.

Then P_E = 12,421/(2 × 48)
$\quad\quad$ = 129 kips

The roof area tributary to column B2 is

$\quad$ A = 24 × 24
$\quad\quad$ = 576 square feet

and the corresponding roof live load is obtained from UBC Table 16–C as

$\quad$ P_R = 576 × 16/1000
$\quad\quad$ = 9 kips

The axial forces in column B2 due to the floor live load and the dead loads are

$\quad$ P_L = 100 kips
$\quad$ P_D = 220 kips

Applying UBC Section 1603.6 the column design loads are

$\quad$ P = $P_D + P_L + P_R$
$\quad\quad$ = 220 + 100 + 9
$\quad\quad$ = 329 kips, with no increase allowed in the permissible stresses,

or $\quad$ P = $P_D + P_L + P_E$
$\quad\quad$ = 220 + 100 + 129
$\quad\quad$ = 449 kips, with the allowable stress increased by 1.33.

Applying UBC Section 1628.7.2, the column design loads are

$\quad$ P = $P_D + 0.8P_L + 3(R_w/8)P_E$
$\quad\quad$ = 220 + 0.8 × 100 + 3(8/8)129
$\quad\quad$ = 687 kips, with the allowable stress increased by 1.7,

or $\quad$ P = $0.85P_D \pm 3(R_w/8)P_E$
$\quad\quad$ = 0.85 × 220 ± 3(8/8)129
$\quad\quad$ = 574 or –200 kips, with the allowable stress increased by 1.7 =====

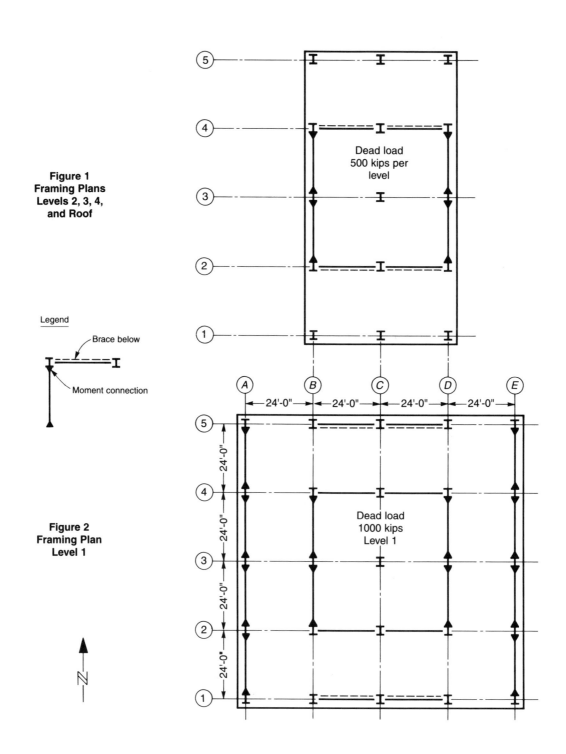

**Figure 1
Framing Plans
Levels 2, 3, 4,
and Roof**

Dead load
500 kips per
level

Legend

Brace below

Moment connection

**Figure 2
Framing Plan
Level 1**

Dead load
1000 kips
Level 1

24'-0" 24'-0" 24'-0" 24'-0"

N

FIGURE 90A−2(i)

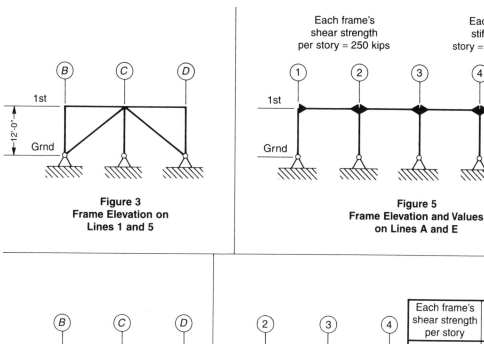

Figure 3
Frame Elevation on
Lines 1 and 5

Figure 5
Frame Elevation and Values
on Lines A and E

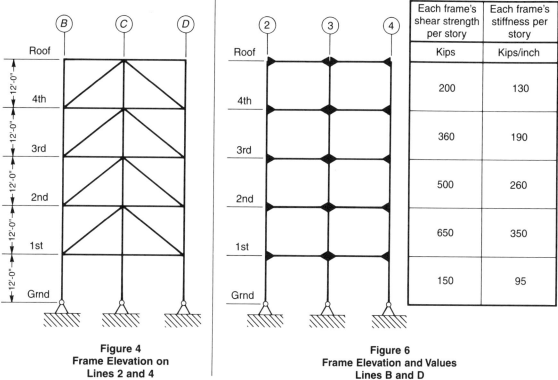

Figure 4
Frame Elevation on
Lines 2 and 4

Figure 6
Frame Elevation and Values
Lines B and D

FIGURE 90A−2(ii)

STRUCTURAL ENGINEER EXAMINATION – 1988

PROBLEM A–2 – WT. 5.0 POINTS

GIVEN: Plan and section of an office building as shown.

CRITERIA: Seismic zone 3, building period $T = 0.56$ seconds, site coefficient $S = 1.50$, importance factor $I = 1.00$, response factor $R_w = 6$. Weight of each floor and roof is 1100 kips each. Non–moment–resisting frame with braces in bays shown. The stiffness of each bracing along B and F is twice that of those along 2 and 3.

REQUIRED: 1. Determine the seismic force at each level.
              2. Determine the maximum seismic forces in members of the frame on line F. Neglect vertical loads.
              3. Check the adequacy of member Z.

SOLUTION

DEAD LOAD

The seismic dead load, assuming no storage occupancies, partition loads, snow loads, or permanent equipment loads, is given by

$$W = 3 \times 1100$$
$$= 3300 \text{ kips}$$

BASE SHEAR

The force coefficient is given by UBC Formula (28–2) as

$$C = 1.25S/T^{2/3}$$

where S = 1.5 for the site coefficient, as given

and T = 0.56 for the structure period, as given

Thus $C = 1.25 \times 1.5/(0.56)^{2/3}$
             = 2.77
             > 2.75 hence use the upper limit of
        C = 2.75

The seismic base shear is given by UBC Formula (28–1) as

$$V = (ZIC/R_w)W$$

where Z = 0.3 for zone 3 from UBC Table 16–I
          I = 1.0 for standard occupancy from UBC Table 16–K
          R_w = 6 for the response factor, as given
          W = 3300 kips for the total structure dead load, as calculated
          C = 2.75 for the force coefficient, as calculated

Then   V $= (0.3 \times 1 \times 2.75/6)3300$
                    $= 0.1375 \times 3300$
                    $= 454$ kips

1. BASE SHEAR DISTRIBUTION

The base shear is distributed over the height of the structure[9,11] in accordance with UBC Formula (28–8)

$$F_x = (V - F_t)\, w_x h_x / \Sigma w_i h_i$$

where F_x = the lateral force at level x
          V = the base shear of 454 kips, as calculated
          F_t = the additional concentrated force at roof level
             = 0, as T is less than 0.7 seconds
         w_x = the structure dead load at level x
         h_x = the height in feet above the base to level x
     $\Sigma w_i h_i$ = the sum of the product $w_x \times h_x$ at all levels
            = 92,400 kips feet, from the Table below

Then F_x = $V w_x h_x / \Sigma w_i h_i$
               = $454 w_x h_x / 92{,}400$
               = $0.00491 w_x h_x$ kips

In the Table below,
         F_x = the lateral force at level x
and V_x = the shear at story x

Level	w_x	h_x	$w_x h_x$	F_x	V_x
Roof	1100	42	46,200	227	–
3rd floor	1100	28	30,800	151	227
2nd floor	1100	14	15,400	76	378
1st floor	–	–	–	–	454
Total	3300	–	92,400	454	–

2. MEMBER FORCES

The roof and floor diaphragms may be considered rigid and, in determining the distribution of the base shear to the braced frames, torsional effects must be considered. Since the structure is symmetrical in plan, the center of mass and the center of rigidity coincide. However, to comply with UBC Section 1628.5 accidental eccentricity shall be allowed for by assuming that the center of mass is displaced a distance equal to five percent of the building dimension perpendicular to the direction of the seismic force. The maximum forces are produced in the frame on line F by a north–south seismic load when the center of mass is displaced to the east. Torsional irregularity does not exist and, in accordance with UBC Section 1628.6, further amplification of the torsional effects is not required.

Then, the accidental eccentricity is given by

$$e = 0.05 \times \text{east-west dimension}$$
$$= 0.05 \times 150$$
$$= 7.5 \text{ feet}$$

The polar moment of inertia of the frames is given by the sum of the products of the relative rigidity of each frame and its squared distance from the center of rigidity. Thus,

$$\Sigma r^2 R = 2 \times 2 \times 60^2 + 4 \times 1 \times 15^2$$
$$= 15,300 \text{ square feet}$$

The torsion acting on the structure at level x is given by

$$T_x = F_x e$$
$$= 7.5 F_x \text{ kip feet}$$

The torsional shear force acting on the frame on line F at level x is

$$F_T = T_x r_F R_F / \Sigma r^2 R$$

where
- r_F = the distance, 60 feet, of the frame on Line F from the center of rigidity
- R_F = the relative rigidity of the frame on line F = 2 as given
- $\Sigma r^2 R$ = the polar moment of inertia = 15,300 as calculated
- T_x = the applied torsion = $7.5 F_x$ as calculated

Then
$$F_T = 2 \times 60 \times 7.5 F_x / 15,300$$
$$= 0.0588 F_x \text{ kips}$$

The in-plane shear force acting on the frame on line F at level x is

$$F_S = F_x R_F / \Sigma R_y$$

where
- F_x = the lateral force at level x
- R_F = the relative rigidity of the frame on line F = 2 as given
- ΣR_y = sum of braced frame rigidities for frames in the north-south direction
$$= 2 + 2$$
$$= 4$$

Then
$$F_S = 2 F_x / 4$$
$$= 0.5 F_x \text{ kips}$$

The total force in the braced frame on line F is given by

$$F_F = F_S + F_T$$
$$= 0.5 F_x + 0.0588 F_x$$
$$= 0.5588 F_x$$

In the Table below,

- F_F = the lateral force at the indicated level
- V = the shear at the indicated story
- M = the moment at the indicated story
$$= \Sigma V H$$

where
- H = the height of the indicated story and the summation $\Sigma V H$ extends from the indicated story to all stories above.

126

The forces in the members of the frame on line F are obtained by resolving forces at the joints and assuming that the lateral force F_F is applied symmetrically on either side of the frame. The force in a beam is equal to half the corresponding story shear. The force in a diagonal equals the beam force times $\cos\theta$, where θ is the angle of inclination of the diagonal. The force in a column equals the corresponding story moment divided by the frame width. These values are given in the Table below and indicated in Figure 88A-2(ii).

Level	F_F	V	VH	M	V/2	$(V/2)\cos\theta$	M/B
Roof	127	–	–	–	–	–	–
3rd floor	84	127	1778	–	63.5	87	–
2nd floor	43	211	2954	1778	105.5	144	59
1st floor	–	254	3556	4732	127.0	174	158
Foundation	–	–	–	8288	–	–	276

3. CHECK MEMBER Z

MEMBER PROPERTIES

The section properties of structural tubing size $7 \times 7 \times \frac{1}{4}$ are given in AISC Table 3–41

$$
\begin{aligned}
A &= 6.59 \text{ square inches} \\
r &= 2.74 \text{ inches} \\
F_y &= 46 \text{ kips per square inch}
\end{aligned}
$$

The unbraced length of member Z, using center line dimensions, is

$$
\begin{aligned}
\ell &= H/\sin\theta \\
&= 14/\sin(43.03)° \\
&= 20.52 \text{ feet}
\end{aligned}
$$

The effective length factor for the brace, assuming hinged ends, is given by AISC Table C–C2.1 item (d) as

$$
K = 1.0
$$

BRACING CONFIGURATION

In accordance with UBC Section 2211.8.4 chevron bracing in an ordinary concentrically braced frame shall be designed for 1.5 times the prescribed seismic force. The required design force is, then

$$
\begin{aligned}
P_R &= 1.5 \times 87 \\
&= 131 \text{ kips}
\end{aligned}
$$

STRESS INCREASE

In accordance with UBC Section 1603.5, a one-third increase is allowed in stresses when designing for seismic forces. The equivalent design force, based on the allowable stress, is

$$
\begin{aligned}
P_E &= P_R/1.33 \\
&= 131/1.33 \\
&= 98 \text{ kips}
\end{aligned}
$$

The equivalent member stress, based on the gross sectional area, is

$$
\begin{aligned}
f &= P_E/A \\
&= 98/6.59 \\
&= 14.9 \text{ kips per square inch}
\end{aligned}
$$

and this stress may be either compressive or tensile.

TENSILE CAPACITY

Assuming that the brace connections are welded, the gross area of the member governs the tension capacity and the allowable tensile stress is given by UBC Section 2251 D1 as

$$
\begin{aligned}
F_t &= 0.6F_y \\
&= 0.6 \times 46 \\
&= 27.6 \text{ kips per square inch} \\
&> 14.9
\end{aligned}
$$

and the brace is adequate for the applied tensile force.

SLENDERNESS

As the building is located in seismic zone 3, the slenderness ratio of the brace is limited, by UBC Section 2211.8.2, to a maximum value of

$$
\begin{aligned}
\ell/r &= 720/(F_y)^{1/2} \\
&= 720/(46)^{1/2} \\
&= 106
\end{aligned}
$$

The actual slenderness ratio is

$$
\begin{aligned}
\ell/r &= 20.52 \times 12/2.74 \\
&= 90 \\
&< 106
\end{aligned}
$$

and the slenderness ratio of the brace is satisfactory.

LOCAL BUCKLING

The width to thickness ratio of square structural tubing is limited, by UBC Section 2251.7 Table B5.1, to a maximum value of

$$b/t = 238/(F_y)^{1/2}$$
$$= 238/(46)^{1/2}$$
$$= 35$$

The actual width to thickness ratio is

$$b/t = 7/0.25$$
$$= 28$$
$$< 35$$

and the width to thickness ratio of the brace is satisfactory.

DISTRIBUTION OF LATERAL FORCE

UBC Section 2211.8.2 stipulates that neither the sum of the horizontal components of the compressive member forces nor the sum of the horizontal components of the tensile member forces, along a line of bracing, shall exceed seventy percent of the applied lateral force. In each story of the building, both braces resist the lateral force equally and the braces are, therefore, satisfactory.

COMPRESSIVE CAPACITY

The effective slenderness ratio is given by

$$K\ell/r = 1.0 \times 90$$
$$= 90$$

The axial compressive stress permitted by UBC Section 2251 E2 is obtained from AISC Table C-50 as

$$F_a = 16.94 \text{ kips per square inch}$$

The stress reduction factor for a bracing member resisting seismic forces is determined from UBC Formula (11-5) as

$$\beta = 1/[1.0 + ([K\ell/r]/2C_c)]$$

where C_c = 111.6 as given by UBC Division IX Table 4

Hence
$$\beta = 1/[1.0 + 90/(2 \times 111.6)]$$
$$= 0.71$$

Then, the allowable stress for a bracing member resisting seismic forces in compression is determined from UBC Formula (11-4) as

$$F_{as} = \beta F_a$$
$$= 0.71 \times 16.94$$
$$= 12.07$$
$$< 14.9$$

and the brace is inadequate for the applied compressive force.

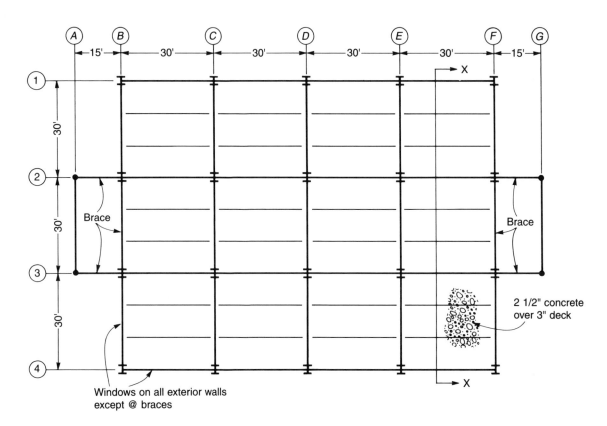

Plan of Building, 1" = 30'

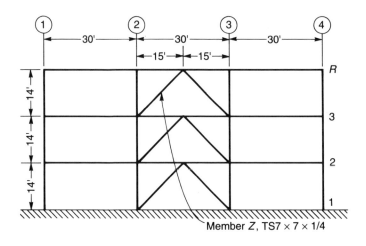

Section X—X, 1" = 30'

FIGURE 88A-2(i)

130

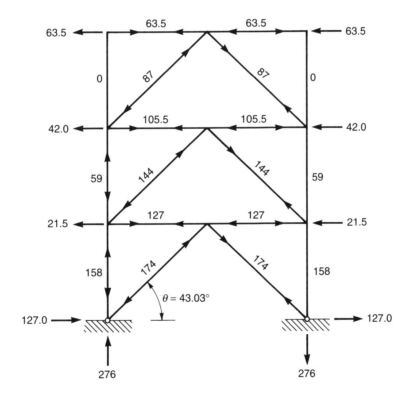

Member Forces

FIGURE 88A–2(ii)

131

SEISMIC PRINCIPLES EXAMINATION – APRIL 1989 ══════

301B. WT. 10 POINTS. (One Hour)

SITUATION: The figures show a building with exterior concrete masonry shear walls and flexible diaphragms.

CRITERIA: Roof dead load (including tributary loads from the walls) = 40 psf
Floor dead load (including tributary loads from the walls) = 60 psf
Importance factor I = 1.0
Seismic zone 4
Force factor C = 2.75
Response factor R_w = 6

REQUIREMENTS: (a) Wt. 2 pts.
Calculate the design base shear in the east–west direction.
(b) Wt. 4 pts.
Calculate the lateral force distribution to each level in the east–west direction.
(c) Wt. 4 pts.
Calculate the maximum diaphragm shear in level 2 in the east–west direction.

SOLUTION

DEAD LOAD

Assuming that there are no partition or storage loads, the dead loads at each level are given by

Roof load   = 40(120 × 60 + 110 × 20)/1000
= 40(7200 + 2200)/1000
= 376 kips

2nd floor load   = 60(7200 + 2200)/1000
= 564 kips

The total load W   = 376 + 564
= 940 kips

a. BASE SHEAR

The base shear is given by UBC Formula (28–1) as

	V	$= (ZIC/R_w)W$
where	Z	$= 0.4$ for zone 4 from UBC Table 16–I
	I	$= 1.0$ as given
	C	$= 2.75$ as given
	R_w	$= 6.0$ as given
Thus	V	$= (0.4 \times 1 \times 2.75/6) \, 940$
		$= 0.183 \times 940$
		$= 172$ kips

b. BASE SHEAR DISTRIBUTION

From UBC Section 1625, the structure period is given by Formula (28–3) as

	T	$= C_t(h_n)^{3/4}$
where	C_t	$= 0.02$ for all buildings other than moment-resisting frames and eccentric braced frames
	h_n	$=$ the roof height of thirty two feet
Thus	T	$= 0.02(32)^{3/4}$
		$= 0.27$ seconds
		< 0.7 seconds

Hence, in accordance with UBC Section 1628.4, the additional concentrated force at roof level is

$$F_t = 0$$

The base shear is distributed over the height of the structure[9,11] in accordance with UBC Formula (28–8)

	F_x	$= (V - F_t)w_x h_x / \Sigma w_i h_i$
where	F_x	$=$ the lateral force at level x
	V	$=$ the base shear of 172 kips, as calculated
	F_t	$= 0$ as calculated
	w_x	$=$ the structure dead load at level x
	h_x	$=$ the height in feet above the base to level x
	$\Sigma w_i h_i$	$=$ the sum of the product $w_x \times h_x$ at all levels
		$= 21{,}056$ kip feet, from the Table below
Then	F_x	$= V w_x h_x / \Sigma w_i h_i$
		$= 172 w_x h_x / 21{,}056$
		$= 0.00817 w_x h_x$

and the force F_x at each level, and the shear V_x at each story is shown in the Table below

Level	W_x	h_x	$w_x h_x$	F_x	V_x
Roof	376	32	12,032	98.29	–
2nd floor	564	16	9,024	73.71	98.29
1st floor	–	–	–	–	172.00
Total	940	–	21,056	172.00	–

c. DIAPHRAGM SHEAR

The diaphragm force at level x is given by UBC Formula (31–1) as

$$F_{px} = w_{px} \Sigma F_i / \Sigma w_i$$

where
F_{px} = the diaphragm force at level x
w_{px} = weight of the diaphragm and tributary elements at level x
ΣF_i = the sum of the lateral forces at level x and above
Σw_i = the sum of the structure dead load w_x at level x and above

and
$$F_{px} \leq 0.75 \, ZIw_{px} = 0.30 w_{px}$$
$$F_{px} \geq 0.35 \, ZIw_{px} = 0.14 w_{px}$$

Assuming that the weights of the walls are negligible compared with the diaphragm weight at each level, the weight tributary to each diaphragm is identical with the structure dead load at that level and, hence,

$$w_{px} = w_x$$

At the second floor level, the diaphragm force is

$$F_{p2} = w_{p2} \Sigma F_i / \Sigma w_i$$

where
F_{p2} = the diaphragm force at level 2
w_{p2} = w_2 = 564 kips, the structure dead load at the second floor
ΣF_i = V, the base shear
Σw_i = W, the total structure dead load

Then
$$F_{p2} = w_2 V / W$$
$$= 0.183 \times 564$$

and this lies within the allowable limits of

$$F_{p2} \leq 0.30 w_2$$
and
$$\geq 0.14 w_2$$
Thus
$$F_{p2} = 0.183 \times 564$$
$$= 103 \text{ kips}$$

134

EAST–WEST SEISMIC

The total diaphragm force at the second floor, for east–west seismic loads, is distributed to the floor areas A and B in proportion to their relative areas, as shown in Figure 89SA-2.

The area A is given by

$$A_A = 120 \times 60$$
$$= 7200 \text{ square feet}$$

and the associated diaphragm force is given by

$$F_{pA} = F_{p2} \times A_A/A_T$$

where A_T = total floor area = 9400 square feet

Then
$$F_{pA} = 103 \times 7200/9400$$
$$= 79 \text{ kips}$$

The diaphragm force associated with area B is given by

$$F_{pB} = 103 - 79$$
$$= 24 \text{ kips}$$

The reaction of the diaphragm force at the north shear wall is given by

$$R_N = F_{pA}/2 + F_{pB} \times 65/120$$
$$= 79/2 + 24 \times 65/120$$
$$= 52.6 \text{ kips}$$

and the unit shear in the diaphragm at the north wall is given by

$$v_N = 52.6 \times 1000/80$$
$$= 658 \text{ pounds per linear foot}$$

The reaction of the diaphragm force at the south shear wall is given by

$$R_S = F_{p2} - R_N$$
$$= 103 - 52.6$$
$$= 50.4 \text{ kips}$$

and the unit shear in the diaphragm at the south wall is given by

$$v_S = 50.4 \times 1000/60$$
$$= 840 \text{ pounds per linear foot}$$

and this value governs for east–west seismic loads at the second floor.

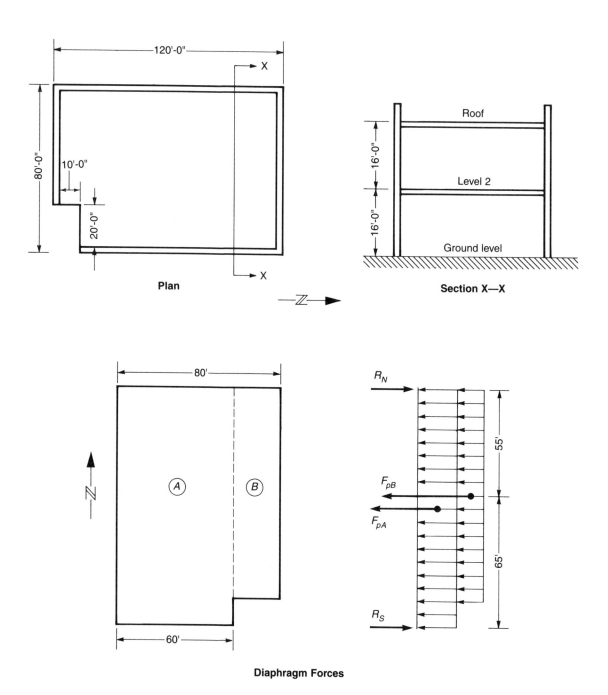

FIGURE 89SA–2

136

STRUCTURAL ENGINEER REGISTRATION

CHAPTER 1

GENERAL STRUCTURAL PRINCIPLES AND SEISMIC DESIGN

SECTION 1.6

STRUCTURAL ELEMENTS

PROBLEMS:

October 1989 Seismic Principles Examination 301

SEISMIC PRINCIPLES EXAMINATION – OCTOBER 1989 ═══════════

301 WT. 10 POINTS (One Hour)

SITUATION: The roof framing plan of a single story warehouse facility is illustrated in Figure 301A. Section A, in Figure 301B, is taken through the west wall.

CRITERIA: Seismic zone 3

Wall C_p	= 0.75
Parapet C_p	= 2.0

GIVEN:

Roof live load	= 20 psf
Roof dead load	= 11.0 psf
Wall weight	= 120 psf

(For purposes of this problem, wall openings are ignored.)

Reinforcing Steel	= ASTM A615, Grade 60
Masonry strength f'_m	= 1500 psi, solid grouted with special inspection

REQUIREMENTS:

(a) Determine the seismic loading to the roof diaphragm for both the east-west and north-south directions. Wt. 3.0 pts.

(b) Determine the required anchorage force between the west wall (Section A) and the roof for east-west seismic forces. Assume anchorage points at 4 feet on center. Wt. 2.0 pts.

(c) The wall anchor for east-West seismic forces is not shown in Section A. Provide a sketch to indicate an appropriate positive connection for the west wall anchorage. Wt. 1.0 pt.

(d) For seismic forces, determine:
1. The diaphragm boundary nailing at south wall. Wt. 1.0 pt.
2. The ledger anchor bolt requirements at west wall. (Neglect the reaction of beams located 20 feet on center). Wt. 2.0 pts.
3. The maximum chord force and minimum required chord reinforcement. Wt. 1.0 pt.

SOLUTION

DEAD LOAD

The wall dead load, tributary to the roof diaphragm, is obtained by taking moments about the top of the ground floor slab[15]. The dead loads are given by

Roof diaphragm	$= 11 \times 240 \times 120/1000$	= 316.8 kips
North wall	$= 120 \times 120 \times 24^2/(2 \times 20.67 \times 1000)$	= 200.6 kips
South wall		= 200.6 kips
East wall	$= 120 \times 240 \times 24^2/(2 \times 20.67 \times 1000)$	= 401.2 kips
West wall		= 401.2 kips

LATERAL FORCE

From UBC Section 1625, the structure period is given by UBC Formula (28-3) as

$$T = C_t(h_n)^{3/4}$$

where C_t = 0.02 for a shear wall type structure
and h_n = the diaphragm height of 20.67 feet
Then T = $0.02 (20.67)^{3/4}$
= 0.194 seconds

The force coefficient is given by UBC Formula (28-2) as

$$C = 1.25S/T^{2/3}$$

where S = 1.5 from UBC Table 16-J for a site with unknown soil properties
Then C = $1.25 \times 1.5/(0.194)^{2/3}$
= 5.6
> 2.75 hence use the upper limit of
C = 2.75

The seismic force acting on the structure is given by UBC Formula (28-1) as

$$V = (ZIC/R_w)W$$

Z = 0.3 for zone 3 from UBC Table 16-I
I = 1.0 from UBC Table 16-K for a standard occupancy structure.
C = 2.75 as calculated
R_w = 6 from UBC Table 16-N for a bearing wall system
W = structure dead load
Then V = $(0.3 \times 1 \times 2.75/6)W$
= 0.1375W

a. DIAPHRAGM SHEAR

The panelized wood roof system may be analyzed as a flexible diaphragm[16].

NORTH-SOUTH SEISMIC

The dead load tributary to the roof diaphragm in the north-south direction is due to the north wall, south wall and the roof diaphragm and is given by

$$W = 316.8 + 200.6 + 200.6$$
$$= 718 \text{ kips}$$

The seismic force on the roof diaphragm in the north-south direction is given by

$$V = 0.1375W$$
$$= 0.1375 \times 718$$
$$= 98.7 \text{ kips}$$

The unit shear in the diaphragm along the east and west walls is given by

$$v = V/(2L)$$
$$= 98.7 \times 1000/(2 \times 240)$$
$$= 206 \text{ pounds per linear foot}$$

EAST–WEST SEISMIC

The dead load tributary to the roof diaphragm in the east-west direction is due to the east wall, west wall, and the roof diaphragm and is given by

$$W = 316.8 + 401.2 + 401.2$$
$$= 1119.2 \text{ kips}$$

The seismic force on the roof diaphragm is the east-west direction is given by

$$V = 0.1375W$$
$$= 0.1375 \times 1119.2$$
$$= 154 \text{ kips}$$

The unit shear in the diaphragm along the north and south walls is given by

$$v = V/(2B)$$
$$= 154/(2 \times 120)$$
$$= 642 \text{ pounds per linear foot}$$

b. ANCHORAGE FORCE

The anchorage force is due to that portion of the wall dead load which is tributary to the roof diaphragm. This is obtained by taking moments about the top of the ground floor slab and is given by

$$W_p = 120 \times 24^2/(2 \times 20.67)$$
$$= 1672 \text{ pounds per linear foot}$$

The horizontal force factor C_p, to be used in determining the anchorage force, is that value applicable to the total wall system, and the value for a parapet only is not relevant[17]. Hence, the applicable value of the horizontal force factor is obtained from UBC Table 16-O, item 1.1.b. as

$$C_p = 0.75$$

The anchorage force between the west wall and the roof diaphragm is given by UBC Formula (30–1) as

$$F_p = ZI_pC_pW_p$$

where
$$Z = 0.3 \text{ as determined}$$
$$I_p = 1.0 \text{ as determined}$$
$$C_p = 0.75 \text{ as determined}$$
$$W_p = 1672 \text{ as calculated}$$

Then $\qquad F_p \qquad = 0.3 \times 1 \times 0.75 \times 1672$
$$= 376 \text{ pounds per linear foot}$$

and this value governs as it exceeds the minimum value of 200 pounds per linear foot specified in UBC Section 1611.

The force on each wall anchor, for a spacing of four feet, is given by

$$P \qquad = 4F_p$$
$$= 4 \times 376$$
$$= 1504 \text{ pounds}$$

c. ANCHORAGE DETAIL

The anchorage must provide a positive direct connection between the roof diaphragm and the wall. A typical detail utilizes a twelve gauge steel strap embedded in the grouted wall and attached to the subpurlin. The strap is hooked around the chord reinforcement in the wall and may be nailed to the top of the subpurlin, as indicated in NAVFAC Figure 5–33, or bolted to the side of the subpurlin[16]. The latter method is more suitable in this case as the $2\times$ subpurlin may tend to split with top nailing.

The allowable load for a ¾ inch diameter bolt in single shear, loaded parallel to the grain of a Douglas Fir–Larch member, 1.5 inches thick with a 0.105 inch thick steel side plate, is given by UBC Section 2336 Equation (36–3) as

$$Z_{\|} \qquad = k_1 D t_s F_{es}/3.6K_\theta$$
$$= 714 \text{ pounds}$$

(The allowable load tabulated in UBC Table 23–III–K for a ¼ inch steel side plate is $Z_{\|} = 800$ pounds.)

UBC Table 23–III–E, for two bolts in a row, gives a value for the group factor of

$$C_G \qquad = 1.00$$

UBC Section 2304.3.4, for seismic loads, gives a load duration factor of

$$C_D \qquad = 1.33$$

Hence, the number of ¾ inch diameter bolts required is

$$n \qquad = P/(Z_{\|} \times C_D \times C_G)$$
$$= 1504/(714 \times 1.33 \times 1.00) = 2 \text{ bolts.}$$

The required bolt spacing, in accordance with UBC Table 23–III–I, is four times diameter, or three inches, the required end distance, in accordance with UBC Table 23–III–H, is seven times diameter, or 5.25 inches, and the required edge distance, in accordance with UBC Section 2336.5.3, is 1.5 times diameter, or 1.125 inches. The angle of inclination of the strap must be such that the vertical component of the seismic force in the anchor does not exceed the available dead load, in this instance approximately $7°$.

The required anchorage detail is shown in Figure 89SO–1(i).

(d)1. NAILING REQUIREMENTS

The nail spacing required, at the south wall for east–west seismic forces, to provide a shear capacity of 652 pounds per linear foot using ½ inch Structural I plywood and ten penny nails with 1.625 inch penetration, may be obtained from UBC Table 23–I–J–1. For a case 2 loading condition, all edges blocked and nominal three inch wide framing members in the end bay, as indicated, the required nail spacing is

at diaphragm boundaries	= 2.5 inches
at all other edges	= 4 inches
at intermediate framing members	= 12 inches

and this provides a shear capacity of 720 pounds per liner foot.

(d)2. LEDGER ANCHOR BOLTS

The loading acting on the ledger at the west wall is due to the vertical dead load and live load on the roof diaphragm and the horizontal load due to the north–south seismic forces. In accordance with UBC Section 1603.6, the two loading combinations, dead load plus seismic load and dead load plus live load, must be investigated. For both the ledger and the masonry wall, a stress increase of one–third is allowed for the first loading combination, in accordance with UBC Section 1603.5. For the ledger, a twenty–five percent increase in stresses is allowed for the roof live load, in accordance with UBC Section 2304.3.4.

COMBINATION DEAD PLUS SEISMIC LOAD

The dead load acting on the ledger is

$$w = 11 \times 8/2$$
$$= 44 \text{ pounds per linear foot.}$$

The seismic load acting on the ledger is

$$v = 206 \text{ pounds per linear foot}$$

The resultant load is given by

$$r = (w^2 + v^2)^{0.5}$$
$$= (44^2 + 206^2)^{0.5}$$
$$= 211 \text{ pounds per linear foot, and this acts at an angle to the grain of}$$
$$\theta = \tan^{-1}(w/v)$$
$$= \tan^{-1}(44/206)$$
$$= 12°$$

The allowable loads on a ¾ inch diameter anchor bolt in the three and a half inch wide Douglas fir ledger are obtained from UBC Table 23–I–F and UBC Section 2311.2 as half the double shear value for a seven inch wide member and are

p = Allowable load parallel to grain = 3220/2 = 1610 pounds
q = Allowable load perpendicular to grain = 2050/2 = 1025 pounds

From UBC Section 2336.2.1, the allowable load when inclined at an angle to the grain of $\theta°$, is obtained from Hankinson's Formula as

$$F_n = pq/(p\sin^2\theta + q\cos^2\theta)$$
$$= 1610 \times 1025/\{1610 \times \sin^2(12°) + 1025 \times \cos^2(12°)\}$$
$$= 1571 \text{ pounds}$$

Taking into account the allowable one–third stress increase, the required bolt spacing, determined by ledger stresses is given by

$$s = 1.33F_n/r$$
$$= 1.33 \times 1571/211$$
$$= 9.90 \text{ feet}$$

The allowable shear capacity of a ¾ inch diameter anchor bolt in a masonry wall with a specified compressive strength of 1500 pounds per square inch is obtained from UBC Table 21–F as

$$B_v = 1780 \text{ pounds}$$

Taking into account the allowable one–third stress increase, the required bolt spacing, determined by masonry stresses is given by

$$s = 1.33B_v/r$$
$$= 1.33 \times 1780/211$$
$$= 11.22 \text{ feet}$$

COMBINATION DEAD PLUS LIVE LOAD

The dead plus live load acting on the ledger is

$$w = (11 + 20) \times 8/2$$
$$= 124 \text{ pounds per liner foot}$$

The required bolt spacing, determined by masonry stresses is given by

$$s = B_v/w$$
$$= 1780/124$$
$$= 14.35 \text{ feet}$$

Taking into account the allowable twenty–five percent stress increase, the required bolt spacing, determined by ledger stresses is given by

$$s = 1.25q/w$$
$$= 1.25 \times 918/124$$
$$= 9.25 \text{ feet}$$

Hence, the governing bolt spacing is

$$s = 9.25 \text{ feet}$$

(d)3. CHORD FORCE

The maximum chord force is produced in the east–west walls by the east–west seismic force. As shown in Figure 89S0–1(ii), the maximum bending moment in the roof diaphragm is given by

$$M \quad = VL/8$$
$$= 154 \times 240/8$$
$$= 4620 \text{ kip feet}$$

The corresponding chord force is given by

$$F \quad = M/B$$
$$= 4620/120$$
$$= 38.5 \text{ kips}$$

The allowable stress in Grade 60 reinforcement is obtained from UBC Section 2107.2.11 and is given by UBC Formula (7–25) as

$$F_s \quad = 0.5f_y$$
$$= 0.5 \times 60$$
$$= 30 \text{ kips per square inch}$$
$$> 24 \text{ hence use the upper limit of}$$
$$F_s \quad = 24 \text{ kips per square inch}$$

Taking into account the allowable one–third stress increase, the required reinforcement area is given by

$$A_s \quad = F/F_s$$
$$= 38.5/(24 \times 1.33)$$
$$= 1.21 \text{ square inches}$$

Use two No. 6 bars which provide an area of

$$A_s \quad = 1.24 \text{ square inches}$$

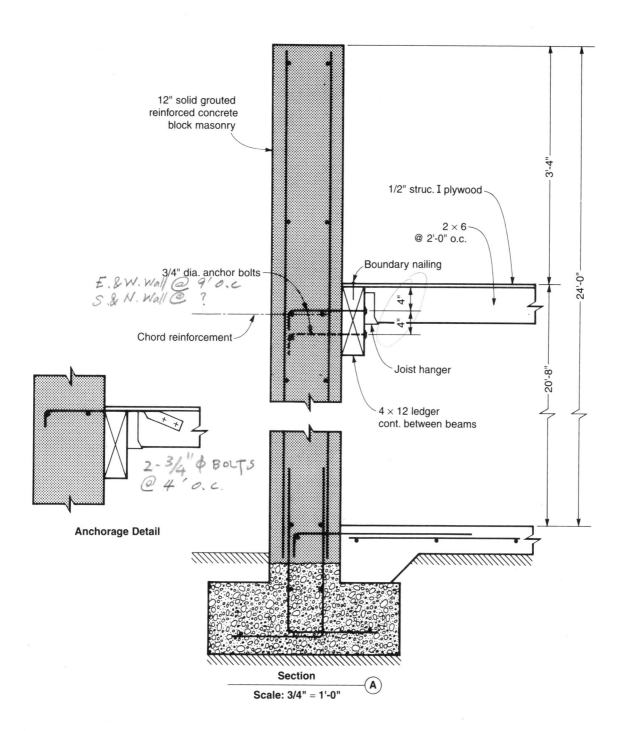

12" solid grouted reinforced concrete block masonry

1/2" struc. I plywood

2 × 6 @ 2'-0" o.c.

Boundary nailing

3'-4"

24'-0"

3/4" dia. anchor bolts

E. & W. Wall @ 9' o.c
S. & N. Wall @ ?

Chord reinforcement

4"

4"

Joist hanger

20'-8"

4 × 12 ledger cont. between beams

Anchorage Detail

2 - 3/4" ⌀ BOLTS @ 4' o.c.

Section

Scale: 3/4" = 1'-0"   Ⓐ

FIGURE 89SO-1(i)

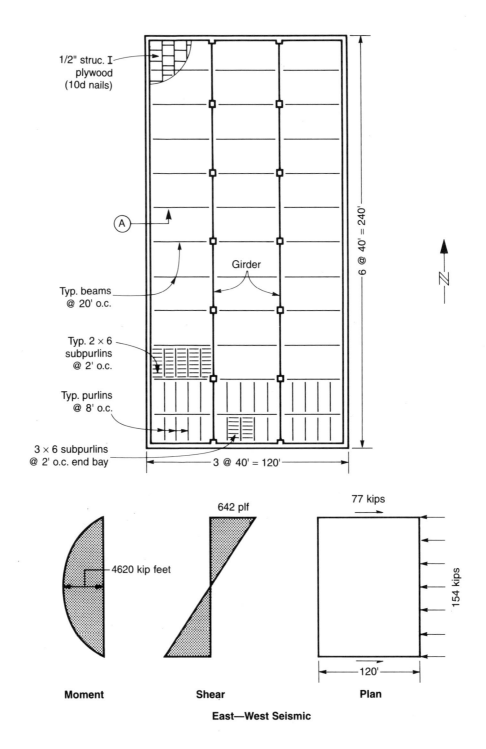

FIGURE 89SO–1(ii)

STRUCTURAL ENGINEER REGISTRATION

CHAPTER 1

GENERAL STRUCTURAL PRINCIPLES AND SEISMIC DESIGN

SECTION 1.7

ROOF FRAMING

PROBLEMS:

1989 A–1
1987 A–3

STRUCTURAL ENGINEER EXAMINATION – 1989

PROBLEM A-1 – WT. 5.0 POINTS

GIVEN: Framing for a hip roof is shown below. Dead load on the projected area including self–weight of all members is 15 psf.

CRITERIA:
- Roof is continuously supported vertically by perimeter walls.
- Roof plan is symmetrical about both major axes.
- Consider joist loads to the main members as uniformly distributed loads, not point loads.

Assumptions:
- Neglect lateral loads and unbalanced live loads.
- Assume the roof structural system is stable, with no lateral loads or unbalanced live loads.
- No lateral restraint at top of walls.

REQUIRED:
1. For the different framing systems listed below, sketch free–body diagrams of a typical hip "AB" and ridge "BC", showing applied (dead + live) loads, reactions, and maximum moments.
   Case 1a: Hips/ridge intersections "B" and "C" are supported by interior posts.
   Case 1b: Hips/ridge intersections "B" and "C" are designed as moment connections (without interior posts).
   Case 1c: The roof is supported by walls at the perimeter, with no moment connection or interior post supports at "B" and "C".
2. For Case 1c, calculate the corner restraint forces at "A".

SOLUTION

LOADING

The area tributary to each hip beam is given by

$$A_H = 10 \times 10/2$$
$$= 50 \text{ squarc feet}$$

The applicable live load for a roof slope of four inches per foot and a tributary area of less than 200 square feet is obtained from UBC Table 16-C as

$$w_L = 16 \text{ pounds per square foot on the projected horizontal plane.}$$

The dead load, on the projected area, is given in the problem statement as

$$w_D = 15 \text{ pounds per square foot.}$$

148

The total load acting on each hip beam is given by

$$
\begin{aligned}
W_H &= (w_D + w_L)A_H \\
&= (15 + 16) \times 50 \\
&= 31 \times 50 \\
&= 1550 \text{ pounds.}
\end{aligned}
$$

The roof joists span from the perimeter beams to each hip beam and produce a triangular load on the hip beam with a maximum value at the apex of

$$
\begin{aligned}
w_H &= 2W_H/L_P
\end{aligned}
$$

where L_P = the projected horizontal length of the hip beam of 14.14 feet

Then w_H = $2 \times 1550/14.14$

            = 219 pounds per foot

The area tributary to the ridge beam is given by

$$
\begin{aligned}
A_R &= 10 \times 20/2 \\
&= 100 \text{ square feet}
\end{aligned}
$$

The uniformly distributed load on the ridge is

$$
w_R = 31 \text{ pounds per square foot}
$$

The total load acting on the ridge beam is

$$
\begin{aligned}
W_R &= w_R A_R \\
&= 31 \times 100 \\
&= 3100 \text{ pounds}
\end{aligned}
$$

The applied loading, acting on the ridge and on each hip beam, is shown in Figure 89A-1.

1a. INTERIOR POSTS

The interior posts provide independent supports to the ridge and the hips. Neglecting the strains produced by the axial forces in the members, only vertical reactions are produced. The maximum moment produced in the hip is given by

$$
\begin{aligned}
M_H &= 0.128W_HL_P \\
&= 0.128 \times 1550 \times 14.14 \\
&= 2805 \text{ pounds feet}
\end{aligned}
$$

and this occurs at a horizontal distance from the eaves of

$$
\begin{aligned}
x &= 0.5774L_P \\
&= 8.16 \text{ feet}
\end{aligned}
$$

The maximum moment produced in the ridge is given by

$$
\begin{aligned}
M_R &= 0.125 \, W_RL_R \\
&= 0.125 \times 3100 \times 10 \\
&= 3875 \text{ pounds feet}
\end{aligned}
$$

The reactions and moments are shown in Figure 89A-1.

1b. RIGID JOINTS

The forces in the ridges and the hips may be determined by the moment distribution procedure[1,2]. Due to the symmetry of the structures and the applied loading, no sway occurs and only joint rotations need be considered. Assuming that the moment of inertia of each hip beam is equal to that of the ridge beam, the equivalent inertia of members AB and CD, which consist of two hip beams combined, is twice that of member BC. The distribution may be carried out in the plane of the ridge with all moments occurring in the hip beams resolved in the direction of the ridge. For the loading on each hip, the initial fixed–end moment at joint B, allowing for the hinged end at A, is given by

$$M_{BA}^F = 2W_H L_P/15$$
$$= 2 \times 1550 \times 14.14/15$$
$$= 2922 \text{ pounds feet}$$

Resolving the fixed–end moment in each hip in the plane of the ridge gives the initial equivalent fixed–end moment at B of

$$M_E^F = 2922 \times 1.414$$
$$= 4132 \text{ pounds feet}$$

For the loading on the ridge beam, the initial fixed–end moment at joint B is given by

$$M_{BC}^F = -W_R L_R/12$$
$$= -3100 \times 10/12$$
$$= -2583 \text{ pounds feet}$$

In the distribution, advantage may be taken of the hinged supports and of the symmetry to modify the stiffness factors, as shown in the distribution table, so that distribution is required in only half of the structure and there is no carry–over between the two halves or to the hinged ends. The distribution proceeds, for the left half of the frame, as shown in the Table, and the final moments in the right half of the frame are equal and of opposite sense to those obtained for the left–half. The convention adopted is that clockwise moments acting from the joint on the member are considered positive.

Joint	A	B	B
Member	AB	BA	BC
Relative EI/L	2/14.53	2/14.53	1/10
Modified stiffness	8/14.53	6/14.53	2/10
Distribution factors	1	0.674	0.326
Fixed-end moments	0	4132	−2583
Distribution	0	−1044	− 505
Final moments, lb.ft.	0	3088	−3088

To determine the forces in a hip beam, the equivalent moment at B must be resolved into the plane of the hip. Then the actual moment in the hip at B is given by

$$M_{BA} = 3088 \times 0.707$$
$$= 2183 \text{ pounds feet, hogging}$$

The vertical reaction at joint A is given by

$$V_A = W_H + W_R/4$$
$$= 1550 + 3100/4$$
$$= 2325 \text{ pounds}$$

The horizontal reaction at joint A is obtained by taking moments about joint B for member AB and is given by

$$H_A = (M_{BA} + V_A L_P - W_H L_P/3)/3.33$$
$$= (2183 + 2325 \times 14.14 - 1550 \times 14.14/3)/3.33$$
$$= 8334 \text{ pounds}$$

The maximum sagging moment in a hip beam is approximately given by

$$M = M_H - M_{BA} \times 0.67$$
$$= 2805 - 2183 \times 0.67$$
$$= 1342 \text{ pounds feet sagging}$$

To determine the forces in the ridge beam, the horizontal forces in each hip beam must be resolved into the plane of the ridge. Then the horizontal thrust in the ridge is given by

$$H_B = H_A \times 1.414$$
$$= 8334 \times 1.414$$
$$= 11784 \text{ pounds}$$
$$V_B = W_R/2$$
$$= 3100/2$$
$$= 1550 \text{ pounds}$$
$$M_{BC} = 3088 \text{ pounds feet, hogging}$$

The maximum sagging moment in the ridge is given by

$$M = M_R - M_{BC}$$
$$= 3875 - 3088$$
$$= 787 \text{ pounds feet, sagging}$$

The reactions and moments are shown in Figure 89A-1.

1c. PINNED JOINTS

The introduction of hinges at joints B and C convert the structure into a mechanism which is, however, stable under symmetrical loading.

To determine the forces in a hip beam, the vertical reaction at joint A is first obtained as

$$V_A = W_H + W_R/4$$
$$= 1550 + 3100/4$$
$$= 2325 \text{ pounds}$$

The horizontal reaction at joint A is obtained by taking moments about joint B for member AB and is given by

$$H_A = (V_A L_P - W_H L_P/3)/3.33$$
$$= (2325 \times 14.14 - 1550 \times 14.14/3)/3.33$$
$$= 7678 \text{ pounds}$$

The maximum moment produced in a hip beam is, as previously determined in part 1a, given by

$$M_H = 2805 \text{ pounds feet}$$

To determine the forces in the ridge beam, the horizontal forces in each hip beam must be resolved into the plane of the ridge to give

$$H_B = H_A \times 1.414$$
$$= 7678 \times 1.414$$
$$= 10857 \text{ pounds}$$
$$V_B = 1550 \text{ pounds}$$
$$M_R = 3875 \text{ pounds feet}$$

2. CORNER RESTRAINT

Corner restraint is provided by the tensile forces in the perimeter beams. This tensile force is obtained by resolving the horizontal reaction at joint A and is given by

$$R_P = H_A \times 0.707$$
$$= 7678 \times 0.707$$
$$= 5428 \text{ pounds}$$

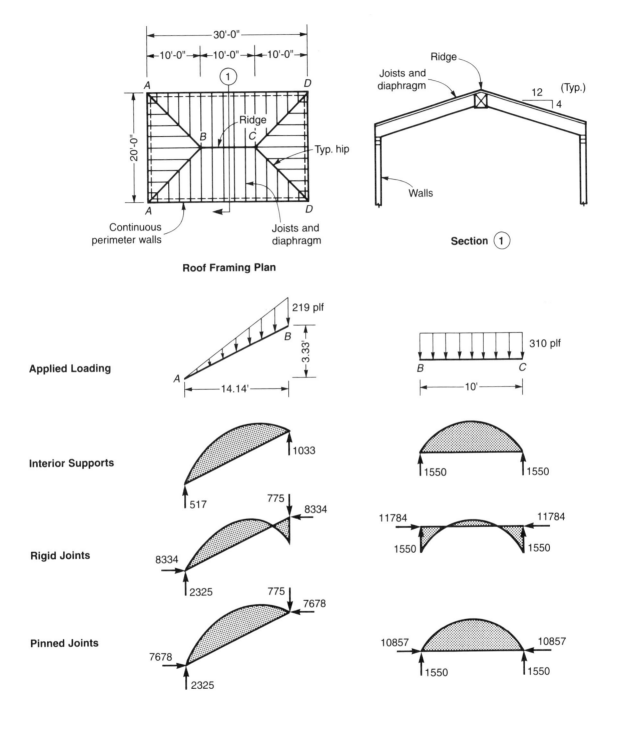

FIGURE 89A-1

153

STRUCTURAL ENGINEER EXAMINATION – 1987 ═══════════════

PROBLEM A–3 – WT. 6.0 POINTS

GIVEN: The roof framing plan, elevation of hip beams, and detail "A".

CRITERIA: Roof dead load = 21 psf (on projected horizontal plane)
Use live load per UBC Section 1605. Neglect snow loads.
All steel is Grade A–36. All welding is E70XX.
Hip beams are pin connected at the locations shown.
Assume the perimeter beams are fully supported in such a way that they do not carry any vertical loads.
Neglect seismic and wind loads.

REQUIRED: 1) Determine the total load to be used, and draw a free body diagram of the hip beams.
2) Find the reactions from the free body diagram (from static loads only).
3) Calculate the forces transmitted from the hip beam thrust to the perimeter struts using the reactions calculated from item 2.
4) Check the W8 × 21 connection shown in detail "A" and comment if it is adequate to resist the strut forces.

SOLUTION

1. LOADING

The area tributary to a hip beam is one-eighth of the total roof area and is given by

$$A = 59 \times 59/8$$
$$= 435 \text{ square feet}$$

The applicable live load for a roof slope of six inches per foot and a tributary area of between 201 and 600 square feet is obtained from UBC Table 16–C as

$$w_L = 14 \text{ pounds per square foot on the projected horizontal plane.}$$

The dead load, which is assumed to include the self-weight of the hip beam, is given in the problem statement as

$$w_D = 21 \text{ pounds per square foot}$$

The total load acting on the hip beam is given by

$$W = (w_D + w_L)A$$
$$= (21 + 14) \times 435/1000$$
$$= 15.23 \text{ kips}$$

The roof joists span from the perimeter beams to the hip beams and produce a triangular load on the hip beam with a maximum value at the apex of

$$w = 2W/L$$
$$= 2 \times 15.23/41.72$$
$$= 0.73 \text{ kips per foot}$$

Figure 87A-3(ii)shows the applied loading on a free body diagram of a hip beam.

2. HIP BEAM REACTIONS

The reactions acting on the hip beam are shown on the free body diagram. By equating vertical forces, the vertical reaction acting at the eaves is given as

$$R_v = W$$
$$= 15.23 \text{ kips}$$

By taking moments about the eaves, the horizontal reaction acting at the apex is given as

$$R_H = 27.81W/14.75$$
$$= 27.81 \times 15.23/14.75$$
$$= 28.72 \text{ kips}$$

By equating horizontal forces, the horizontal reaction at the eaves is given as

$$R_H = 28.72 \text{ kips.}$$

3. PERIMETER BEAM FORCES

The horizontal component of the force in the hip beam R_H is transmitted to the perimeter beams meeting at each corner as shown in Figure 87A-3(ii). Resolving forces in the East-West direction, the force in each perimeter beam is given as

$$R_P = R_H \cos 45°$$
$$= 28.72 \times 0.7071$$
$$= 20.31 \text{ kips}$$

4. PERIMETER BEAM CONNECTION

PLATE TO BEAM WELD

The allowable force on the E70 grade 3/16 inch fillet weld is governed by the tensile strength of the weld metal and is obtained from UBC Section 2251 Table J2.5 as

$$q = 3 \times 0.928$$
$$= 2.78 \text{ kips per inch}$$

The total allowable force on the twelve inch long weld is

$$2.78 \times 12 = 33.36 \text{ kips}$$

This exceeds the applied force and is satisfactory.

PLATE STRENGTH

The gross plate area is given by

$$A_g = Bt$$

where
B = 6 inches = plate width
t = 0.5 inches = plate thickness

Then
A_g = 6 × 0.5
= 3 square inches

Based on the gross area, the allowable plate strength in tension is defined by UBC Section 2251.D1 and is given by

$$P_t = 0.6F_yA_g$$
$$= 0.6 \times 36 \times 3$$
$$= 64.8 \text{ kips.}$$

The maximum permissible value of the effective net area is defined by UBC Section 2251.B3 as

$$A_e = 0.85A_g$$
$$= 0.85 \times 3$$
$$= 2.55 \text{ square inches}$$

For tensile forces, the actual effective net area of the section is defined by UBC Section 2251.B3 and is given by

$$A_e = t(B - nd_e)$$

where
n = 2 = the number of bolts, of diameter d_b, in one vertical row
d_e = effective diameter of the bolt hole as given by UBC Section 2251.B2
$= d_b + 0.125$
$= 0.875$ inches

Then
A_e = 0.5(6 – 2 × 0.875)
= 2.125 square inches
< 2.55

Hence the applicable effective net area is

$$A_e = 2.125 \text{ square inches}$$

Based on the effective net area, the allowable plate strength is defined by UBC Section 2251.D1 and is given by

$$P_t = 0.5F_uA_e$$
$$= 0.5 \times 58 \times 2.125$$
$$= 61.6 \text{ kips}$$
$$< 64.8$$

Hence the allowable plate strength, as governed by the effective net area, is

$$P_t = 61.6 \text{ kips}$$

This exceeds the applied force and is satisfactory.

WEB TEAR-OUT

The net effective area resisting shear is given by AISC page 4-9 as

$$A_v = 2t_w(3 + L_e - 1.5d_h)$$

where
$$t_w = \text{web thickness of the W8} \times 21$$
$$= 0.25 \text{ inches}$$
$$L_e = \text{minimum bolt end distance given by UBC Section 2251.J3.7}$$
$$= 1.5d_b$$
$$= 1.5 \times 0.75$$
$$= 1.125 \text{ inches}$$
$$d_h = \text{nominal diameter of a standard hole in shear as given by UBC Section 2251 Table J3.1}$$
$$= d_b + 0.0625$$
$$= 0.8125 \text{ inches}$$

and
$$A_v = 2 \times 0.25(3 + 1.125 - 1.5 \times 0.8125)$$
$$= 1.45 \text{ square inches}$$

The net effective area resisting tension is given by

$$A_t = t_w(3 - d_h)$$
$$= 0.25(3 - 0.8125)$$
$$= 0.55 \text{ square inches}$$

The resistance to shearing rupture is determined from UBC Section 2251.J4 as

$$P_r = 0.3F_uA_v + 0.5F_uA_t$$
$$= 0.3 \times 58 \times 1.45 + 0.5 \times 58 \times 0.55$$
$$= 41.18 \text{ kips ... exceeds the applied force and is satisfactory.}$$

WEB BEARING

The allowable load on a 0.75 inch diameter bolt in a 0.25 inch thick web is obtained from AISC Table I-E as 13.1 kips. The allowable load on four bolts is given by

$$P_b = 4 \times 13.1$$
$$= 52.4 \text{ kips ... exceeds the applied force and is satisfactory.}$$

BOLT SHEAR

The allowable shear force on a 0.75 inch diameter A307 bolt is obtained from AISC Table I-D as 4.4 kips. The allowable load on four bolts is given by

$$P_s = 4 \times 4.4$$
$$= 17.6 \text{ kips}$$

This is less than the applied force and is unsatisfactory. In order to carry the applied force, the A307 bolts must be replaced with A325 bolts.

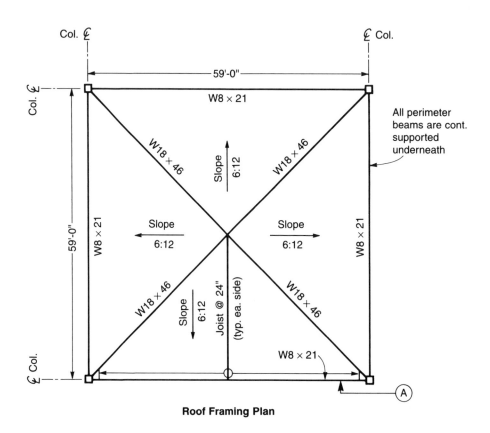

Roof Framing Plan

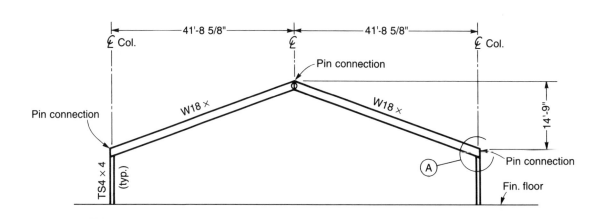

Hip Beam Elevation

FIGURE 87A–3(i)

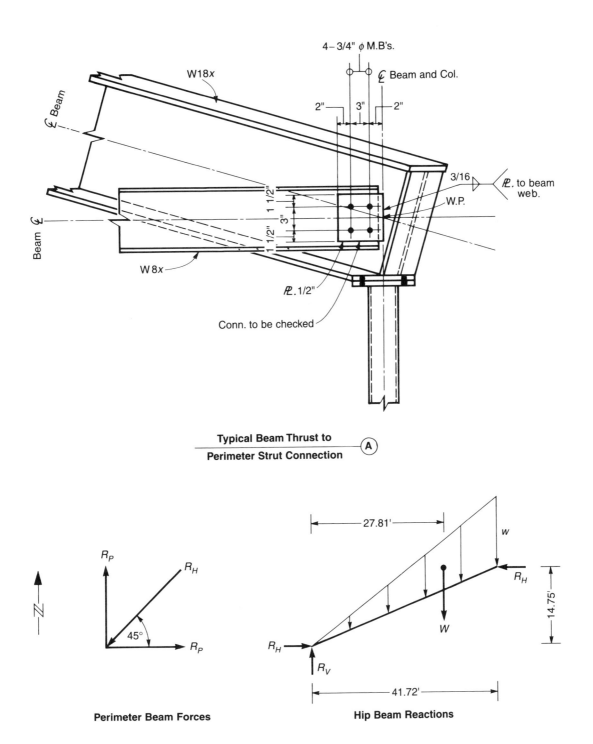

Typical Beam Thrust to
Perimeter Strut Connection (A)

Perimeter Beam Forces

Hip Beam Reactions

FIGURE 87A–3(ii)

159

CHAPTER 1

REFERENCES

1. Cross, H. The analysis of continuous frames by distributing fixed end moments. *Trans. Am. Soc. Civil Eng.*, Vol 96, 1932, pp. 1–156.

2. Williams A. *The analysis of indeterminate structures.* Hart Publishing Company, Inc., New York, 1968, pp. 114–184.

3. Hsu, T.H. *Stress and strain data handbook.* Gulf Publishing Company, Houston, TX, 1986. Table 3–1.

4. Tuma, J.J. *Structural Analysis.* Table P–6.14. McGraw–Hill, New York, 1969.

5. Bush, V.R. *Handbook to the Uniform Building Code.* International Conference of Building Officials, Whittier, CA, 1988, pp. 156–163.

6. Structural Engineers Association of Washington. *Wind commentary to the Uniform Building Code.* Seattle, WA, 1995.

7. American Society of Civil Engineers. *Minimum design loads for buildings and other structures.* New York, 1990.

8. Mehta, K.C. et al. *Guide to the use of the wind load provisions of ASCE 7–88.* American Society of Civil Engineers, New York, 1991.

9. American Plywood Association. *Diaphragms.* APA Design/Construction Guide, Tacoma, WA, 1989, pp. 9–23.

10. Paz, M. *Structural dynamics.* Van Nostrand Reinhold, New York, 1991.

11. Williams, A. *Seismic design of buildings and bridges.* Engineering Pess, Inc., San Jose, CA, 1995.

12. Corps. of Engineers. *Seismic design guidelines for essential buildings.* NAVFAC, Technical Manual P–335.1, Washington, 1986.

13. Ramasco, R. An iterative method for the calculation of natural vibration frequencies in plane frames. *Il Cemento*, Vol. 3, No. 4, 1968, pp. 66–69.

14. Clough, R.W. and Penzien, J. *Dynamics of structures.* McGraw–Hill, New York, 1975.

15. Sheedy, P. Anchorage of concrete and masonry walls. *Building Standards*, October 1983. International Conference of Building Officials, Whittier, CA.

16. International Conference of Building Officials. *Illustrative guide to the 1988 Uniform Building Code*, Whittier, CA, 1989, pp. 123–128.

17. Sheedy, P. Discussion on reference 15. *Building Standards*, April 1984. International Conference of Building Officials, Whittier, CA.

2

STRUCTURAL STEEL DESIGN

SECTION 2.1

MEMBER DESIGN

PROBLEMS:

1990 B–1
1989 B–2
1988 B–2

STRUCTURAL ENGINEER EXAMINATION – 1990 ═══════════

PROBLEM B–1 – WT. 6.0 POINTS

GIVEN: The roof truss as shown.
- Dead load = 8 psf (top chord loads only – neglect the dead load of the truss)
- Live load = 12 psf
- Wind uplift = 20 psf

CRITERIA: Materials:
- Steel beams Grade A36, F_y = 36 ksi
- E70XX Electrodes

Assumptions:
- Trusses are spaced at 20 feet on center.
- All loads are applied to the top chord of the truss, i.e., no ceiling loads.
- Purlins are spaced at 5 feet on center.
- Bottom chord of the truss is braced at 20 feet on center.
- Use steel stitch plates along the length of the double angle members.

REQUIRED: 1. Select the lightest 4 inch tee section for the bottom chord member "3–4".
             2. Select the lightest double angle for the diagonal member "2–10".
             3. Check the WT 6 × 17.5 for top chord member "10–11".
             4. For a force of 16 kips (6DL + 10LL) tension or compression, design the welded connection at the lower end of member "2–10" using the least amount of 3/16" fillet weld. Complete detail 1.

SOLUTION

LOADING

The loads applied to the interior panel points of the truss are given by

Dead load W_D	= 8 × 10 × 20/1,000	= 1.6 kips
Live load W_L	= 12 × 10 × 20/1,000	= 2.4 kips
Wind load W_w	= 20 × 10 × 20/1,000	= 4.0 kips
Maximum downward load W_T	= $W_D + W_L$	= 4.0 kips
Maximum upward load W_U	= $W_w - W_D$	= 2.4 kips

The loads applied to the outer panel points are one–half of these values.

MEMBER FORCES

For a panel point load of W, the support reaction at each end of the truss is given by

$$R = 3W$$

Applying the method of sections to panel 3–4–11–10, and taking moments about joint 3, gives the force in the top chord member 10–11 as

$$
\begin{aligned}
P_{10\text{-}11} &= (20R - 10W - 20W/2)/5 \\
&= (60 - 10 - 10)W/5 \\
&= 8W
\end{aligned}
$$

The force in member 10–11 due to dead load plus live load is

$$P_{10\text{-}11} = 8 \times W_T = 32 \text{ kips compression}$$

The force in member 10–11 due to dead load plus wind load is

$$P_{10\text{-}11} = 8 \times W_U = 19.2 \text{ kips tension}$$

Taking moments about joint 11 gives the force in the bottom chord member 3–4 as

$$
\begin{aligned}
P_{3\text{-}4} &= (30R - 10W - 20W - 30W/2)/5 \\
&= (90 - 10 - 20 - 15)W/5 \\
&= 9W
\end{aligned}
$$

The force in member 3–4 due to dead load plus live load is

$$P_{3\text{-}4} = 9 \times W_T = 36 \text{ kips tension}$$

The force in member 3–4 due to dead load plus wind load is

$$P_{3\text{-}4} = 9 \times W_U = 21.6 \text{ kips compression}$$

Applying the method of sections to panel 2–3–10–9, and resolving forces vertically, gives the force in the diagonal member 2–10 as

$$
\begin{aligned}
P_{2\text{-}10} &= (R - W - W/2)5^{0.5} \\
&= (3 - 1 - 0.5)W5^{0.5} \\
&= 3.35W
\end{aligned}
$$

The force in member 2–10 due to dead load plus live load is

$$P_{2\text{-}10} = 3.35 \times W_T = 13.42 \text{ kips compression}$$

The force in member 2–10 due to dead load plus wind load is

$$P_{2\text{-}10} = 3.35 \times W_U = 8.05 \text{ kips tension}$$

1. BOTTOM CHORD MEMBER 3-4

To comply with UBC Section 2251.B7, the slenderness ratio, for a compression member, is limited to

$$K\ell/r \leq 200$$

where
K = effective length factor
ℓ = actual unbraced length
r = radius of gyration

The out-of-plane unbraced length of the bottom chord is given in the problem statement as twenty feet and the effective length, with respect to the Y-axis is

$$K\ell = 1.0 \times 20$$
$$= 20 \text{ feet}$$

As indicated in AISC Section C-C2, the effective length of a truss member, in the plane of the truss, is normally assumed equal to the actual distance between panel points. Hence, the effective length, with respect to the X-axis, is

$$K\ell = 1.0 \times 10$$
$$= 10 \text{ feet}$$

In accordance with UBC Section 1603.5, a one-third increase is allowed in permissible stresses when designing for wind forces. The equivalent design force in compression, based on the basic allowable stress is, then

$$P_E = P_{3-4}/1.33$$
$$= 21.6/1.33$$
$$= 16.24 \text{ kips}$$

From AISC Table 3-105, a WT 4 × 12 section has an allowable axial load of

$$P_A = 24 \text{ kips}$$
$$> 16.24$$

The member is adequate in compression since the $K\ell/r$ values about both axes are less than 200.

The allowable tensile force in the WT 4 × 12 section is governed by the gross area and is given by UBC Section 2251.D1 as

$$P_A = 0.6F_yA_g$$
$$= 21.6 \times 3.54$$
$$= 76.5 \text{ kips}$$
$$> 36$$

Hence the WT 4 × 12 section is also adequate in tension.

2. DIAGONAL MEMBER 2–10

The compressive force of 13.42 kips, due to dead load plus live load, governs. The effective length, with respect to both the X–axis and the Y–axis, is

$$K\ell = 1.0 \times 11.18$$
$$= 11.18 \text{ feet}$$

From AISC Table 3–74, a double angle $3 \times 2\frac{1}{2} \times 3/16$ with long legs back to back has an allowable axial load of

$$P_A = 15.8 \text{ kips}$$
$$> 13.42$$

and this is the lightest double angle which is adequate in compression.

The governing slenderness ratio of the double angle section is

$$K\ell/r_x = 1.0 \times 11.18 \times 12/0.954$$
$$= 141$$

Using two stitch plates along the length of the double angle member, the local slenderness ratio of the individual angles is given by

$$K\ell/r_z = 1.0 \times 11.18 \times 12/(3 \times 0.533)$$
$$= 84$$
$$< 141 \times 0.75$$

Hence the compound section satisfies UBC Section 2251.E4 which requires that the slenderness ratio of the individual angles, between fasteners, shall not exceed three–fourths times the governing slenderness ratio of the build–up member.

3. TOP CHORD MEMBER 10–11

The properties of the 6×17.5 structural tee are obtained from AISC Table 1–64 as

$$A = 5.17 \text{ inches}^2$$
$$S_x = 3.23 \text{ inches}^3$$
$$r_x = 1.76 \text{ inches}$$
$$r_y = 1.54 \text{ inches}$$
$$Q_s = \text{No value}$$

The axial forces in member 10–11 were derived on the basis of the loads being applied at the panel points of the roof truss. In fact, the loads are applied to the top boom by the purlins and the top boom is subjected to local bending in addition to the axial forces.

When reversal under wind load occurs, the axial force in the top boom changes to tension and, in addition, the bending from the purlins, is reduced by forty percent. Hence it is unnecessary to check the top boom under the applied upward load.

Due to the downward load, the sagging moment produced in the span adds to the axial compression in the top of the flange of the section. Conversely, the hogging moment at the supports adds to the axial compression in the toe of the stem, as shown in Figure 90B-1. Thus the appropriate modulus about the X-axis must be used in determining the stresses. Since the minimum section modulus applies to the toe of the stem, it is unnecessary to consider the span moment as the support moment gives the critical combined stress.

The top boom may be considered as a beam continuous over six spans with a concentrated load applied in the center of each span. The support moment for span 10-11 is obtained from tabulated coefficients for a continuous beam, ignoring skip loading, and is given by

$$M = 0.119 \, W_pL$$

where W_p = applied purlin load of two kips due to dead plus live load

L = span length of ten feet

Then M = $0.119 \times 2 \times 10 \times 12$

= 28.56 kip inches

The compressive stress produced in the toe of the stem by this moment is given by

$$f = M/S_x$$

where S_x = Minimum section modulus about the X-axis, as tabulated

Then f_b = 28.56/3.23

= 8.84 kips per square inch

The value of L_c is given by UBC Section 2251.F1 as

$$L_c = 20{,}000A_f/dF_y$$

$$= 20{,}000 \times 6.56 \times 0.52/(6.25 \times 36)$$

$$= 303 \text{ inches}$$

or $L_c = 76b_f/(F_y)^{0.5}$

$$= 76 \times 6.56/6$$

$$= 83 \text{ inches} \ldots \text{ governs}$$

The actual unbraced length is

$$L_b = 60 \text{ inches}$$

$$< L_c$$

Hence the allowable bending stress is given by UBC Section 2251 Equation (F1-1) as

$$F_b = 0.66F_y$$

$$= 23.76 \text{ kips per square inch}$$

The out-of-plane unbraced length of the top chord members is equal to the distance of five feet between purlins and the in-plane unbraced length is equal to the distance of ten feet between panel points. Then, the slenderness ratios about the Y-axis and the X-axis are given by

$$Kl/r_y = 1.00 \times 5 \times 12/1.54$$

$$= 38.96$$

166

and
$$K\ell/r_x = 1.00 \times 10 \times 12/1.76$$
$$= 68.18 \ldots > 38.96$$

Hence, the slenderness ratio about the X-axis governs and the allowable compressive stress is obtained from AISC Table C-36 as

$$F_a = 16.62 \text{ kips per square inch}$$

The actual compressive stress in member 10-11, due to dead load plus live load, is given by

$$f_a = P_{10-11}/A$$
$$= 32/5.17$$
$$= 6.19 \text{ kips per square inch} \ldots > 0.15 \, F_a$$

Hence, in checking the adequacy of the member in combined axial compression and bending, in accordance with UBC Section 2251.H1, Equations (H1-1) and (H1-2) are applicable. Equation (H1-1) is

$$f_a/F_a + C_m f_b/F_b(1 - f_a/F'_e) \leq 1.0$$

where C_m = the bending coefficient for a continuous member braced against joint translation which is given by UBC Section 2251.H1 and AISC Section C-H1 as 0.85.

and F'_e = the factored Euler stress for a slenderness ratio, $K\ell_b/r_b$ with respect to the axis of bending, of 68.18 which is obtained from AISC Table 8 as 32.2 kips per square inch.

Using calculator program A.4.2 in Appendix A, the left hand side of Equation (H1-1) is 0.76.

Equation (H1-2) is

$$f_a/0.6F_y + f_b/F_b \leq 1.0$$

and using calculator program A.4.2, the left hand side is 0.66.

Hence both Equations are satisfied and the 6 × 17.5 structural tee section is satisfactory.

4. WELDED CONNECTION

The capacity of the E70XX grade 3/16 inch fillet weld is obtained from UBC Section 2251 Table J2.5 as

$$q = 3 \times 0.928$$
$$= 2.78 \text{ kips per inch}$$

Balanced welds are not required as the connection is not subjected to fatigue loading. The length of each of the four sections of weldment at the joint is given by

$$\ell = P/4q$$
$$= 16/(4 \times 2.78)$$
$$= 1.4 \text{ inches}$$

The completed details are shown on Figure 90B-1.

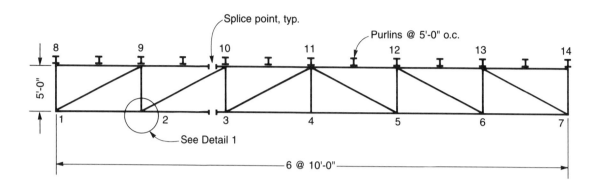

Elevation

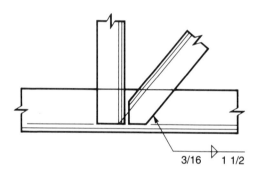

Detail 1

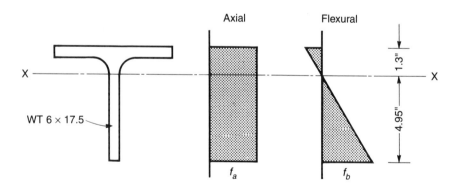

Combined Stresses

Figure 90B-1

168

STRUCTURAL ENGINEER EXAMINATION – 1989 =========

PROBLEM B–2 – WT. 8.0 POINTS

GIVEN: A portal frame with a truss girder is shown. Applied loads are as follows:
$P_{(dead + live)}$ = 10 kips
$V_{(seismic)}$ = 20 kips

CRITERIA: Materials:
- Grade A36 steel
- Grade A325–SC bolts, ¾ inch diameter

Assumptions:
- Each column resists half of the applied lateral force.
- The truss girder is rigid for lateral deflection calculation.
- All top and bottom chord members are laterally braced at the panel points only.
- Effective length factors for the columns are K_x = 1.8, K_y = 1.0.

REQUIRED: 1. Size the lightest W14 column for the portal frame for both stress and deflection, using dead, live and lateral loads. The allowable deflection shall be 0.005 times the story height. Show all stress calculations and draw the shear and moment diagrams for the column.
2. Size the lightest WT5 top chord for the truss girder.
3. Size the lightest 6" double channel bottom chord for the truss girder.
4. With only the forces given in Section "A", compute the number of bolts required at the connection between the vertical double angle and the gusset plate. The forces are for combined dead, live and lateral loads.

SOLUTION

MEMBER FORCES

(i) DUE TO LATERAL LOAD

The horizontal reactions at the supports are given by

$$H_1 = H_2 = 20/2$$
$$= 10 \text{ kips}$$

The vertical reactions at the supports are obtained by taking moments about each support in turn and are given by

$$V_1 = V_2 = 20 \times 28/32$$
$$= 17.5 \text{ kips}$$

Applying the method of sections to panel 3-4-7-6, and taking moments about joint 3, gives the force in the top chord member 67 as

$$P_{67} = (8V + 20H_1)/8$$
$$= 45 \text{ kips compression}$$

Taking moments about joint 7 gives the force in the bottom chord member 34 as

$$P_{34} = (28H_1 - 8V_1)/8$$
$$= 17.5 \text{ kips tension}$$

Applying the method of sections to panel 3-4-8-7, and taking moments about joint 4, gives the force in the top chord member 78 as

$$P_{78} = (8V + 20H_1 - 16V_1)/8$$
$$= 10 \text{ kips compression}$$

Similarly $\quad P_{89} = 25 \text{ kips tension}$

The forces acting on column 2-5-9, and the corresponding shear force and bending moment diagrams, are shown in Figure 89B-2(ii). The maximum moment occurs at joint 5 and is given by

$$M_5 = 20H_2$$
$$= 20 \times 10$$
$$= 200 \text{ kip feet}$$

(ii) DUE TO VERTICAL LOADS

The two-hinged frame is one degree redundant, and it is convenient to consider the horizontal reaction R at the hinges as the external redundant. The cut-back structure is produced by removing this redundant, and the applied loads and a unit virtual load replacing R are applied, in turn, to the cut-back structure, as shown in Figure 89B-2(i). The redundant reaction is obtained from the virtual work method[1,2] as

$$R = -(\Sigma PuL/AE + \int Mm\,ds/EI)/(\Sigma u^2L/AE + \int m^2\,ds/EI)$$

where P and u are the member forces in the cut-back structure due to the applied loads and unit virtual load, respectively, and M and m are the bending moments at any point in a member of the cut-back structure due to the applied loads and the unit virtual load, respectively. Since M is zero for all members

$$\int Mm\,ds/EI = 0$$

The summations are obtained in the Table below from the relevant values of P and u and by using an initial estimate of the cross sectional areas of the members. Only the top chord and bottom chord members need be included in the Table since the force u is zero for all other members. Tensile forces are considered positive.

The allowable lateral displacement at the top of the frame is

$$\delta = 0.005H$$

where H = story height of 28 feet

then, δ = $0.005 \times 28 \times 12$

$$= 1.68 \text{ inches}$$

Since the truss girder is assumed to be rigid for lateral displacement, no rotation occurs at the tops of the columns which act as vertical cantilevers. Then the minimum allowable moment of inertia for each column, so as not to exceed the allowable displacement, is given by

$$I = WL^3/3E\delta$$

where I = required moment of inertia of one column

            W = applied lateral force of 10 kips on one column

            L = column height of 240 inches

Then I = $10 \times (240)^3/(3 \times 29{,}000 \times 1.68)$

$$= 946 \text{ inches}^4$$

The lightest W14 column, which satisfies this requirement, is a W14 × 90 which has a moment of inertia of 999 inches⁴.

The term $\int m^2 ds/EI$ is applicable to the columns only with

$$m_3 = m_5 = 1.0 \times 20 \times 12$$
$$= 240 \text{ kip inches}$$

and I = 999 in⁴, as calculated

Using the volume integration technique,

$$\int m^2 ds/EI = 2 \times (1/3)m_3^2(L_{13} + L_{36})/EI$$
$$= 2 \times (1/3) \times 240^2(20 + 8)/999E$$
$$= 1076/E$$

Member	P	u	L	A	PuL/A	u²L/A	P+uR
34	10	−3.5	16	4.8	−117	41	9.00
45	10	−3.5	16	4.8	−117	41	9.00
67	0	2.5	8	3.2	0	16	0.74
78	−10	2.5	16	3.2	−125	32	−9.30
89	0	2.5	8	3.2	0	16	0.74
Total	–	–	–	–	−359	146	–

Then the redundant reaction is given by

$$R \quad = 359/(146 + 1076)$$
$$= 0.294 \text{ kips}$$

The final forces in the actual structure are obtained from the expression $(P + uR)$ and are shown in the Table. The forces acting on column 2-5-9, and the corresponding shear force and bending moment diagrams, are shown in Figure 89B-2(ii). The maximum moment occurs at joint 5 and is given by

$$M_5 \quad = 20R$$
$$= 20 \times 0.294$$
$$= 5.88 \text{ kip feet}$$

1. COLUMN MEMBER 2-5-9

The properties of a W14 $\times$ 90 column are:

$$
\begin{aligned}
F_y &= 36 \text{ kips per square inch} \\
F_y' &= 40.4 \text{ kips per square inch} \\
I_x &= 999 \text{ inches}^4 \\
S_x &= 143 \text{ inches}^3 \\
A &= 26.5 \text{ inches}^2 \\
r_x &= 6.14 \text{ inches} \\
r_y &= 3.70 \text{ inches} \\
L_c &= 15.3 \text{ feet} \\
L_u &= 34.0 \text{ feet (for } C_b = 1)
\end{aligned}
$$

The design of the column is governed by the combination of vertical load plus lateral load and the column is subjected to combined axial compression and bending. The forces acting on the column are given by

$$
\begin{aligned}
P &= 17.5 + 20 \\
&= 37.5 \text{ kips} \\
M &= 200 + 5.88 \\
&= 205.88 \text{ kip feet}
\end{aligned}
$$

The compressive stress produced in the column flange by the bending moment is

$$
\begin{aligned}
f_b &= M/S_x \\
&= 205.88 \times 12/143 \\
&= 17.3 \text{ kips per square inch}
\end{aligned}
$$

172

The section is compact since

$$F_y' > F_y$$

and the allowable bending stress is dependent on the value of the maximum unbraced length of the compression flange which is assumed to be

$$\ell = 20 \text{ feet}$$

The bending coefficient is given by UBC Section 2251 F1.3 as

$$C_b = 1.75 + 1.05(M_1/M_2) + 0.3(M_1/M_2)^2$$

where M_1 = smallest bending moment in the column = 0

M_2 = largest bending moment in the column = 202.88 kip feet

Hence $C_b = 1.75$

> 1.00

and the actual value of L_u the maximum unbraced length of the compression flange at which the allowable bending stress may be taken as $0.6F_y$, is greater than 34 feet.

Hence $L_c < 1 < L_u$

and the allowable bending stress is given by

$$F_b = 0.6F_y$$
$$= 21.6 \text{ kips per square inch}$$

The slenderness ratios about the Y-axis and the X-axis are given by

$$K_y \, l/r_y = 1 \times 20 \times 12/3.7$$
$$= 64.86$$

and $$K_x \, l/r_x = 1.8 \times 20 \times 12/6.14$$
$$= 70.36$$
$$> 64.86$$

Hence, the slenderness ratio about the X-axis governs and the allowable compressive stress is obtained from AISC Table C-36 as

$$F_a = 16.39 \text{ kips per square inch}$$

The actual compressive stress in the column, due to dead load plus live load, is given by

$$f_a = P/A$$
$$= 37.5/26.5$$
$$= 1.42 \text{ kips per square inch}$$
$$< 0.15F_a$$

Hence in checking the adequacy of the column in combined axial compression and bending, in accordance with UBC Section 2251.H1, Equation (H1-3) is applicable. Equation (H1-3), after allowing for the one-third increase in permissible stresses when designing for seismic loads, is given by

$$f_a/F_a + f_b/F_b \leq 1.00 \times 1.33$$

The left hand side of the Equation is evaluated as

$$
\begin{aligned}
1.42/16.39 + 17.3/21.6 \quad &= 0.09 + 0.80 \\
&= 0.89 \\
&< 1.33
\end{aligned}
$$

In addition, for seismic zones 3 and 4 and in accordance with UBC Section 2211.5, the column must satisfy the Equation

$$1.0P_{DL} + 0.7P_{LL} + 0.375R_wP_E \leq 1.7F_aA$$

Assuming the given vertical loads are equally due to dead load and live load, and obtaining the value of R_w from UBC Section 1628 Table 16-N for a special moment-resisting frame, the left hand side is evaluated as

$$1.0 \times 10 + 0.7 \times 10 + (3 \times 12/8)17.5 = 95.75$$

The right hand side is evaluated as

$$1.7 \times 16.39 \times 26.5 = 738 > 95.75$$

Hence the W14 × 90 section is satisfactory.

2. TOP CHORD MEMBER 67

Member 67 is the most highly loaded top chord member with a compressive force, due to vertical load plus lateral load, of

$$
\begin{aligned}
P_{67} \quad &= 45 - 0.74 \\
&= 44.26 \text{ kips}
\end{aligned}
$$

In accordance with UBC Section 1603.5, a one-third increase is allowed in permissible stresses when designing for seismic forces. The equivalent design force in compression, based on the basic allowable stress is, then

$$
\begin{aligned}
P_E \quad &= P_{67}/1.33 \\
&= 44.26/1.33 \\
&= 33.28 \text{ kips}
\end{aligned}
$$

Member 67 is braced laterally at its ends and the effective length, with respect to the Y-axis, is

$$
\begin{aligned}
K\ell \quad &= 1.0 \times 8 \\
&= 8 \text{ feet}
\end{aligned}
$$

The effective length, with respect to the X–axis, is specified by AISC Section C–C2 as equal to the actual length and is given by

$$K\ell = 1.0 \times 8$$
$$= 8 \text{ feet}$$

From AISC Table 3–103, a WT 5 × 11 is the lightest suitable section and has an allowable axial load of

$$P_A = 50 \text{ kips}$$
$$> 33.28$$

and complies with UBC Section 2251.B7 since the $K\ell/r$ values about both axes are less than 200.

2. TOP CHORD MEMBER 78

Member 78 is more lightly loaded than member 67 but, as it has a greater effective length, must also be checked. The compressive force due to vertical load plus lateral load is

$$P_{78} = 10 + 9.3$$
$$= 19.3 \text{ kips}$$

and the equivalent design force is

$$P_E = P_{78}/1.33$$
$$= 14.5 \text{ kips}$$

The effective length of member 78 with respect to both the X and Y–axes is

$$K\ell = 1.00 \times 16$$
$$= 16 \text{ feet}$$

From AISC Table 3–103, a WT 5 × 11 with an effective length of 16 feet has an allowable load of

$$P_A = 23 \text{ kips}$$
$$> 14.5$$

Hence the WT 5 × 11 is adequate.

2. TOP CHORD MEMBER 89

The tensile force due to vertical load plus lateral load is

$$P_{89} = 25 + 0.74$$
$$= 25.74 \text{ kips}$$

and the equivalent design force is

$$P_E = P_{89}/1.33$$
$$= 19.4 \text{ kips}$$

By inspection, the WT 5 × 11 is adequate. Hence, the WT 5 × 11 is adequate for all top chord members.

3. BOTTOM CHORD MEMBER 34

Member 34 is subjected to a tensile force, due to vertical load plus lateral load, of

$$P_{34} = 17.5 + 9$$
$$= 26.5 \text{ kips tension}$$

The equivalent design force, allowing for the seismic load, is

$$P_E = P_{34}/1.33$$
$$= 26.5/1.33$$
$$= 19.9 \text{ kips tension}$$

The allowable tensile stress is governed by the gross area and is given by UBC Section 2251.D1 as

$$F_t = 0.6 \, F_y$$
$$= 21.6 \text{ kips per square inch}$$

The required sectional area is given by

$$A_g = P_E/F_t$$
$$= 19.9/21.6$$
$$= 0.92 \text{ square inches}$$

From AISC Table 1–40 a C6 × 8.2 channel is the lightest suitable section and provides a sectional area of

$$A = 2 \times 2.4$$
$$= 4.8 \text{ square inches}$$
$$> 0.92$$

Hence, a double C6 × 8.2 member is adequate for the bottom chord tensile forces. However, when the seismic force reverses, member 34 is subjected to a compressive force of

$$P_{34} = 17.5 - 9$$
$$= 8.5 \text{ kips compression}$$

The equivalent design force, allowing for the seismic load, is

$$P_E = P_{34}/1.33$$
$$= 8.5/1.33$$
$$= 6.4 \text{ kips compression}$$

The relevant properties of a single C6 × 8.2 channel are

$$A = 2.4 \text{ inches}^2$$
$$I_y = 0.693 \text{ inches}^4$$
$$\underline{x} = 0.511 \text{ inches}$$
$$r_y = 0.537 \text{ inches}$$
$$r_x = 2.34 \text{ inches}$$

The relevant properties of a double C6 × 8.2, back–to–back with a ⅜ inch gusset plate are obtained as

$$I_y = 2[0.693 + 2.4(0.511 + 0.375/2)^2]$$
$$= 3.73 \text{ inches}^4$$
$$r_y = [3.73/(2 \times 2.4)]^{0.5}$$
$$= 0.88 \text{ inches}$$
$$r_x = 2.34 \text{ inches}$$

The effective length of member 34 with respect to both the X and Y-axis is

$$K\ell = 1.0 \times 16$$
$$= 16 \text{ feet}$$

The slenderness ratio about the Y–axis governs and is given by

$$K\ell/r_y = 16 \times 12/0.88$$
$$= 218$$

This is less than the limit of 300 specified in UBC Section 2251.B7 for tension members, but exceeds the limit of 200 specified for compression members. However, the limit is exceeded by only 9 percent and the compressive stress in the member is low. Also, as specified in UBC Section 2251.B7, members which have been designed as tension members, but experience some compression loading need not satisfy the compression slenderness limit. The allowable compressive stress in member 34 is given by UBC Equation (E2–2) as

$$F_a = 12\pi^2 E/23(K\ell/r)^2$$
$$= 12 \times 3.14^2 \times 29,000/)(23 \times 218^2)$$
$$= 3.14 \text{ kips per square inch.}$$

The actual stress in member 34 due to the combined loading is

$$f_a = P_E/A$$
$$= 6.4/4.8$$
$$= 1.33 \text{ kips per square inch}$$
$$< 3.14$$

Hence, the double C6 × 8.2 section is satisfactory for the bottom chord members.

4. BOLTED GUSSET PLATE

Assuming a total of seven bolts, the applied loading on the bolt group is given by:

Vertical shear V	$= 0.707 \times 40$
	$= 28.3 \text{ kips}$
Horizontal shear H	$= 0.707 \times 40 + 20$
	$= 48.3 \text{ kips}$
Eccentricity moment M	$= 20(14/2 + 2.25)$
	$= 185 \text{ kip inches}$

The geometrical properties of the bolt group are obtained by applying the unit area method. Then, the inertia of the bolt group about its centroid is

$$I = 2(3^2 + 6^2 + 9^2)$$
$$= 252 \text{ inches}^4$$

The modulus of an outer bolt is

$$S = 252/9$$
$$= 28 \text{ inches}^3$$

The co-existent forces on an outer bolt are:

Vertical force dues to vertical shear	$= V/7$
	$= 28.3/7$
	$= 4.04$ kips
Horizontal force due to horizontal shear	$= H/7$
	$= 48.3/7$
	$= 6.90$ kips
Horizontal force due to eccentricity moment	$= M/S$
	$= 185/28$
	$= 6.61$ kips
Total horizontal force	$= 6.90 + 6.61$
	$= 13.51$ kips

The resultant force on an outer bolt is given by

$$R = (4.04^2 + 13.41^2)^{1/2}$$
$$= 14.10 \text{ kips}$$

Allowing for the one-third increase in permissible stress when designing for seismic loads, the allowable double shear value of a ¾ inch diameter A325-SC bolt is obtained from AISC Table 1-D as

$$P_s = 15 \times 1.33$$
$$= 20 \text{ kips}$$

This exceeds the applied force of 14.10 kips and is satisfactory. ===================

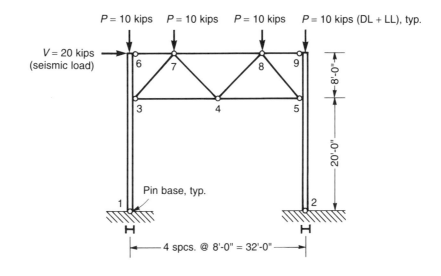

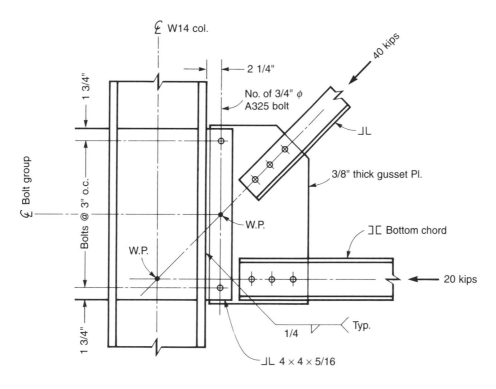

Section "A"

Figure 89B–2(i)

179

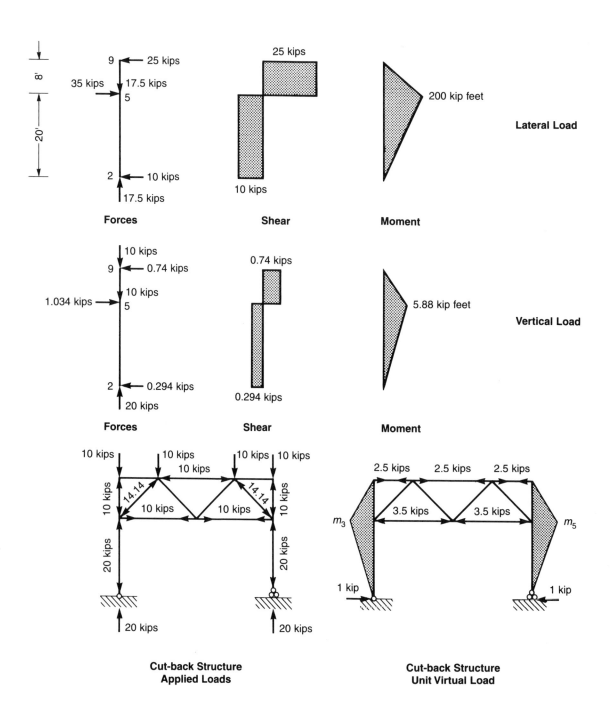

Figure 89B–2(ii)

STRUCTURAL ENGINEER EXAMINATION – 1988 ══════════════

PROBLEM B–2 – WT. 5.0 POINTS

GIVEN: Jib crane system as shown, consisting of a vertical pipe column, horizontal I–beam monorail, and diagonal rod.

CRITERIA:
Hoist weight	= 250 lbs.
Max. lifted load	= 3000 lbs.
Impact factor	= 1.25 (on lifted load).

Materials:
- All members Grade A36 steel
- Rod = ¾ inch diameter.
- Beam = S5 × 10 (I beam).
- Column = 8" nominal diameter, standard weight pipe.

Assumptions:
- Neglect the rod and beam weight
- Neglect deflections
- Assume all connections are adequate
- Neglect eccentricity of all connections
- Assume support conditions shown as "(PIN)" or "(FIXED)"

REQUIRED:
1. Maximum design axial stress in the rod, and its adequacy to support the design vertical loads.
2. Maximum design axial and bending stresses in the beam, and its adequacy to support the design vertical loads.
3. Maximum design axial and bending stresses in the column, and its adequacy to support the design vertical loads.

SOLUTION:

APPLIED LOAD

The maximum applied load due to the hoist weight and the lifted load is given by

$$W = (250 + 1.25 \times 3000)/1000$$
$$= 4 \text{ kips}$$

1. TIE ROD DESIGN STRESS

The loading case which produces the maximum rod force is shown in Figure 88B-2. The maximum tensile force in the rod is given by

$$P_{34} = W/\sin\theta$$
$$= 4/\sin 33.69°$$
$$= 7.21 \text{ kips}$$

The rod area A is 0.4418 square inches and

$$f_t = P_{34}/A$$
$$= 7.21/0.4418$$
$$= 16.32 \text{ kips per square inch}$$

The allowable tensile stress, for a pin-connected member, is given by UBC Section 2251.D1 as

$$F_t = 0.60F_y$$
$$= 0.60 \times 36$$
$$= 21.60 \text{ kips per square inch}$$
$$> 16.32$$

Hence the ¾ inch diameter rod is satisfactory.

2. BEAM DESIGN STRESS

The loading case which produces the maximum beam stress is shown in Figure 88B-2. The beam is subjected to combined axial compression and bending and the forces acting on the beam are

$$\text{Mid-span moment} = M_5 = WL/4$$
$$= 4 \times 6 \times 12/4$$
$$= 72 \text{ kip inches}$$
$$\text{Axial compression} = P_{24} = W/(2\tan\theta)$$
$$= 4/(2\tan 33.69°)$$
$$= 3 \text{ kips}$$

The relevant properties of the S5 × 10 beam are

$$
\begin{aligned}
F_y &= 36 \text{ kips per square inch} \\
F_y' &= \text{No value} \\
S_x &= 4.92 \text{ inches}^3 \\
A &= 2.94 \text{ inches}^2 \\
r_y &= 0.643 \text{ inches} \\
L_c &= 3.2 \text{ feet} \\
L_u &= 9.1 \text{ feet (for } C_b = 1)
\end{aligned}
$$

The compressive stress produced in the beam flange by the bending moment is

$$
\begin{aligned}
f_b &= M_5/S_x \\
&= 72/4.92 \\
&= 14.63 \text{ kips per square inch}
\end{aligned}
$$

Since no value is listed for F_y', the section is compact and the allowable bending stress is dependent on the value of the maximum unbraced length of the compression flange which is assumed to be

$$
\ell = 6 \text{ feet}
$$

The bending coefficient is obtained from UBC Section 2251.F1.3 as

$$
C_b = 1.00
$$

Hence
$$
L_c < \ell < L_u
$$

and the allowable bending stress is given by

$$
\begin{aligned}
F_b &= 0.6F_y \\
&= 0.6 \times 36 \\
&= 21.6 \text{ kips per square inch}
\end{aligned}
$$

The slenderness ratio about the Y-axis governs and is given by

$$
\begin{aligned}
K\ell/r_y &= 1.00 \times 6 \times 12/0.643 \\
&= 112
\end{aligned}
$$

The allowable compressive stress is obtained from AISC Table C-36 as

$$
F_a = 11.4 \text{ kips per square inch}
$$

The actual compressive stress in the beam is given by

$$
\begin{aligned}
f_a &= P_{24}/A \\
&= 3/2.94 \\
&= 1.02 \text{ kips per square inch} \\
&< 0.15F_a
\end{aligned}
$$

Hence in checking the adequacy of the column in combined axial compression and bending, in accordance with UBC Section 2251.H1, Equation (H1-3) is applicable.

Equation (H1–3) is given by

$$f_a/F_a + f_b/F_b \leq 1$$

The left hand side of the Equation is evaluated as

$$
\begin{aligned}
1.02/11.4 + 14.63/21.6 \quad &= 0.09 + 0.68 \\
&= 0.77 \\
&< 1.0
\end{aligned}
$$

Hence the S5 × 10 beam is satisfactory.

3. COLUMN DESIGN STRESS

The loading case, which produces the maximum column stress, is shown in Figure 88B–2 together with the bending moment in the column. The column is subjected to combined axial compression and bending and the forces acting are

$$
\begin{aligned}
\text{Bending moment } M_2 \quad &= W \times 6 \times 12 \\
&= 288 \text{ kip inches}
\end{aligned}
$$

and, allowing for the column self weight,

$$
\begin{aligned}
\text{Axial compression } P_{13} \quad &= W + 11 \times 0.0286 \\
&= 4.31 \text{ kips}
\end{aligned}
$$

The relevant properties of the eight inch nominal diameter standard pipe are

$$
\begin{aligned}
F_y \quad &= 36 \text{ kips per square inch} \\
S \quad &= 16.8 \text{ inches}^3 \\
D \quad &= 8.625 \text{ inches} \\
t \quad &= 0.322 \text{ inches}
\end{aligned}
$$

The compressive stress produced in the pipe by the bending moment is

$$
\begin{aligned}
f_b \quad &= M_2/S \\
&= 288/16.8 \\
&= 17.14 \text{ kips per square inch}
\end{aligned}
$$

The diameter–thickness ratio of the pipe is

$$
\begin{aligned}
D/t \quad &= 8.625/0.322 \\
&= 26.79 \\
&< 3300/F_y
\end{aligned}
$$

Hence the section is compact in accordance with UBC Section 2251.B5 Table B5.1 and the allowable bending stress is given by

$$F_b = 0.66F_y$$
$$= 23.8 \text{ kips per square inch}$$

The effective length factor for the column is given by AISC Table C–C2.1 as

$$K = 2.1 \text{ and the effective length is}$$
$$K\ell = 2.1 \times 11$$
$$= 23.1 \text{ feet}$$

The allowable axial load is obtained from AISC Table 3–36 as

$$P_a = 115 \text{ kips}$$

The load ratio is

$$P_{13}/P_a = 4.31/115$$
$$= 0.037$$
$$< 0.15$$

Hence Equation (H1–3) is applicable and

$$f_a/F_a + f_b/F_b = P_{13}/P_a + f_b/F_b$$
$$= 0.037 + 17.14/23.8$$
$$= 0.757$$
$$< 1.0$$

Hence the pipe column is adequate.

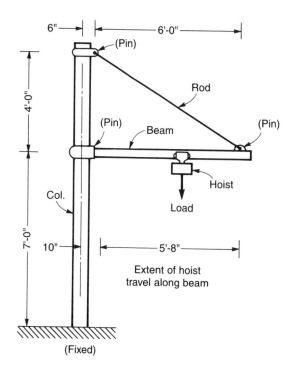

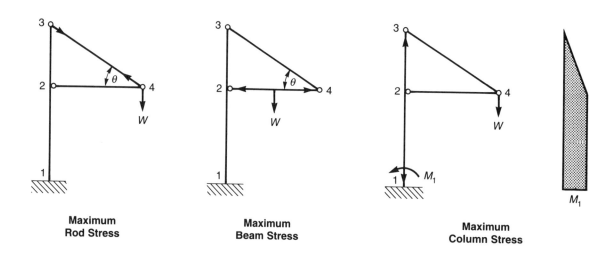

Maximum
Rod Stress

Maximum
Beam Stress

Maximum
Column Stress

Figure 88B–2

186

CHAPTER 2

STRUCTURAL STEEL DESIGN

SECTION 2.2

CONNECTION DESIGN

PROBLEMS

STRUCTURAL ENGINEER EXAMINATION – 1990 ══════════════════

PROBLEM B–3 – WT. 6.0 POINTS

GIVEN: The partial elevation of steel framing in a building.
- Dead load   = as shown on the framing elevation
- Live load   = none

CRITERIA: Materials:
- A325–SC (slip critical) bolts
- Steel beams Grade A36, F_y = 36 ksi
- E70XX Electrodes

Assumptions:
- Use of Tables in steel manual is acceptable.
- The dead load of framing members is included in the loads shown, i.e., neglect dead load of members.
- Adequate lateral support is provided only at the ends of members.

REQUIRED:
1. Select the lightest W14× beam to support an up or down load of 73 kips. Neglect shear and deflection.
2. Select the lightest double angle hanger for the axial load. Indicate the size and spacing of any required spacer plates. Show the capacity of the double angles per AISC tables.
3. Select the number of ¾ inch diameter A325–SC bolts for the connection of the double angles to the W14× beam. Show capacity of the bolts per AISC Tables.
4. Determine the length "L_4" of the WT8 × 22.5 hanger connection using AISC table "Preliminary Selection Table for Hanger Type Connection". Note that the beam stiffener plate is adequate as shown.
5. Select a frame beam connection with A325–SC bolts with short–slotted holes. Determine the thinnest 4 × 3 double angles, the number of ¾ inch diameter bolts required, and the length L_5. Show the capacity of the connection per AISC tables.
6. Select a welded, stiffened, seated beam connection. Determine the smallest "W_6" and "L_6" dimensions, the thinnest plate thickness t_{p1} and t_{p2}, and the thinnest weld t_w. Show the capacity of the connection per AISC tables.

SOLUTION

1. BEAM SELECTION

The maximum bending moment in the beam is given by

$$M = 73 \times 7 \times 3/10 = 153.3 \text{ kip feet}$$

The unbraced length of the beam is

$$\ell = 10 \text{ feet}$$

From AISC Table 2-11, a suitable W14 section may be selected.

A W14 $\times$ 53 is unsatisfactory as its value for the maximum unbraced length at which the allowable bending stress may be taken as $0.66F_y$ is

$$L_c = 8.5 \text{ feet} < \ell$$

A W14 $\times$ 61 is satisfactory as it has a value of

$$L_c = 10.6 \text{ feet} > \ell$$

and its resisting moment is

$$M_R = 183 \text{ kip feet} > M$$

2. HANGER SELECTION

The design of the hanger is governed by the maximum applied compressive load, with an impact allowance of 33 percent as specified by UBC Section 2251.A4.2. Hence the design load is

$$P = 1.33 \times 73 \times 3/10 = 29.13 \text{ kips}$$

The effective length factor, assuming pinned ends, is given by AISC Table C-C2.1 as

$$K = 1.0$$

and the effective length about both axes is

$$K\ell = 1.0 \times 14 = 14 \text{ feet}$$

From AISC Table 3-71, neglecting eccentricity, a double angle hanger $4 \times 3\frac{1}{2} \times \frac{1}{4}$, with long legs back to back, has an allowable axial load of

$$\begin{aligned} P_a &= 31 \text{ kips} \\ &> P \end{aligned}$$

and this is the lightest double angle which can support the applied load.

To conform to the requirements of AISC Table 3-71, a spacing of $\frac{3}{8}$ inch must be provided between the angles. A suitable size for spacer plates is $4 \times 4 \times \frac{3}{8}$.

The governing slenderness ratio of the double angle section is

$$\begin{aligned} K\ell/r_x &= 1.0 \times 14 \times 12/1.27 \\ &= 132 \end{aligned}$$

Using two spacer plates at the third points of the double angles, the local slenderness ratio of the individual angles is given by

$$\begin{aligned} K\ell/r_z &= 1.0 \times 14 \times 12/(3 \times 0.734) \\ &= 76 \\ &< 132 \times 0.75 \ldots \text{satisfactory} \end{aligned}$$

Hence, the compound section satisfies UBC Section 2251.E4 which requires that the slenderness ratio of the individual angles, between fasteners, shall not exceed three-fourths times the governing slenderness ratio of the build-up member. In addition, to comply with AISC 3-53, a minimum of two connectors are required for adequate shear transfer.

3. SELECTION OF BOLTS

The maximum load on the bolts is

$$P = 29.13 \text{ kips}$$

From AISC Table IIA, the double shear value of two $\frac{3}{4}$ inch diameter A325-SC bolts using $\frac{1}{4}$ inch angles is

$$\begin{aligned} P_s &= 30 \text{ kips} \\ &> P \ldots \text{satisfactory} \end{aligned}$$

Since the beam is coped, web tear out must be checked. Using two $\frac{3}{4}$ inch diameter bolts with edge distances of

$$\ell_v = \ell_h = 1.5 \text{ inches}$$

AISC Table 1–G indicates the value of the coefficients

$$C_1 = 1.2$$
$$C_2 = 0.33$$

Hence the block shear value is

$$
\begin{aligned}
P_r &= (C_1 + C_2)F_u t \\
&= (1.2 + 0.33)58 \times 0.375 \\
&= 33.3 \text{ kips} \\
&> P
\end{aligned}
$$

Hence two ¾ inch diameter A325–SC bolts are adequate.

4. HANGER CONNECTION

The relevant properties of the WT8 $\times$ 22.5 are

$$t_w = 0.375 \text{ inches}$$
$$t_f = 9/16 \text{ inches}$$

The design of the hanger connection is governed by the maximum applied downward load which is

$$P = 29.13 \text{ kips}$$

Assuming four bolts are used which are located on the beam gage of 4 inches, to provide for wrench clearance, the distance from bolt centerline to face of the section stem is

$$
\begin{aligned}
b &= (4 - t_w)/2 \\
&= (4 - 0.345)/2 \\
&= 1.83 \text{ inches.}
\end{aligned}
$$

From AISC Table 4–89, using this value of b and the flange thickness of 9/16 inches, the permissible load in kips per inch, allowing for prying action, is obtained as

$$q = 3.14 \text{ kips per inch}$$

Hence the required length of the hanger connection is

$$
\begin{aligned}
L_4 &= P/q \\
&= 29.13/3.14 \\
&= 9.5 \text{ inches}
\end{aligned}
$$

5. FRAMED BEAM CONNECTION

The maximum reaction on the connection is

$$R = 21.9 \times 4/14 + 4 \times 14/2$$
$$= 34.26 \text{ kips}$$

From AISC Table IIB, the double shear value of three ¾ inch diameter A325–SC bolts with short–slotted holes in $4 \times 3 \times$ ¼–inch angles is

$$P_s = 39.8 \text{ kips}$$
$$> R \text{ ...satisfactory}$$

To allow for wrench clearance, using the normal angle gage, holes in the angles must be staggered and the length required for the angles is given in AISC Table IIB as

$$L_5 = 10 \text{ inches}$$

6. STIFFENED, SEATED BEAM CONNECTION

The relevant properties of the W14 $\times$ 61 beam, obtained from AISC Table 2–65, are

$$R_1 = 32 \text{ kips}$$
$$R_2 = 8.91$$
$$R_3 = 37.6$$
$$R_4 = 3.6$$
$$t_w = \text{⅜ inch}$$

The maximum reaction on the connection is

$$R = 21.9 \times 10/14 + 4 \times 14/2$$
$$= 43.64 \text{ kips}$$

From AISC 2–36, the length of bearing required to prevent local web yielding is

$$N_y = (R - R_1)/R_2$$
$$= (43.64 - 32)/8.91$$
$$= 1.30 \text{ inches}$$

The length of bearing required to prevent web crippling is

$$N_c = (R - R_3)/R_4$$
$$= (43.64 - 37.6)/3.6$$
$$= 1.68 \text{ inches}$$

Adopting a practical length of 3.5 inches for the length of bearing and using a 0.5 inch setback gives a seat width of

$$W_6 = 3.5 + 0.5$$
$$= 4 \text{ inches}$$

192

The web thickness of the W12 $\times$ 65 column is $\frac{3}{8}$ inch and the minimum allowable fillet weld size is given by the UBC Section 2251 Table J2.4 as

$$t_w = 3/16 \text{ inch}$$

The stiffener plate thickness required to develop the welds is

$$
\begin{aligned}
t_{p2} &= 2 \times t_w \\
&= 2 \times 3/16 \\
&= \tfrac{3}{8} \text{ inch}
\end{aligned}
$$

and this is satisfactory as it is not less than the thickness of the column web.

A suitable thickness for the seat plate is

$$t_{p1} = \tfrac{3}{8} \text{ inch}$$

Adopting a length for the stiffener plate of

$$L_6 = 11 \text{ inches}$$

the capacity of the connection is obtained from AISC Table VIII as

$$
\begin{aligned}
R_w &= 47.9 \text{ kips} \\
&> R
\end{aligned}
$$

and the connection is adequate.

The minimum length of weld required between the stiffener plate and the edge of the seat plate is

$$
\begin{aligned}
L_w &= 0.2L_6 \\
&= 0.2 \times 11 \\
&= 2.2 \text{ inches}
\end{aligned}
$$

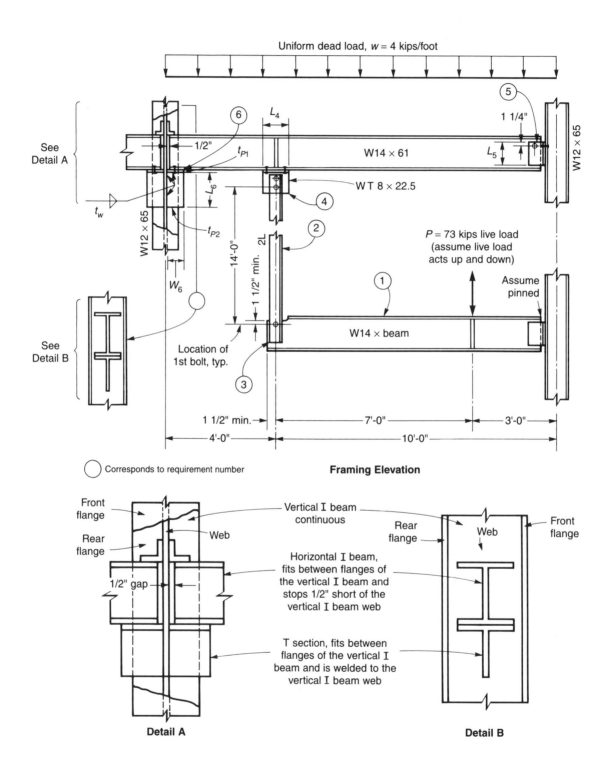

Uniform dead load, $w = 4$ kips/foot

See Detail A

t_w

W12 × 65

t_{P1}

t_{P2}

W_6

See Detail B

Location of 1st bolt, typ.

1/2"

L_6

14'-0"

1 1/2" min. 2L

6

4

WT 8 × 22.5

2

1

3

W14 × 61

L_4

L_5

1 1/4"

W12 × 65

5

$P = 73$ kips live load (assume live load acts up and down)

Assume pinned

W14 × beam

1 1/2" min.   7'-0"   3'-0"

4'-0"   10'-0"

⬭ Corresponds to requirement number

Framing Elevation

Front flange

Rear flange

1/2" gap

Web

Vertical I beam continuous

Horizontal I beam, fits between flanges of the vertical I beam and stops 1/2" short of the vertical I beam web

T section, fits between flanges of the vertical I beam and is welded to the vertical I beam web

Rear flange

Web

Front flange

Detail A

Detail B

FIGURE 90B–3

194

STRUCTURAL ENGINEER EXAMINATION – 1989

PROBLEM B-1 – WT. 5.0 POINTS

GIVEN: Shop drawings have been returned to you for consideration. The fabricator has substituted different connections from the ones originally prepared by your office.

CRITERIA: Materials:
- Grade A36 steel
- A325–SC bolts
- E70XX welding rods

Assumptions:
- All shop welds

REQUIRED:
1. Check the proposed detail connection "A" for the following design requirements:
   - A. Tension and compression force on the stiffener plates.
   - B. Weld of stiffener plates to column web.
   - C. Shear weld at beam to column.
2. Check the proposed detail connection "B" for the following design requirements:
   - A. Brace and beam connection to the column.
   - B. Beam connection on the left side of the column.
3. Identify two areas of concern regarding each of the proposed detail connections "A" and "B".

SOLUTION

1A. STIFFENER PLATE REQUIREMENTS

The relevant properties of the W12 × 26 beam are

Yield stress F_{yb}	= 36 kips per square inch
Flange width b	= 6.49 inches
Flange thickness t_b	= 0.38 inches

The relevant properties of the W8 × 40 column are

Yield stress F_{yc}	= 36 kips per square inch
Web thickness t	= 0.36 inches
Fillet depth k	= 1.0625 inches
Section depth d	= 8.25 inches
Flange thickness t_f	= 0.56

The relevant properties of the stiffeners are

Yield stress F_{yst} = 36 kips per square inch
Thickness t_{st} = ⅜ inch

The force in the beam flange is defined as

$$
\begin{aligned}
P_{bf} &= A_{fb}F_{yb} \\
&= t_b b F_{yb} \\
&= 6.49 \times 0.38 \times 36 \\
&= 88.8 \text{ kips}
\end{aligned}
$$

The required stiffener area, at the bottom compression flange, is given by UBC Section 2251 Equation (K1–9) as

$$A_{st} \geq [P_{bf} - F_{yc}t(t_b + 5k]/F_{yst}$$

Using calculator program A4.4 in Appendix A, the value is

$$A_{st} = 0.42 \text{ square inches}$$

The required stiffener area at the top tension flange is given by

$$
\begin{aligned}
A_{st} &= [P_{bf} - F_{yc}t(t_b + 2.5k)]/F_{yst} \\
&= [88.8 - 36 \times 0.36(0.38 + 2.5 \times 1.0625)]/36 \\
&= 1.37 \ldots \text{ governs}
\end{aligned}
$$

Hence, stiffeners are required and it is unnecessary to check UBC Equations (K1–1) and (K1–8) since the stiffeners are provided over the full depth of the column web.

To conform to UBC Section 2251.K1.8, the minimum thickness required for the stiffeners is

$$
\begin{aligned}
t_{st} &= t_b/2 \\
&= 0.38/2 \\
&= 0.19 \text{ inches} \\
&< 0.375
\end{aligned}
$$

Hence the thickness provided is satisfactory.

The minimum width required for the stiffeners is

$$
\begin{aligned}
b_{st} &= b/3 - t/2 \\
&= 6.49/3 - 0.36/2 \\
&= 1.98 \text{ inches}
\end{aligned}
$$

Provide a stiffener plate with a width of two and a half inches.

The stiffener area, after allowing for a half inch corner snip is

$$
\begin{aligned}
A_{st} &= 2 \times 0.375 \times 2 \\
&= 1.5 \text{ square inches} \\
&> 1.37
\end{aligned}
$$

Hence the stiffener provided satisfies UBC Section 2251 Equation (K1–9).

The maximum width–thickness ratio is given by UBC Section 2251 Table B5.1 as

$$95/(F_y)^{0.5} = 15.8$$

The width–thickness ratio provided is $2 \times 8/3 = 5.3$ ... satisfactory.

1B. STIFFENER WELD REQUIREMENTS

The web thickness of the W8 × 40 column is ⅜ inch and the minimum allowable fillet weld size is given by UBC Section 2251 Table J2.4 as

$$t_w \qquad = 3/16 \text{ inch}$$

and this is also satisfactory for the ⅜ inch stiffener plate.

The welds connecting the stiffeners to the column web must develop the required stiffener force

$$
\begin{aligned}
P_{st} \quad &= A_{st}F_y \\
&= 1.37 \times 36 \\
&= 49.32 \text{ kips}
\end{aligned}
$$

The allowable force on a 1/16 inch fillet weld is governed by the strength of the weld metal and is obtained from UBC Section 2251 Table J2–5 as

$$q_u \qquad = 0.928 \text{ kips per inch per } 1/16 \text{ inch weld}$$

The minimum size of the two lengths of single fillet required is

$$
\begin{aligned}
D \quad &= P_{st}/(2q_u)(d - 2t_f - 1.0) \\
&= 49.32/(2 \times 0.928)(8.25 - 2 \times 0.56 - 1.0) \\
&= 4.33
\end{aligned}
$$

Hence a 5/16 weld is required.

1C. BEAM WEB CONNECTION

The beam reaction is

$$R \qquad = 32 \text{ kips}$$

The minimum length of a single 3/16 inch fillet weld required for the web connection is

$$
\begin{aligned}
\ell_w \quad &= R/q \\
&= 32/(3 \times 0.928) \\
&= 11.5 \text{ inches}
\end{aligned}
$$

The length provided almost equals this. However, the flange thickness of the W8 × 40 column is 9/16 inch and the minimum allowable fillet weld size is given by UBC Section 2251 Table J2.5 as

$$t_w \qquad = \text{¼ inch}$$

Hence the 3/16 inch fillet weld provided is unsatisfactory.

2A. CONNECTION ON RIGHT SIDE OF COLUMN

The relevant properties of the W8 × 15 column are

Yield stress F_y	= 36 kips per square inch
Depth of section d	= 8.11 inches
Flange thickness t	= 5/16 inches

The relevant properties of the W12 × 26 beam are

Yield stress F_y	= 36 kips per square inch
Depth of section d	= 12.22 inches

(i) SHEAR PLATE AND WELD

The flange thickness of the W8 × 15 column is 5/16 inch and the minimum allowable fillet weld size is given by UBC Section 2251 Table J2.4 as

$$t_w \quad = 3/16 \text{ inch}$$

The 3/16 inch weld indicated is therefore satisfactory in this respect.

The geometrical properties of the single weld are obtained by assuming unit size of weld. Then, the length of the weld is

$$
\begin{aligned}
L \quad &= 12.22/2 + 3 + 4 + 1.25 \\
&= 14.36 \text{ inches}
\end{aligned}
$$

The weld area is

$$
\begin{aligned}
A \quad &= 1.0 \times L \\
&= 14.36 \text{ inches}^2 \text{ per inch}
\end{aligned}
$$

The modulus at the extreme ends of the weld is

$$
\begin{aligned}
S \quad &= 1.0 \times L^2/6 \\
&= 14.36^2/6 \\
&= 34.37 \text{ inches}^3 \text{ per inch}
\end{aligned}
$$

The applied loading on the weld is given by

Vertical shear V	= 48 – 10 – 12
	= 26 kips
Horizontal shear H	= 13 kips
Moment M	= (48 – 10) × 8.11/2 – 12(8.11 + 3 – 1.25)
	+ 13 (14.36/2 – 3 – 1.25)
	= 73.86 kip inches

198

The co-existent forces acting at the extreme ends of the unit weld are

Vertical force due to vertical shear $= V/A$
$= 26/14.36$
$= 1.81$ kips per inch

Horizontal force due to horizontal shear $= H/A$
$= 13/14.36$
$= 0.91$ kips per inch

Horizontal force due to eccentricity moment $= M/S$
$= 73.86/34.37$
$= 2.15$ kips per inch

Total horizontal force $= 0.91 + 2.15$
$= 3.06$ kips per inch

The resultant force at the ends of the weld is given by

$R \quad = (1.81^2 + 3.06^2)^{0.5}$
$= 3.55$ kips per inch

The allowable force on 1/16 inch E70XX grade fillet weld is obtained from UBC Section 2251 Table J2.5 as

$q \quad = 0.928$ kips per inch per 1/16 inch

The required weld size in 1/16 inch is

$D \quad = 3.55/0.928$
$= 3.83$

Hence a single ¼ inch fillet weld is required and the 3/16 inch weld indicated is unsatisfactory.

The geometrical properties of the shear plate are

Area $A \quad = 14.36 \times 3/16$
$= 2.69$ square inches ... at the column flange

Modulus $S \quad = 34.37 \times 3/16$
$= 6.44$ inches^3 ... at the column flange

Area $A' \quad = (14.36 - 3 \times 11/16)3/16$
$= 2.31$ square inches ... at the line of bolts

The shear stress in the plate is

$f_v \quad = V/A'$
$= 26/2.31$
$= 11.28$ kips per square inch ... at the line of bolts

The allowable shear stress is given by UBC Section 2251.F4 as

$F_v \quad = 0.4F_y$
$= 14.4$ kips per square inch
> 11.28

The flexural stress in the plate is

$$f_b = M/S$$
$$= 73.86/6.44$$
$$= 11.47 \text{ kips per square inch ... at the column flange}$$

The allowable flexural stress is given by UBC Section 2251.F2 as

$$F_b = 0.6F_y$$
$$= 21.6$$
$$> 11.47$$

The tensile stress in the plate is

$$f_t = H/A$$
$$= 13/2.69$$
$$= 4.83 \text{ kips per square inch ... at the column flange}$$

The combined tensile and flexural stress must satisfy UBC Section 2251 Equation (H2–1)

$$f_a/0.6F_y + f_b/F_b \leq 1.0$$
$$4.83/21.6 + 11.47/21.6 = 0.75$$
$$< 1.0$$

Hence the 3/16 inch plate is satisfactory.

(ii) BLOCK SHEAR IN THE BEAM WEB

The block shear capacity is given by UBC Section 2251.J.4 as

$$R_{BS} = 0.3A_vF_u + 0.5A_tF_u$$
$$= 0.3 \times 58 \times 1.43 + 0.5 \times 58 \times 0.21$$
$$= 30.93 \text{ kips}$$
$$> 12 \text{ ... satisfactory}$$

(iii) BOLT GROUP

The geometrical properties of the bolt group are obtained by applying the unit area method. The distance between the top and bottom bolt is

$$h = 14.36 - 2.5$$
$$= 11.86 \text{ inches}$$

The height of the bolt group centroid is

$$d = (11.86 + 6)/3$$
$$= 5.95 \text{ inches}$$

The inertia of the bolt group about its centroid is

$$I = 5.95^2 + 0.05^2 + 5.91^2$$
$$= 70.33 \text{ inches}^4$$

The modulus of the bottom bolt is

$$S = 70.33/5.95$$
$$= 11.82 \text{ inches}^3$$

200

The applied loading acting on the bolt group centroid is

Vertical shear V = 26 kips

Horizontal shear H = 13 kips

Eccentricity moment M = $38(8.11/2 + 1.75) - 12(8.11 + 2 \times 1.75)$
$+ 13(5.95 - 3)$
$= 119.62$ kip inches

The co-existent forces on the bottom bolt are

Vertical force due to vertical shear = V/3
= 26/3
= 8.67 kips

Horizontal force due to horizontal shear = H/3
= 13/3
= 4.33 kips

Horizontal force due to eccentricity moment = M/S
= 119.62/11.82
= 10.12 kips

Total horizontal force = 4.33 + 10.12
= 14.45 kips

The resultant force on the bottom bolt is given by

R $= (8.67^2 + 14.45^2)^{0.5}$
$= 16.85$ kips

The allowable single shear value of a ⅝ inch diameter A325-SC bolt is obtained from AISC Table I-D as

P_s = 5.22 kips

This is less than the applied force of 16.85 kips and the ⅝ inch diameter bolts are inadequate. To resist the applied forces, the number of bolts used and their diameter must be increased.

(iv) WELDING OF BRACE

The allowable force on the E70XX Grade 3/16 inch fillet weld is governed by the shear strength of the base metal and is obtained from UBC Section Table J2.5 as

q $= 3 \times 0.928$
$= 2.78$ kips per inch

The total allowable load on the four inch length of weld provided is

P_a $= 2.78 \times 4$
$= 11.1$ kips

This is less than the applied force of twenty kips and an eight inch length of weld must be provided by welding to both sides of the gusset plates. However, to comply with UBC Section 2251.J2.2b, the weld length in each run must not be less than the width of the 3 × 3 tubular section. Hence, four lengths of weld metal are required, each three inches long.

2B. CONNECTION ON LEFT SIDE OF COLUMN

(i) SHEAR PLATE AND WELD

The geometrical properties of the single weld are obtained by assuming unit size of weld. Then the length of the weld is

$$L = 6 + 2 \times 1.25$$
$$= 8.5 \text{ inches}$$

The weld area is

$$A = 1.0 \times 8.5$$
$$= 8.5 \text{ inches}^2 \text{ per inch}$$

The modulus at the ends of the weld is

$$S = 1.0 \times L^2/6$$
$$= 8.5^2/6$$
$$= 12.04 \text{ inches}^3 \text{ per inch}$$

The applied loading on the weld is given by

Vertical shear V	$= 12 \text{ kips}$
Horizontal shear H	$= 13 \text{ kips}$
Eccentricity moment M	$= 12 \times 1.75$
	$= 21 \text{ kip inches}$

The co-existent forces acting at the ends of the weld are

Vertical force due to vertical shear	$= V/A$
	$= 12/8.5$
	$= 1.41 \text{ kips per inch}$
Horizontal force due to horizontal shear	$= H/A$
	$= 13/8.5$
	$= 1.53 \text{ kips per inch}$
Horizontal force due to eccentricity moment	$= M/S$
	$= 21/12.04$
	$= 1.74 \text{ kips per inch}$
Total horizontal force	$= 1.53 + 1.74$
	$= 3.27 \text{ kips per inch}$

The resultant force at the ends of the weld is given by

$$R = (1.41^2 + 3.27^2)^{0.5}$$
$$= 3.57 \text{ kips per inch}$$

202

The required weld size in 1/16 inch is

$$D = 3.57/0.928$$
$$= 3.85$$

Hence a single ¼ inch fillet weld is required and the 3/16 inch weld indicated is unsatisfactory.

The geometrical properties of the shear plate are

Area A $= 8.5 \times 3/16$
$= 1.59$ square inches
Modulus S $= 23.04 \times 3/16$
$= 2.25$ inches^3

The shear stress in the plate is

$$f_v = V/A$$
$$= 12/1.59$$
$$= 7.55 \text{ kips per square inch}$$

The allowable shear stress is given by UBC Section 2251.F4 as

$$F_v = 0.4F_y$$
$$= 14.4 \text{ kips per square inch}$$
$$> 7.55$$

The flexural stress in the plate is

$$f_b = M/S$$
$$= 21/2.25$$
$$= 9.33 \text{ kips per square inch}$$

The allowable flexural stress is given by UBC Section 2251.F2 as

$$F_b = 0.6F_y$$
$$= 21.6 \text{ kips per square inch}$$
$$> 9.33$$

The tensile stress in the plate is

$$f_t = H/A$$
$$= 13./1.59$$
$$= 8.18 \text{ kips per square inch}$$

The combined tensile and flexural stress must satisfy the UBC Section 2251 Equation (H2-1)

$$f_a/0.6F_y + f_b/F_b \leq 1.0$$
$$8.18/21.6 + 9.33/21.6 = 0.81$$
$$< 1.0$$

Hence the 3/16 inch plate is satisfactory

(ii) BOLT GROUP

For a flexible bolted connection, in accordance with AISC page 4–9, it is permissible to ignore the eccentricity produced by the vertical reaction at the center line of the bolt group. The co-existent forces on the bolts are

$$
\begin{aligned}
\text{Vertical force due to vertical shear} \quad &= V/2 \\
&= 12/2 \\
&= 6 \text{ kips} \\
\text{Horizontal force due to horizontal shear} \quad &= H/2 \\
&= 13/2 \\
&= 6.5 \text{ kips}
\end{aligned}
$$

The resultant force on the bottom bolt is given by

$$
\begin{aligned}
R \quad &= (6^2 + 6.5^2)^{0.5} \\
&= 8.85 \text{ kips}
\end{aligned}
$$

The allowable single shear value of a ⅝ inch diameter A325–SC bolt is obtained from AISC Table 1D as

$$
P_s \quad = 5.22 \text{ kips}
$$

This is less than the applied force of 8.85 kips and the ⅝ inch diameter bolts are inadequate. To resist the applied forces, three ¾ inch diameter A325–SC bolts are required.

3. AREAS OF CONCERN

In the case of connection "A", it is unnecessarily expensive to use a groove weld to connect the stiffeners to the column flange. In addition, this detail is liable to cause lamella tearing in the column flange.

Welding the beam web to the column flange requires very close tolerances in cutting the beam to size. The use of a shear plate is more practical.

In the case of connection "B", the bolted shear plate connections and the welded brace connection are inadequate.

The indicated single-sided fillet welds of the stiffener plates to the column flanges are unable to sustain accidental transverse loads.

No connection is indicated for the 7 × 4 × 3/16 gusset plate to the beam. ════════

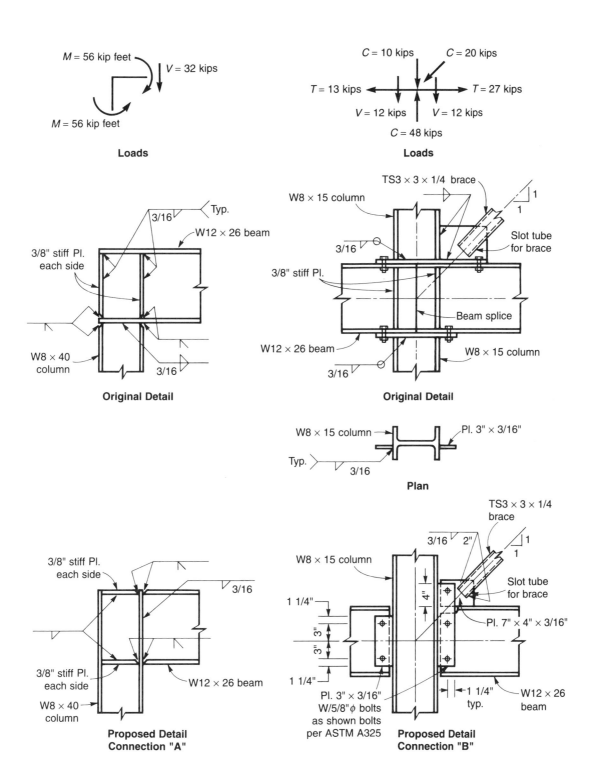

Loads

Loads

Original Detail

Original Detail

Plan

Proposed Detail Connection "A"

Proposed Detail Connection "B"

FIGURE 89B–1

205

STRUCTURAL ENGINEER EXAMINATION – 1989

PROBLEM B–3 – WT. 5.0 POINTS

GIVEN: Structural steel framing supported on concrete corbels and walls as shown in the building cross–section. Loadings: Vertical loads on member "C" and a moving load on member "A" are shown in the cross–section.

CRITERIA: Materials:
- Grade A36 steel
- E70XX welding rods

Assumptions:
- Do not consider weight of steel members.
- Allowance for impact and lateral loads is not required.
- Ignore deflections.

REQUIRED:
1. Determine the lightest section for the depths shown of members "A", and "B" and "C".
2. Provide calculations and complete the sketch in the workbook of the welded connection for Detail "1" for the joint at the center line of member "C". Indicate on this sketch any welds and/or additional plates.

SOLUTION

1A. MEMBER "A"

The maximum moment occurs in member "A" with the twenty kip moving load at the limit of its operating length. The maximum moment, at the center of the beam, is

$$M \quad = 20 \times 5$$
$$= 100 \text{ kip feet}$$

The unbraced length of the beam is

$$\ell \quad = 5 \text{ feet}$$

From AISC Table 2–12, a suitable section is a W12 × 40. This has no value listed for F'_y, indicating that the section is compact. The maximum unbraced length at which the allowable bending stress may be taken as $0.66F_y$ is

$$L_c = 8.4 \text{ feet}$$
$$> \ell$$

Hence the indicated value of 103 kip feet for the resisting moment is applicable and exceeds the applied moment of 100 kip feet.

1B. MEMBER "B

The maximum moment in member "B" is

$$M = 100 \text{ kip feet}$$

and this occurs simultaneously with a tensile force of

$$T = 20 \text{ kips}$$

The unbraced length of the member is

$$\ell = 5 \text{ feet}$$

AISC Table 2–12 indicates that a W12 × 45, which has a L_c value in excess of ℓ, satisfies UBC Section 2251.H2 as the resisting moment is

$$M_R = 115 \text{ kip feet}$$
$$> M$$

The tensile stress in the member is

$$f_t = T/A$$
$$= 20/13.2$$
$$= 1.51 \text{ kips per square inch}$$

The combined tensile and flexural stress must satisfy the UBC Equation (H2–1)

$$f_t/0.6F_y + f_b/F_b \leq 1.0$$
or $$f_t/0.6F_y + M/M_R \leq 1.0$$
$$1.51/21.6 + 100/115 = 0.94$$
$$< 1.0$$

Hence the W12 × 45 section is satisfactory.

1C. MEMBER "C"

The moments and shears induced in member "C" by the applied loads are indicated in Figure 89B-3(ii). The maximum sagging moment in member "C" is given by

$$M = 45 \times 10 - 30 \times 3.33$$
$$= 350 \text{ kip feet}$$

The unbraced length of the member is

$$\ell = 6.67 \text{ feet}$$

AISC Table 2-11 indicates that a W24 × 76, which has a L_c value in excess of ℓ, is satisfactory in flexure as the resisting moment is

$$M_R = 348 \text{ kip feet}$$
$$\approx M$$

The maximum end shear on member "C" is 45 kips. AISC Table 2-55 indicates that a W24 × 76 has an allowable web shear of 152 kips. Hence the W24 × 76 is satisfactory.

2. JOINT DETAIL

The relevant properties of the W12 × 45 member "B" are

Yield stress F_{yb}	= 36 kips per square inch
Web thickness t_w	= 0.335 inches
Web depth d_w	= 10.91 inches
Flange width b	= 8.045 inches
Flange thickness t_b	= 0.575 inches
Depth d_B	= 12.06 inches
Area A_B	= 13.20 square inches

The relevant properties of the W24 × 76 member "C" are

Yield stress F_{yc}	= 36 kips per square inch
Web thickness t	= 0.44 inches
Fillet depth k	= 1.4375 inches
Web depth between fillets d_c	= 21 inches
Flange thickness t_f	= 0.680 inches
Depth d	= 23.92 inches

FLANGE CONNECTION

In order to develop the full flexural and axial capacity of the flanges of member "B", it is necessary for the flanges to be connected to member "C" with full penetration groove welds as detailed in SEAOC Section 4F.1.

WEB CONNECTION

No shear stress is developed in the web of member "B". However, the fillet weld connection between the web of member "B" and the flange of member "C" must develop the applied tensile force of

$$
\begin{aligned}
T_w &= TA_w/A_B \\
&= T \times t_w \times d_w/A_B \\
&= 20 \times 0.335 \times 10.91/13.2 \\
&= 5.54 \text{ kips}
\end{aligned}
$$

The tensile force developed in each flange of member "B" is

$$
\begin{aligned}
T_f &= (T - T_w)/2 \\
&= (20 - 5.54)/2 \\
&= 7.23 \text{ kips}
\end{aligned}
$$

The flange thickness of member "C" is 0.680 inches and the minimum allowable fillet weld size is given by UBC Section 2251 Table J2.4 as ¼ inch. The allowable force on a ¼ inch fillet weld is governed by the strength of the weld metal and is obtained from UBC Section 2251 Table J2.5 as

$$
\begin{aligned}
q &= 4 \times 0.928 \\
&= 3.71 \text{ kips per inch}
\end{aligned}
$$

The strength of the double ¼ inch fillet weld in tension is

$$
\begin{aligned}
P_w &= 2qd_w \\
&= 2 \times 3.71 \times 10.91 \\
&= 81 \text{ kips} \\
&> 5.54
\end{aligned}
$$

Hence the ¼ inch fillet weld is satisfactory.

PANEL ZONE SHEAR REQUIREMENTS

In member "C" the shear capacity of the web must be compared with the maximum shear force applied at the connection. This must take into account the tensile force applied[3], the forces in the flanges of member "B" due to the applied moment, and the co-existent shears in member "C". From AISC Section C-E6, the forces in the tension flange P_t and in the compression flange P_c of member "B" are given by

$$
\begin{aligned}
P_t &= M/(0.95d_B) + T_f \\
&= 100 \times 12/(0.95 \times 12.06) + 7.23 \\
&= 111.97 \\
P_c &= M/(0.95d_B) - T_f \\
&= 97.51
\end{aligned}
$$

The co-existent shears in member "C", on the left hand side of the connection V_L and on the right hand side of the connection V_R are given by

$$V_L = 15 \text{ kips}$$
$$V_R = 5 \text{ kips}$$

The resultant forces acting on the connection are shown in Figure 89-3(ii)

The maximum shear force acting on the web of member "C" is given by AISC Equation (C-E6-1) as

$$\Sigma F = P_c + V_R$$
$$= 97.51 + 5$$
$$= 102.51 \text{ kips}$$

From UBC Section 2251.F4 the web capacity is

$$F_w = 0.4 F_{yc} \times t \times d$$
$$= 0.4 \times 36 \times 0.44 \times 23.92$$
$$= 151.56$$
$$> \Sigma F$$

Hence the web of member "C" need not be reinforced.

STIFFENER PLATE REQUIREMENTS

The factored force in the flange of member "B" is defined in UBC Section 2251.K1.2 as

$$P_{bf} = 1.67 P_t$$
$$= 1.67 \times 111.97$$
$$= 187 \text{ kips}$$

The required stiffener area is given by UBC Section 2251 Equation (K1-9) as

$$A_{st} \geq \{P_{bf} - F_{yc}t(t_b + 5k)\}/F_{yst}$$

Using calculator program A4.4 in Appendix A, the value is

$$A_{st} = 1.78 \text{ square inches}$$

Hence stiffeners are required and, since the force is applied to only one flange of member "C", the stiffeners need not exceed one half the depth of member "C". A pair of stiffeners is required, opposite both the compression and the tension flanges, and these must be welded to the flange of member "C" as the applied moment is reversible.

To conform to UBC Section 2251.K1.8 the minimum thickness required for the stiffeners is

$$t_{st} = t_b/2$$
$$= 0.575/2$$
$$= 0.283 \text{ inches}$$

210

The minimum width required for the stiffeners is

$$b_{st} = b/3 - t/2$$
$$= 8.045/3 - 0.44/2$$
$$= 2.46 \text{ inches}$$

Using a pair of stiffener plates $3/8 \times 3.5 \times 11.5$ inches long provides a width–thickness ratio of

$$3.5 \times 8/3 = 9.33$$
$$< 95/(F_y)^{0.5}$$
$$= 15.8$$

as required by UBC Section 2251 Table B5.1

The stiffener area provided, after allowing for a one–inch corner snip, is

$$A_{st} = 2 \times 0.375 \times 2.5$$
$$= 1.875 \text{ square inches}$$
$$> 1.78$$

Hence the stiffeners provided satisfy UBC Equation (K1–9)

STIFFENER WELD REQUIREMENTS

From UBC Section 2251 Table J2.4, the minimum allowable fillet weld size to the 0.68 inch flange is ¼ inch and to the 0.44 inch web is 3/16 inch.

The force delivered to the weld at the compression flange is determined from UBC Section 2251 Equation (12–1) as

$$P_{st} = A_{st}F_{yst}$$
$$= 1.78 \times 36$$
$$= 64.08 \text{ kips}$$

The force delivered to the weld at the tension flange is determined by UBC Equation (K1–1) as

$$P_{st} = P_{bf} - t_f^2 F_{yc}/0.16$$
$$= 187 - 0.68^2 \times 36/0.16$$
$$= 82.96 \text{ kips ... governs}$$

The non–factored load on each stiffener is

$$P = 82.96/(2 \times 1.67)$$
$$= 24.84 \text{ kips}$$

The allowable force on a 1/16 inch fillet weld is governed by the strength of the weld metal and is obtained from UBC Section 2251 Table J2.5 as

$$q = 0.928 \text{ kips per inch per 1/16 inch}$$

The strength of a single 3/16 inch fillet weld for the full length of the stiffener is

$$
\begin{aligned}
P_w &= 3q(\ell_{st} - 1.0) \\
&= 3 \times 0.928 \times (11.5 - 1.0) \\
&= 29.23 \text{ kips} \\
&> P
\end{aligned}
$$

The strength of a double 5/16 inch fillet weld for the full width of the stiffener is

$$
\begin{aligned}
P_w &= 2 \times 5q(b_{st} - 1.0) \\
&= 2 \times 5 \times 0.928 \times (3.5 - 1.0) \\
&= 23.20 \text{ kips} \\
&> 53.6/(2 \times 1.67)
\end{aligned}
$$

The shear stress in the stiffener base metal is

$$
\begin{aligned}
f_v &= P/t_{st}(\ell_{st} - 1.0) \\
&= 19.19/0.375(11.5 - 1.0) \\
&= 4.87 \text{ kips per square inch}
\end{aligned}
$$

The allowable shear stress is obtained from UBC Section 2251.F4 as

$$
\begin{aligned}
F_v &= 0.4F_{yst} \\
&= 0.4 \times 36 \\
&= 14.4 \text{ kips per square inch} \\
&> f_v
\end{aligned}
$$

The shear stress in the beam web is

$$
\begin{aligned}
f_v &= 2P/t(\ell_{st} - 1.0) \\
&= 2 \times 19.19/0.44(11.5 - 1.0) \\
&= 8.31 \text{ kips per square inch} \\
&< F_v
\end{aligned}
$$

Hence the stiffener plates and welding are adequate and the completed detail is shown in Figure 89B-3(ii)

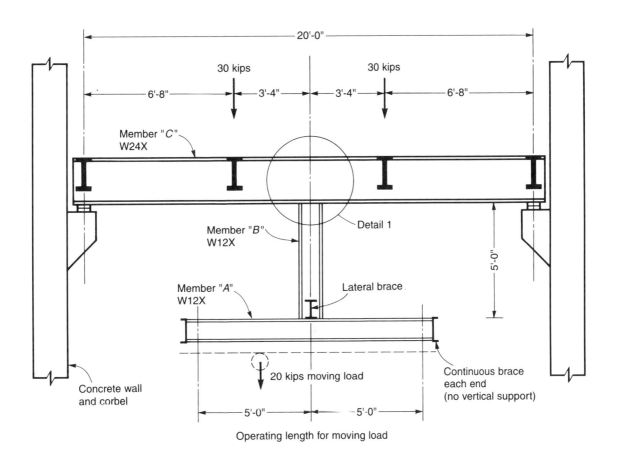

Building Cross Section

Figure 89B–3(i)

213

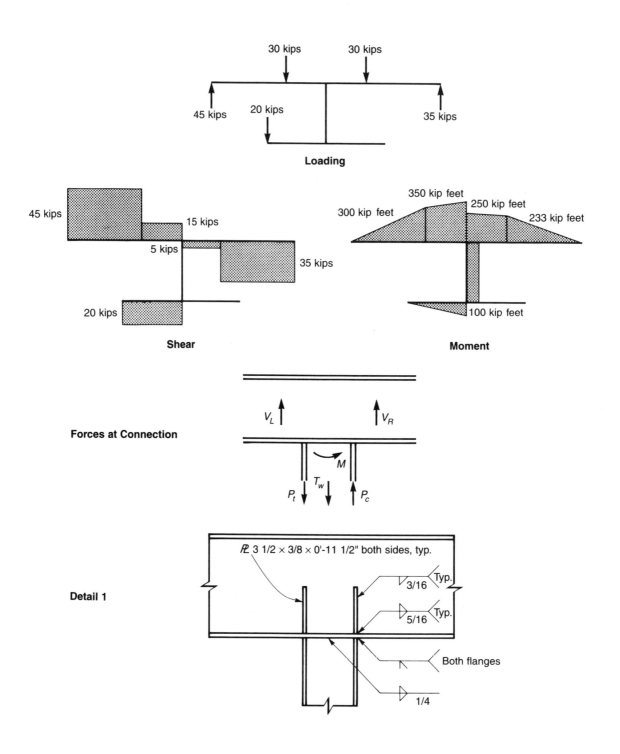

Figure 89B-3(ii)

STRUCTURAL ENGINEER EXAMINATION – 1987

PROBLEM B–2 – WT. 3.0 POINTS

GIVEN: A structural tube is supporting the loads as shown.

CRITERIA: Weld Electrodes are E70XX. Assume the steel plate is adequately connected to the concrete wall.

REQUIRED: Determine the maximum resultant stress on the weld shown in isometric view and determine the minimum size of fillet weld for the structural tube to steel plate connection.

SOLUTION

WELD PROPERTIES

Using the elastic vector analysis technique and assuming unit size of weld, the weld properties are given by

Length of weld L	$= 2(12 + 6)$
	$= 36$ inches
Area of weld A	$= 1.0 \times L$
	$= 36$ inches^2 per inch
Inertia about x-axis I_x	$= 2(12^3/12 + 6 \times 6^2)$
	$= 720$ inches^4 per inch
Inertia about y-axis I_y	$= 2(6^3/12 + 12 \times 3^2)$
	$= 252$ inches^4 per inch
Polar inertia I_o	$= I_x + I_y$
	$= 720 + 252$
	$= 972$ inches^4 per inch

APPLIED LOAD

The moment acting about the y–axis of the weld is

$$M_y \quad = P_H \times 4$$
$$= 20 \times 4$$
$$= 80 \text{ kip inch}$$

The moment acting about the x–axis of the weld is

$$M_x \quad = P_V \times 4$$
$$= 10 \times 4$$
$$= 40 \text{ kip inch}$$

The moment acting about the z–axis of the weld is

$$M_z \quad = P_H(5 + 6)$$
$$= 20 \times 11$$
$$= 220 \text{ kip inch}$$

WELD STRESSES

The co–existent forces acting at the top left corner of the weld in the x–direction, y–direction and z–direction are given by

$$f_x \quad = P_H/A + 6M_z/I_o$$
$$= 20/36 + 6 \times 220/972$$
$$= 0.556 + 1.358$$
$$= 1.914 \text{ kips per inch}$$
$$f_y \quad = -P_V/A + 3M_z/I_o$$
$$= -10/36 + 3 \times 220/972$$
$$= -0.278 + 0.679$$
$$= 0.401 \text{ kips per inch}$$
$$f_z \quad = -3M_y/I_y - 6M_x/I_x$$
$$= -3 \times 80/252 - 6 \times 40/720$$
$$= -0.952 - 0.333$$
$$= -1.285 \text{ kips per inch}$$

The resultant force at the top left corner is

$$f_r \quad = (f_x^2 + f_y^2 + f_z^2)^{0.5}$$
$$= 2.34 \text{ kips per inch}$$

216

Similarly, at the top right corner of the weld, the forces are

$$f_x = 0.556 + 1.358$$
$$= 1.914 \text{ kips per inch}$$
$$f_y = -0.278 - 0.679$$
$$= -0.957 \text{ kips per inch}$$
$$f_z = 0.952 - 0.333$$
$$= 0.619 \text{ kips per inch}$$
$$f_r = 2.23 \text{ kips per inch}$$

At the bottom left corner of the weld, the forces are

$$f_x = 0.556 - 1.358$$
$$= -0.802 \text{ kips per inch}$$
$$f_y = -0.278 + 0.679$$
$$= 0.401 \text{ kips per inch}$$
$$f_z = -0.957 + 0.333$$
$$= -0.619 \text{ kips per inch}$$
$$f_r = 1.09 \text{ kips per inch}$$

At the bottom right corner of the weld, the forces are

$$f_x = 0.556 - 1.358$$
$$= -0.802 \text{ kips per inch}$$
$$f_y = -0.278 - 0.679$$
$$= -0.957 \text{ kips per inch}$$
$$f_z = 0.952 + 0.333$$
$$= 1.285 \text{ kips per inch}$$
$$f_r = 1.79 \text{ kips per inch}$$

Hence the maximum force, which occurs at the top left corner, is

$$f_r = 2.34 \text{ kips per inch}$$

The allowable force on a 1/16 inch E70XX grade fillet weld is obtained from UBC Section 2251 Table J2.5 as

$$q = 0.928 \text{ kip per inch per 1/16-inch.}$$

Hence the required weld size in 1/16 inch is given by

$$D = f_r/q$$
$$= 2.34/0.928$$
$$= 2.52$$

However the minimum size of weld, required for the ¾ inch plate, is given by UBC Section 2251 Table J2.5 as ¼ inch.

217

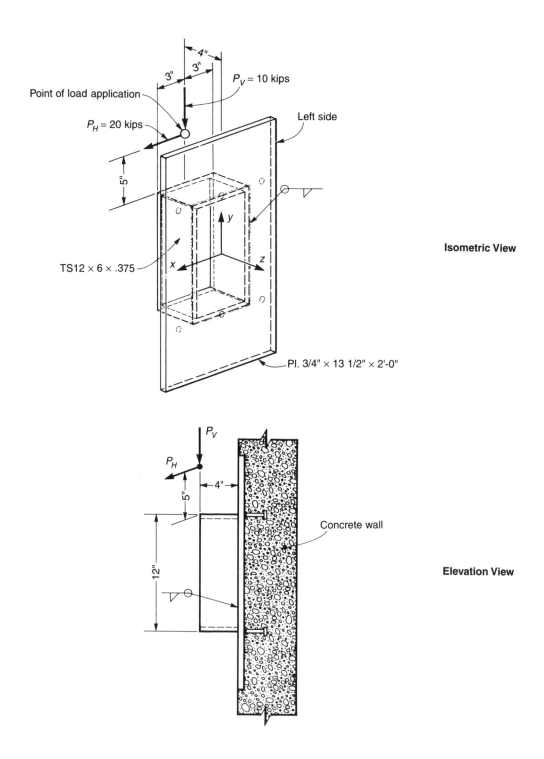

Isometric View

Point of load application

$P_V = 10$ kips

$P_H = 20$ kips

Left side

4"

3" 3"

5"

y

x z

TS12 × 6 × .375

Pl. 3/4" × 13 1/2" × 2'-0"

P_V

P_H

4"

5"

12"

Concrete wall

Elevation View

Figure 87B–2

218

STRUCTURAL ENGINEER REGISTRATION

CHAPTER 2

STRUCTURAL STEEL DESIGN

SECTION 2.3

BRACED FRAMES

PROBLEMS

1990 B–2
1988 B–4

STRUCTURAL ENGINEER EXAMINATION – 1990 ══════════════

PROBLEM B-2 – WT. 7.0 POINTS

GIVEN: The elevation of a metal stud wall, with one "X" braced frame "abcd" to resist seismic load, is shown in Figure 90B-2(i). The seismic load along the length of the wall is 300 pounds per linear foot.

CRITERIA: Seismic zone 4
Response modification factor R_w = 8

Materials:
- Structural tubes F_y = 46 ksi
- Steel beams, Grade A36, F_y = 36 ksi
- Bolts, Grade A325–SC
- All welding Grade E70XX Electrodes

Assumptions:
- The seismic load of 300 plf (reversible) is adequately transferred through a nailer on the top flange of the W14 × 22 beam.
- The braced frame is adequately anchored to the concrete foundation.
- Neglect gravity loads on the columns.
- All joints are pin connected.

REQUIRED:
1. Determine the drag strut force at points "a", "b", and "e".
2. Design and detail the beam to beam bolted connection at point "e" for drag forces only. Use ¾" diameter bolts and complete detail 2.
3. Check the adequacy of the diagonal "ad" assuming an axial force of 16 kips.
4. Using the load shown in detail 1, and assuming that the maximum force that can be transferred to the brace by the system is 60 kips:

   a. Determine the force necessary to design the connection at point "b" on detail 1.
   b. Design and detail the welded connection at "b" and complete detail 2. Calculate the required size of fillet welds for the gusset plate shown.

SOLUTION

1. DRAG FORCES

The shear distribution, net shear, and drag force distribution along the wall are shown in Figure 90B-2(ii). The shear along the length of the nailer is given as

$$q_N = 300 \text{ pounds per linear foot}$$

The total seismic force applied to the wall is

$$
\begin{aligned}
V &= L_N q_N \\
&= 60 \times 300/1{,}000 \\
&= 18 \text{ kips}
\end{aligned}
$$

The shear developed by each brace is

$$
\begin{aligned}
Q_B &= V/2 \\
&= 18/2 \\
&= 9 \text{ kips}
\end{aligned}
$$

The drag forces at "a", "b" and "e" are

$$
\begin{aligned}
F_a &= 10q_N/1{,}000 \\
&= 10 \times 300/1{,}000 \\
&= 3 \text{ and } 6 \text{ kips} \\
F_b &= 40q_N/1{,}000 \\
&= 40 \times 300/1{,}000 \\
&= 12 \text{ and } 3 \text{ kips} \\
F_e &= 39q_N/1{,}000 \\
&= 39 \times 300/1{,}000 \\
&= 11.7 \text{ kips}
\end{aligned}
$$

2. BEAM SPLICE

The splice design load is

$$F_e = 11.7 \text{ kips}$$

Allowing for the one-third increase in permissible stress for seismic loads, in accordance with the UBC Section 1603.5, the equivalent design force is

$$
\begin{aligned}
P_E &= F_e/1.33 \\
&= 11.7/1.33 \\
&= 8.8 \text{ kips}
\end{aligned}
$$

The W14 × 22 beam has a clear web depth between fillets of twelve inches and the splice may be fabricated with six ¾ inch A325-SC bolts and a ¼ × 7 × 9 inch splice plate, as shown in Figure 90B-2(iii).

PLATE STRENGTH IN TENSION

The gross plate area is given by

$$A_g = Bt$$
$$= 9 \times 1/4$$
$$= 2.25 \text{ square inches}$$

based on the gross area, the allowable plate strength is defined by UBC Section 2251.D1 and is given by

$$P_t = 0.6F_yA_g$$
$$= 0.6 \times 36 \times 2.25$$
$$= 48.6 \text{ kips}$$

The maximum allowable value of the effective net area is defined by UBC Section 2251.B3 as

$$A_e = 0.85A_g$$
$$= 0.85 \times 2.25$$
$$= 1.91 \text{ square inches}$$

The actual effective net area is defined by UBC Section 2251.B3 and is given by

$$A_e = t(B-nd_e)$$

where $n = 3 =$ the number of bolts, of diameter d_b, in one vertical row

d_e = effective width of hole as given by Section 2251.B2 and Table J3.1
$$= d_b + 0.125$$
$$= 0.875 \text{ inches}$$

Then
$$A_e = 0.25 (9 - 3 \times 0.875)$$
$$= 1.59 \text{ square inches}$$
$$< 1.91$$

Hence, the applicable effective net area is

$$A_e = 1.59 \text{ square inches}$$

based on the effective net area, the allowable plate strength is defined by UBC Section 2251.D1 and is given by

$$P_t = 0.5F_uA_e$$
$$= 0.5 \times 58 \times 1.59$$
$$= 46.1 \text{ kips}$$

This exceeds the equivalent design force and is satisfactory.

WEB TEAR-OUT

The net effective area resisting shear is given by AISC page 4–9 as

$$A_v = t_w(2 \times 1.5 - d_h)$$

where t_w = web thickness of the W14 × 22
= 0.23 inches

d_h = nominal diameter of a standard hole in shear as given by UBC Section 2251 Table J3.1
= d_b + 0.0625
= 0.8125 inches

and A_v = 0.23(2 × 1.5 − 0.8125)
= 0.50 square inches

The net effective area resisting tension is given by

$$A_t = t_w(6 - 2d_h)$$
$$= 0.23(6 - 2 \times 0.8125)$$
$$= 1.01 \text{ square inches}$$

The resistance to shearing rupture is determined from UBC Section 2251.J4 as

$$P_r = 0.3F_u A_v + 0.5F_u A_t$$
$$= 0.3 \times 58 \times 0.50 + 0.5 \times 58 \times 1.01$$
$$= 38.04 \text{ kips ... exceeds the equivalent design force and is satisfactory.}$$

BOLT SHEAR

The allowable single shear force on a ¾ inch diameter A325-SC bolt is given in AISC Table 1-D as 7.51 kips. The allowable load on three bolts, conforming to the minimum edge distance and spacing criteria, is

$$P_s = 3 \times 7.51$$
$$= 22.53 \text{ kips}$$

This exceeds the equivalent design force and the splice is adequate as shown.

3. BRACE CAPACITY

Design the brace in accordance with the requirements for an ordinary concentrically braced frame. The relevant properties of the 5 × 5 × 3/16 structural tubing are

Yield stress F_y = 46 kips per square inch
Area A = 3.52 square inches
Radius of gyration r = 1.95 inches

STRESS INCREASE

Allowing for the one-third increase in permissible stress for seismic loads, in accordance with UBC Section 1603.5 the equivalent design force is

$$P_E = 16/1.33$$
$$= 12.0 \text{ kips}$$

The equivalent member stress, based on the gross sectional area is

$$f = P_E/A$$
$$= 12/3.52$$
$$= 3.41 \text{ kips per square inch}$$

and this stress may be either tensile or compressive.

TENSILE CAPACITY

Since the brace connections are welded, the gross area of the member governs the tensile capacity and the allowable tensile stress is given by UBC Section 2251.D1 as

$$F_t = 0.6F_y$$
$$= 0.6 \times 46$$
$$= 27.6 \text{ kips per square inch}$$
$$> 3.41$$

and the brace is adequate for the applied tensile force.

SLENDERNESS

Since the building is located in seismic zone 4, the slenderness ratio of the brace is limited, by UBC Section 2211.8.2, to a maximum value of

$$\ell/r = 720/(F_y)^{0.5}$$
$$= 720/(46)^{0.5}$$
$$= 106$$

The distance between lateral supports, for each brace, may be assumed to be half the full theoretical length[4] of the brace, for both in-plane and out-of-plane effects. The actual slenderness ratio is then

$$\ell/r = 17.2 \times 12/(2 \times 1.95)$$
$$= 52.9$$
$$< 106$$

and the slenderness ratio of the brace is satisfactory.

LOCAL BUCKLING

The width to thickness ratio of square structural tubing is limited, by UBC Section 2251.7 Table B5.1 to a maximum value of

$$b/t = 238/(F_y)^{0.5}$$
$$= 238/(46)^{0.5}$$
$$= 35$$

The actual width to thickness ratio is

$$b/t = 5 \times 16/3$$
$$= 26.7 \ldots \text{satisfactory}$$

DISTRIBUTION OF LATERAL FORCE

UBC Section 2211.8.2 stipulates that neither the sum of the horizontal components of the compressive member forces nor the sum of the horizontal components of the tensile member forces, along a line of bracing, shall exceed seventy percent of the applied lateral force. Both braces resist the lateral force equally and are, therefore, satisfactory.

COMPRESSIVE CAPACITY

The effective length factor for a pin-ended member is given by AISC Table C–C2.1 as

$$K = 1.0$$

The effective slenderness ratio of the brace, using center line dimensions, is given by

$$K\ell/r = 1.0 \times 52.9$$
$$= 52.9$$

The axial compressive stress permitted by UBC Section 2251.E2 is obtained from AISC Table C–50 as

$$F_a = 23.90 \text{ kips per square inch.}$$

The stress reduction factor for a bracing member resisting seismic forces is determined from UBC Formula (11–5) as

$$\beta = 1/\{1.0 + [(k\ell/r)/2C_c]\}$$

where $C_c = 111.6$ as given by UBC Section 2251 Table 4

Hence $\beta = 1/[1.0 \times 52.9/(2 \times 111.6)]$
$$= 0.808$$

The allowable stress for a bracing member resisting seismic forces in compression is determined from UBC Section 2211.8.2 Formula (11–4) as

$$F_{as} = \beta F_a$$
$$= 0.808 \times 23.9$$
$$= 19.31 \text{ kips per square inch.}$$
$$> 3.41$$

Hence, the brace satisfies all the requirements of the UBC code.

4a. CONNECTION FORCES

The force applied to the brace is given by

$$P = 18 \times 17.2/(2 \times 10)$$
$$= 15.48 \text{ kips}$$

and this stress may be either tensile or compressive. The brace connection shall be designed, in accordance with UBC Section 2211.8.3 for the lesser of the following three values:

(i)   The tensile strength of the brace,

$$P_{st} = AF_y \text{ from UBC Section 2211.4}$$
$$= 3.52 \times 46$$
$$= 162 \text{ kips}$$

(ii)   The applied force times $0.375R_w$, where R_w is the applicable response factor for a building frame system, given in UBC Table 16-N item 2.4. Hence,

$$P_b = 0.375R_wP$$
$$= 0.375 \times 8 \times 15.48$$
$$= 46.44$$

(iii) The maximum force that can be transferred to the brace by the system

$$P_m = 60 \times 17.2/(2 \times 10)$$
$$= 51.6 \text{ kips}$$

Hence the design load value for the brace is

$$P_b = 46.44 \text{ kips}$$

By resolving forces at the joint, the design loads at the connection are obtained as shown in Figure 90B-2(ii) and are

$$P_c = \text{design load value for the column}$$
$$= 46.44 \times 14/17.2$$
$$= 37.8 \text{ kips}$$
$$P_g = \text{design load value for the beam}$$
$$= 46.44 \times 10/17.2$$
$$= 27.0 \text{ kips}$$

4b. CONNECTION DESIGN

The connection consists of a ⅜ × 9 × 18 gusset plate with a six inch length of weld to the brace and a eight inch length of weld to the column. In accordance with UBC Section 2251 Table J2.4, the minimum size of weld required for the 5/16 beam flange and the ⅜ inch gusset plate is 3/16 inch. The strength capacity of the 3/16 inch fillet weld is given by UBC Section 2211.4 as

$$q = 1.7 \times 3 \times 0.928$$
$$= 4.73 \text{ kips per inch}$$

The strength of the eight inch weld to the column is given by

$$P_w = 4q \times 8$$
$$= 4 \times 4.73 \times 8$$
$$= 151.5 \text{ kips}$$
$$> P_c$$

Hence the column connection is satisfactory.

The strength of the six inch weld to the brace is given by

$$P_w = 4q \times 6$$
$$= 4 \times 4.73 \times 6$$
$$= 113.5 \text{ kips}$$
$$> P_b$$

Hence the brace connection is satisfactory.

The geometrical properties of the gusset plate weld are obtained by assuming unit size of weld.

Weld length L $= 18$ inches
Weld area A $= 1.0 \times 18$
$= 18$ inches² per inch
Plastic modulus Z $= 1.0 \times L^2/4$
$= 1.0 \times 18^2/4$
$= 81$ inches³ per inch

The depth of the W14 × 22 beam is

$$d = 13.74 \text{ inches}$$

and the applied loading on the weld is obtained from Figure 90B-2(ii) as

Horizontal shear H $= P_g$
$= 27$ kips
Eccentricity moment M $= P_g \times d/2$
$= 27 \times 13.74/2$
$= 185.5$ kip inches

The co-existent forces acting at the extreme ends of the unit weld are

Horizontal force $\quad= H/A$
$= 27/18$
$= 1.5$ kips per inch
Vertical force $\quad= M/Z$
$= 185.5/81$
$= 2.29$ kips per inch

The resultant force at the ends of the weld is

$$R \quad= (1.5^2 + 2.29^2)^{0.5}$$
$$= 2.74 \text{ kips per inch}$$

The strength capacity of the double 3/16 inch fillet weld is

$$q_d \quad= 2q$$
$$= 2 \times 4.73$$
$$= 9.46 \text{ kips per inch}$$
$$> R$$

Hence the gusset plate weld is adequate.

The gusset plate shear capacity is given by UBC Section 2211.4 as

$$V_s \quad= 0.55F_y dt$$
$$= 0.55 \times 36 \times 18 \times 3/8$$
$$= 133.7 \text{ kips}$$
$$> P_g$$

The gusset plate moment capacity is

$$M_s \quad= ZF_y$$
$$= (td^2/4)F_y$$
$$= 36 \times 3 \times 18^2/(8 \times 4)$$
$$= 1094 \text{ kip inches}$$
$$> M$$

Hence the gusset plate is satisfactory.

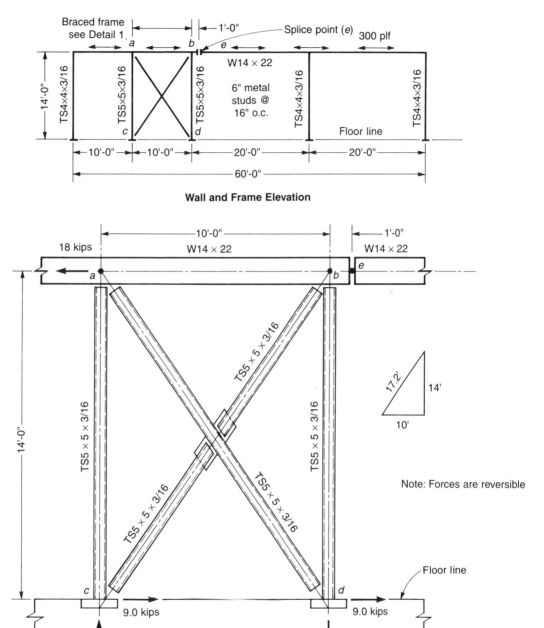

Wall and Frame Elevation

Detail 1

FIGURE 90B-2(i)

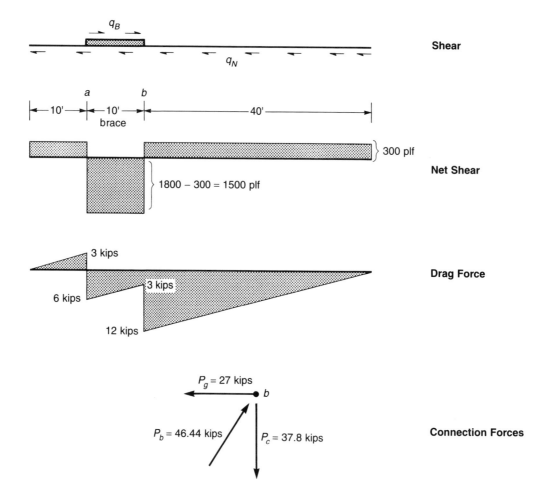

Shear

Net Shear

Drag Force

Connection Forces

FIGURE 90B–2(ii)

230

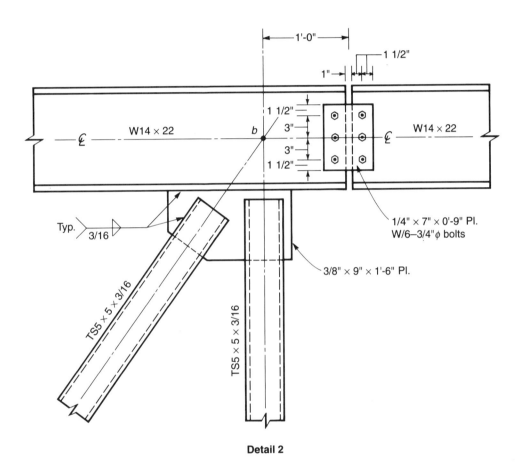

Detail 2

FIGURE 90B–2(iii)

STRUCTURAL ENGINEER EXAMINATION – 1988 ========

PROBLEM B–4 – WT. 4.0 POINTS

GIVEN: The elevation of an industrial building seismic bracing system, as shown.

CRITERIA: All columns are W8 × 31.
All beams are W16 × 26.

Assumptions:
- Neglect the elastic shortening of the beam members.
- All connections are assumed pinned.
- Diagonal members are designed to carry tension and compression loads.
- Neglect gravity loads.

REQUIRED: 1. Compute the member forces and base reactions in braced frame "A".
2. Check the diagonal member in braced frame "B".
3. Find the drift of braced frames "A" and "B".

SOLUTION

STIFFNESS OF FRAME "B"

The stiffness of frame "B" is obtained by applying a unit virtual load to the frame, as shown in Figure 88B–4. The horizontal displacement at the point of application of the load is determined by the virtual work method[1, 2] by evaluating the expression

$$\delta = \Sigma u^2 L / AE$$

where u is the force in a member due to the unit virtual load, L is the length of the member, A is the sectional area of the member, and E is the modulus of elasticity. The summation extends over all the members, with the exception of the beam, which is considered to have negligible elastic shortening. The relevant values are tabulated below.

Member	L	A	u	u^2L/A
Brace 13	22.62	6.59	−1.414	6.86
Column 34	16.00	9.13	1.000	1.75
Total	–	–	–	8.61

The horizontal displacement of point 2 is given by

$$\delta = \Sigma u^2 L/AE$$
$$= 8.61 \times 12/29,000$$
$$= 0.00356 \text{ inches}$$

The lateral stiffness of frame "B", which is defined as the force required to produce unit horizontal displacement, is given by

$$K_B = 1/\delta$$
$$= 1/0.00356$$
$$= 280.7 \text{ kips per inch}$$

STIFFNESS OF FRAME "A"

A unit load at joint 2 produces the forces shown in Figure 88B-4. Since the elastic shortening of the beam may be neglected, only the diagonal members of the frame contribute to the stiffness[5] which is given by

$$K_A = \Sigma AE\cos^2\theta/L$$

where θ is the angle of inclination of the diagonal member to the horizontal and the summation extends over the diagonal members only.

The stiffness of frame "A" is

$$K_A = \Sigma AE\cos^2\theta/L$$
$$= [2 \times 4.27 \times 29,000 \times \cos^2(53.13°)]/(20 \times 12)$$
$$= 371.5 \text{ kips per inch}$$

233

1. FORCES IN FRAME A

The lateral load resisted by frame "A" is given by

$$F_A = 45K_A/(K_A + K_B)$$
$$= 45 \times 371.5/(280.7 + 371.5)$$
$$= 25.6 \text{ kips}$$

The lateral load resisted by frame "B" is

$$F_B = 45 - F_A$$
$$= 45 - 25.6$$
$$= 19.4 \text{ kips}$$

The forces in frame "A" are shown in Figure 88B-4 and are obtained by multiplying the forces produced by unit load by 25.6.

2. FRAME "B" DIAGONAL

The relevant properties of the $7 \times 7 \times \frac{1}{4}$ structural tubing are

Yield stress F_y	= 46 kips per square inch
Area A	= 6.59 square inches
Radius of gyration r	= 2.74 inches

STRESS INCREASE

In accordance with UBC Section 1603.5, a one-third increase is allowed in permissible stresses when designing for seismic forces. The actual force in the diagonal is

$$P = 19.4 \times 1.414$$
$$= 27.43 \text{ kips}$$

The equivalent design force, allowing for the one-third increase and based on the basic allowable stress, is

$$P_E = P/1.33$$
$$= 27.43/1.33$$
$$= 20.62 \text{ kips}$$

The equivalent member stress, based on the gross sectional area, is

$$f = P_E/A$$
$$= 20.62/6.59$$
$$= 3.13 \text{ kips per square inch}$$

and this stress may be either compressive or tensile.

TENSILE CAPACITY

Assuming that the brace connections are welded, the gross area of the member governs the tension capacity and the allowable tensile stress is given by UBC Section 2251.D1 as

$$
\begin{aligned}
F_t &= 0.6F_y \\
&= 0.6 \times 46 \\
&= 27.6 \text{ kips per square inch} \\
&> 3.13
\end{aligned}
$$

and the brace is adequate for the applied tensile force.

SLENDERNESS

Assuming that the frame is an ordinary braced frame located in seismic zone 4, the slenderness ratio of the brace is limited, by UBC Section 2211.8.2, to a maximum value of

$$
\begin{aligned}
\ell/r &= 720/(F_y)^{0.5} \\
&= 720/(46)^{0.5} \\
&= 106
\end{aligned}
$$

The actual slenderness ratio is

$$
\begin{aligned}
\ell/r &= 22.62 \times 12/2.74 \\
&= 99 \\
&< 106
\end{aligned}
$$

and the slenderness ratio of the brace is satisfactory.

LOCAL BUCKLING

The width to thickness ratio of square structural tubing is limited, by UBC Section 2251.7 Table B5.1, to a maximum value of

$$
\begin{aligned}
b/t &= 238/(F_y)^{0.5} \\
&= 238/(46)^{0.5} \\
&= 35
\end{aligned}
$$

The actual width to thickness ratio is

$$
\begin{aligned}
b/t &= 7/0.25 \\
&= 28 \\
&< 35
\end{aligned}
$$

and the width to thickness ratio of the brace is satisfactory.

DISTRIBUTION OF LATERAL FORCE

UBC Section 2211.8.2 stipulates that neither the sum of the horizontal components of the compressive member forces nor the sum of the horizontal components of the tensile member forces, along a line of bracing, shall exceed seventy percent of the applied lateral force. The sum of the horizontal components of the compression brace forces in frames "A" and "B" is

$$\Sigma H = F_A/2 + F_B$$
$$= 25.6/2 + 19.4$$
$$= 32.20$$

The ratio of the sum of the horizontal components to the applied force is

$$\Sigma H/V = 32.2/45$$
$$= 0.72$$
$$> 0.70$$

The distribution of lateral force is unsatisfactory and the brace must be designed to resist $3(R_W/8)$ times the prescribed forces where R_W is the response factor, for a building frame system, given in UBC Table 16-N item 2.4, as 8. Hence, the equivalent design compressive stress in the brace as given by UBC Section 2211.8.2 is

$$f_a = 3(R_W/8)P_E/A$$
$$= 3 \times (8/8) \times 20.62/6.59$$
$$= 9.39 \text{ kips per square inch}$$

COMPRESSIVE CAPACITY

The effective length factor for the pin-ended brace is given by AISC Table C-C2.1 item (d) as

$$K = 1.0$$

The effective slenderness ratio of the brace, using center line dimensions, is given by

$$K\ell/r = 1.0 \times 99$$
$$= 99$$

The axial compressive stress permitted by UBC Section 2251.E2 is obtained from AISC Table C-50 as

$$F_a = 14.94 \text{ kips per square inch}$$
$$> 9.39$$

and the brace is adequate in compression since, in accordance with UBC Section 2211.8.2.3, the strength reduction factor β may be neglected. Hence the brace in frame "B" satisfies all the requirements of the UBC code.

3. FRAME DRIFT

Since the elastic shortening of the beam members may be neglected, the lateral displacements of both frame "A" and frame "B" are identical. Hence, the drift, or lateral displacement due to the seismic load V of 45 kips is given by

$$\delta = V/(K_A + K_B)$$
$$= 45/(371.5 + 280.7)$$
$$= 0.069 \text{ inches}$$

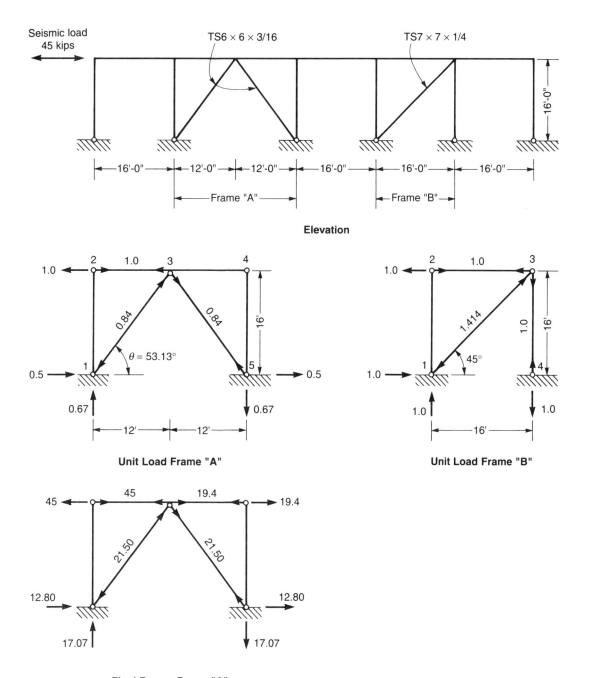

FIGURE 88B-4

STRUCTURAL ENGINEER REGISTRATION

CHAPTER 2

STRUCTURAL STEEL DESIGN

SECTION 2.4

DUCTILE MOMENT FRAMES

PROBLEMS:

1988 B-5

STRUCTURAL ENGINEER EXAMINATION – 1988 =================

PROBLEM B-5 – WT. 6.0 POINTS $\times 9.6^{min}/_{ft} \doteq 57.6^{min}$

GIVEN: Analyze the beam-column joint of the special moment-resisting frame shown in Figure 88B-5(ii).

CRITERIA: Seismic zone 3
          All welds are Grade E70XX Electrodes
          All material is ASTM A36
          Assumptions:
          • All requirements for compact sections are satisfied.
          • Plastic hinges at this joint will not be the last to form.
          • Beam-column shear connections are adequate.
          • Design must adhere to UBC requirements for seismic design of steel.

REQUIRED: 1. Complete the design of the beam-column connection, such that it satisfies seismic ductility requirements.
          2. Complete a sketch showing the results of Part 1. Welding need not be shown or calculated.
          3. Calculate the maximum permissible distance from the column face to a point of lateral support along each beam. Assume both beam deflected shapes are single curvature between the points considered.
          4. Show by calculation that this beam-column represents a strong-column/weak-beam design.

SOLUTION

1. CONNECTION DESIGN

The relevant properties of the W24 × 146 column are

Plastic modulus Z_c	= 418 inches³
Yield stress F_{yc}	= 36 kips per square inch
Depth d_c	= 24.74 inches
Flange thickness t_{cf}	= 1.09 inches
Flange width b_c	= 12.9 inches
Web thickness t	= 0.65 inches
Fillet depth k	= 1.875 inches
Clear depth between fillets T	= 21 inches
Area A	= 43 square inches

The relevant section properties of the W30 × 132 beam 1 are

Plastic modulus Z_1	= 437 inches^3
Yield stress F_{yb}	= 36 kips per square inch
Depth d_{b1}	= 30.31 inches
Flange thickness t_{b1}	= 1.00 inches
Flange width b_1	= 10.55 inches
Minimum radius of gyration r_{y1}	= 2.25 inches

The relevant section properties of the W24 × 68 beam 2 are

Plastic modulus Z_2	= 177 inches^3
Yield stress F_{yb}	= 36 kips per square inch
Depth d_{b2}	= 23.73 inches
Flange thickness t_{b2}	= 0.585 inches
Flange width b_2	= 8.97 inches
Minimum radius of gyration r_{y2}	= 1.87 inches

APPLIED LOADS

From the problem statement, the gravity and seismic moments in beam 1 are

$$
\begin{aligned}
M_{V1} &= (420 - 280)/2 \\
&= 70 \text{ kip feet} \\
M_{E1} &= 420 - 70 \\
&= 350 \text{ kip feet}
\end{aligned}
$$

The gravity and seismic moments in beam 2 are

$$
\begin{aligned}
M_{V2} &= (170 - 80)/2 \\
&= 45 \text{ kip feet} \\
M_{E2} &= 170 - 45 \\
&= 125 \text{ kip feet}
\end{aligned}
$$

The applied forces at the joint, for seismic load acting from left to right, are shown in Figure 88B–5(ii).

PANEL ZONE SHEAR

The panel zone design moments for seismic load acting from left to right, are obtained in accordance with UBC Section 2211.7.2.1 as

$$
\begin{aligned}
M_1 &= M_{V1} + 1.85M_{E1} \\
&= 70 + 1.85 \times 350 \\
&= 717.5 \text{ kip feet} \\
M_2 &= M_{V2} + 1.85M_{E2} \\
&= -45 + 1.85 \times 125 \\
&= 186.25 \text{ kip feet}
\end{aligned}
$$

The panel zone shear is obtained from AISC Formula (C-E6-1) as

$$V_{PZ} = M_1/0.95d_{b1} + M_2/0.95d_{b2} - (M_1 + M_2)/H$$

where H = distance between column midheights
  = 14 feet

Hence V_{PZ} = $12 \times 717.5/(0.95 \times 30.31) + 12 \times 186.25/(0.95 \times 23.73)$
  $-(717.5 + 186.25)/14$
  = 333.6 kips

Similarly, for seismic load acting from right to left, the panel zone design moments and the shear are

$$M_1 = 717.5 - 2 \times 70$$
$$= 577.5 \text{ kip feet}$$
$$M_2 = 186.25 + 2 \times 45$$
$$= 276.25 \text{ kip feet}$$
$$V_{PZ} = 325.7 \text{ kips}$$
$$< 333.6$$

However, in accordance with UBC Section 2211.7.2.1, the panel zone design shear need not exceed the shear produced by 0.8 times the full plastic capacities of the beams. The plastic capacities of the beams are

$$M_{s1} = F_{yb}Z_1$$
$$= 36 \times 437/12$$
$$= 1311 \text{ kip feet}$$
$$M_{s2} = F_{yb}Z_2$$
$$= 36 \times 177/12$$
$$= 531 \text{ kip feet}$$

and these moments produce a panel zone shear in excess of that due to seismic load acting from left to right. Hence, the design panel zone shear is

$$V_{PZ} = 333.6 \text{ kips}$$

In determining the shear capacity of the unreinforced column web, the depth of the deeper beam is used as recommended in the SEAOC Commentary 4F.2.a, and the capacity is given by UBC Section 2211.7 Equation (11-1) as

$$V = 0.55F_{yc}d_c t(1.0 + 3b_c t_{cf}^2/d_{b1}d_c t)$$
$$= 0.55 \times 36 \times 24.74 \times 0.65\,[1 + 3 \times 12.9 \times 1.09^2/(30.31 \times 24.74 \times 0.65)]$$
$$= 348.44$$
$$> V_{PZ}$$

Hence the unreinforced web is adequate and no doubler plate is required.

The minimum panel zone thickness to prevent shear buckling failure is given by UBC Section 2211.7 Equation (11-2) as

$$t_z = (d_z + w_z)/90$$

where $\quad d_z \quad$ = panel zone depth between continuity plates
$$\approx d_{b1} - t_{b1}$$
$$= 30.31 - 1$$
$$= 29.31 \text{ inches}$$

and $\quad w_z \quad$ = panel zone width between flanges
$$= d_c - 2t_{cf}$$
$$= 24.74 - 2 \times 1.09$$
$$= 22.56 \text{ inches}$$

Hence $\quad t_z \quad = (29.31 + 22.56)/90$
$$= 0.58 \text{ inches}$$
$$< t$$

Thus the column web thickness is adequate to prevent shear buckling failure.

STIFFENER REQUIREMENTS

The force in the beam flange for beam 1, for seismic zone 3, is defined as

$$P_{bf} = A_{fb}F_{yb}$$
$$= t_{b1}b_1F_{yb}$$
$$= 1.00 \times 10.55 \times 36$$
$$= 380 \text{ kips}$$

The required stiffener area, to resist the concentrated flange force, is given by UBC Section 2211.7 Equation (K1-9) as

$$A_{st} \geq [P_{bf} - F_{yc}t(t_b + 5k)]/F_{yst}$$

Using calculator program A4.4 in Appendix A, the value is

$$A_{st} = 3.81 \text{ square inches}$$

Hence stiffeners, or continuity plates, are required and it is unnecessary to check Equations (K1-8) and (K1-1) since the stiffeners are provided over the full depth of the column web. Using a pair of stiffener plates ½ inch × 4 inches provides a stiffener area of

$$A_{st} = 2 \times 4 \times 0.5$$
$$= 4.00 \text{ square inches}$$
$$> 3.81$$

The minimum stiffener thickness required is defined by UBC Section 2251.K8 as

$$t_{st} = t_{b1}/2$$
$$= 1.00/2$$
$$= 0.50 \text{ inches ... provided}$$

The minimum stiffener width required is defined by UBC Section 2251.K8 as

$$b_{st} = b_1/3 - t/2$$
$$= 10.55/3 - 0.65/2$$
$$= 3.2 \text{ inches}$$
$$< 4$$

The maximum width–thickness ratio is defined by UBC Section 2251.B5.1 as

$$b_{st}/t_{st} = 95/(F_{yst})^{0.5}$$
$$= 95/(36)^{0.5}$$
$$= 15.8$$

The actual width–thickness ratio provided is

$$b_{st}/t_{st} = 4 \times 2/1$$
$$= 8$$
$$< 15.8$$

Hence the stiffeners provided are adequate.

The force in the beam flange for beam 2 is

$$P_{bf} = t_{b2}b_2F_{yb}$$
$$= 0.585 \times 8.97 \times 36$$
$$= 189 \text{ kips}$$

The required stiffener area, to resist the concentrated flange force, is

$$A_{st} \geq [P_{bf} - F_{yc}t(t_b + 5k)]/F_{yst}$$

and this is evaluated by calculator program A4.4 as

$$A_{st} = -1.22$$

$$\frac{\left[189^{k} - 36 \frac{ksi}{} (.65)(.59'' + 5 \times 1\frac{7}{8}'')\right]}{36\, ksi.} = -1.22$$

and stiffeners are not required.

To prevent buckling opposite the compression flange, UBC Section 2251 Equation (K1–8) requires the provision of stiffeners if the column web depth clear of fillets exceeds the value

$$T = 4100t^3(F_{yc})^{0.5}/P_{bf}$$

and this is evaluated by calculator program A4.4 as

$$T = 35.75 \text{ inches}$$

The actual clear web depth is

$$T = 21 \text{ inches}$$

and stiffeners are not required opposite the compression flange.

To prevent flange bending opposite the tension flange, UBC Section 2251 Equation (K1–1) and Section 2211.7.4 require the provision of stiffeners if the column flange thickness is less than the value

$$t_{cf} = 0.4(1.8P_{bf}/F_{yc})^{0.5}$$

and this is evaluated by calculator program A4.4 as

$$t_{cf} = 1.23 \text{ inches}$$

The actual flange thickness is

$$t_{cf} = 1.09 \text{ inches}$$
$$< 1.23$$

and, hence, stiffeners are required opposite the tension flange and a pair of stiffeners ⅜ inch × 3 inches satisfy the requirements of UBC Section 2251.K8. Details of the connection are shown in Figure 88–B5(ii).

3. BEAM BRACING

In accordance with UBC Section 2211.7.8 both flanges of the beams shall be braced at a maximum distance, between column center lines, given by the expression

$$\ell = 96r_y$$

For beam 1, this length is given by

$$\ell = 96 \times 2.25/12$$
$$= 18 \text{ feet}$$

For beam 2, this length is given by

$$\ell = 96 \times 1.87/12$$
$$= 15 \text{ feet}$$

4. COLUMN–BEAM STRENGTH RATIO

From the problem statement, the maximum column load is

$$P_T = 344 \text{ kips}$$

The corresponding axial stress is

$$f_a = P_T/A$$
$$= 344/43$$
$$= 8 \text{ kips per square inch}$$

To ensure that hinges will form in the beams rather than in the columns, when the panel zone shear strength does not control, UBC Section 2211 Equation (11–3.1) specifies that

$$1.0 < \Sigma Z_c(F_{yc} - f_a)/\Sigma Z_b F_{yb}$$
$$< 2 \times 418(36-8)/36(437 + 177)$$
$$< 1.06$$

Hence, the requirement for a strong–column/weak–beam design is satisfied. ═════

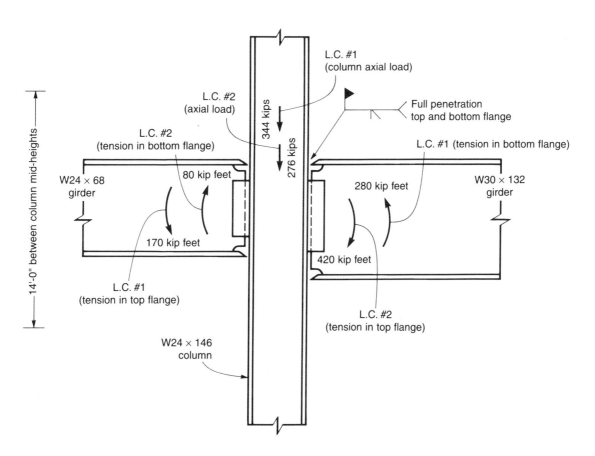

FIGURE 88B–5(i)

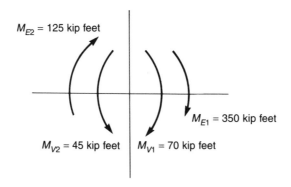

Applied Forces at the Joint

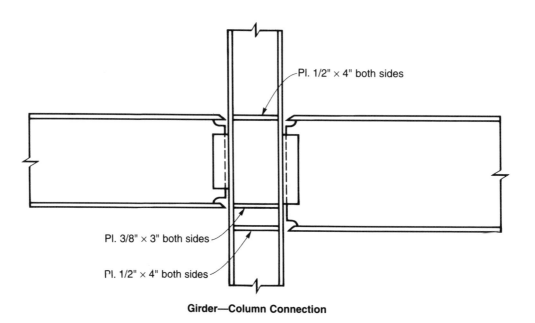

Girder—Column Connection

FIGURE 88B–5(ii)

STRUCTURAL ENGINEER REGISTRATION

CHAPTER 2

STRUCTURAL STEEL DESIGN

SECTION 2.5

NOTCHED BEAM

PROBLEMS:

1990 B–4
1987 B–4

PROBLEM B-4 – WT. 6.0 POINTS

GIVEN: The steel framing shown on detail 1 was erected. Later it was determined that beam BM-1 and the adjacent ducts would have to be raised, resulting in the notched condition shown on detail 2.
- The loading is indicated on detail 1.

CRITERIA: Materials:
- All steel beams are Grade A36 with F_y = 36 ksi
- All welding is Grade E70XX electrodes

Assumptions:
- Neglect skip live load.
- The top elevation of the beam shall not be changed.
- Consider only the loading condition given.
- Do not investigate the web stiffener under the pipe column shown in detail 2.

REQUIRED: 1. Design any required reinforcing to be added to the beam to provide for bearing, shear, and moment at the support. Use 3-1/16 × ½ inch plates for any flange reinforcement and 3 × ⅜ inch plates for any required stiffeners.
2. Determine the extent of the required reinforcing and complete the detail.
3. Design the welding required to attach the reinforcing.

SOLUTION

1. REQUIRED REINFORCING

The relevant properties of the W14 × 38 beam are

Depth d	= 14.1 inches
Web thickness t_w	= 0.31 inches
Flange width b_f	= 6.77 inches
Flange thickness t_f	= 0.515 inches
Flange area A_f	= $b_f \times t_f$
	= 3.487 square inches

APPLIED LOADS

The moment and shear forces occurring at the notch are shown in Figure 90B-4(ii)

MOMENT

The maximum moment at the support is

$$M = 40 \times 12$$
$$= 480 \text{ kip inches}$$

The allowable bending stress is

$$F_b = 0.6F_y$$
$$= 21.6 \text{ kips per square inch}$$

The required section modulus is

$$S_r = M/F_b$$
$$= 480/21.6$$
$$= 21.82 \text{ inches}^3$$

and to provide this modulus, two 3-1/16 × ½ inch plates must be welded to the bottom of the web plate, as shown in Figure 90B-4(ii). This provides a bottom flange area of

$$A_b = 0.5(2 \times 3.0625 + 0.31)$$
$$= 3.218 \text{ square inches}$$

The total area of the section is

$$A = A_f + A_b + t_w(d - 6 - 0.515 - 0.5)$$
$$= 3.487 + 3.218 + 2.196$$
$$= 8.90 \text{ square inches}$$

Hence, comparing the top and bottom flanges, the area variation is given by

$$\delta_A = 100(A_f - A_b)/A$$
$$= 100(3.487 - 3.218)/8.90$$
$$= 3.02 \text{ percent}$$

This exceeds the allowable variation, for W sections, of 2.5 percent as given by AISC Section 1-150. Hence, it is not permissible to consider the section as symmetrical and the section properties must be determined by using the calculator program A.5.1 in Appendix A as tabulated below

Part	A	y	I	Ay	Ay^2
Top flange	3.487	7.84	0.08	–	–
Bottom flange	3.218	0.25	0.07	–	–
Web	2.196	4.04	9.19	–	–
Total	8.90	–	9.34	37.01	250.37

The centroid height is given by

$$\bar{y} = \Sigma Ay/\Sigma A$$
$$= 37.01/8.90$$
$$= 4.16 \text{ inches}$$

The moment of inertia is given by

$$I_x = \Sigma I + \Sigma Ay^2 - \bar{y}^2 \Sigma A$$
$$= 105.79 \text{ inches}^4$$

The section modulus is given by

$$S = I_x/\bar{y}$$
$$= 25.44 \text{ inches}^3$$
$$> S_r$$

Hence the built up member is adequate in flexure.

SHEAR

The maximum shear at the notch is

$$V = 20 \text{ kips}$$

and the shear stress in the web is given by

$$f_v = V/t_w(d - 6)$$
$$= 20/(0.31 \times 8.1)$$
$$= 7.97 \text{ kips per square inch}$$

The allowable shear stress is given by UBC Section 2251.F4 as

$$F_v = 0.4F_y$$
$$= 0.4 \times 36$$
$$= 14.4 \text{ kips per square inch}$$
$$> f_v$$

Hence the built up member is adequate in shear.

SIDESWAY WEB BUCKLING

To prevent sidesway web buckling, when the loaded flange is restrained against rotation, stiffeners must be provided if the concentrated load exceeds the value given by UBC Section 2251 Equation (K1-6) as

$$P = R_5(1.0 + 0.4R_6^3)$$

where $R_6 = d_c b_f / \ell t_w$

d_c = web depth clear of fillets

= 14.1 − 6.0 − 0.5 − 0.1875 − 1.0625 = 6.35 inches

b_f = flange width

= 2 × 3.0625 + 0.31 = 6.44 inches

ℓ = largest unbraced length of either flange

= 20 × 12 = 240 inches

t_w = web thickness = 0.31 inches

and R_6 = 6.35 × 6.44/(240 × 0.31)

= 0.55

< 2.3 ... Equation (K1-6) must be checked

$R_5 = 6800 t_w^3 / h$

h = clear depth between flanges

= 14.1 − 6.0 − 0.5 − 0.515 = 7.09 inches

and R_5 = 6800 × 0.31³/7.09 = 28.59

then P = 28.59(1.0 + 0.4 × 0.55³)

= 29.66 kips

< 32 ... stiffeners are required

The stiffeners need not extend more than one–half the depth of the web in accordance with AISC Commentary K1.5 and it is unnecessary to check for local web yielding or web crippling, in accordance with UBC Section 2251.K1.4. Using a pair of stiffener plates ⅜ × 3 × 6 inches long provides a width–thickness ratio of

b_{st}/t_{st} = 3 × 8/3 = 8

< 95/$(F_y)^{0.5}$ = 15.8

as required by UBC Section 2251.B5.1, and the stiffeners extend approximately to the edge of the flange plates.

The capacity of a 1/16 inch fillet weld is given by UBC Section 2251.J2.4 as

q = 0.928 kips per inch

Allowing for ¾ inch corner snips, the capacity of the 3/16 inch weld provided to the web is

P_w = 4 × 3 × 0.928(6 − 0.75)

= 58.46 kips

> R ... satisfactory

Allowing for ¾ inch corner snips, the capacity of the ¼ inch weld provided to the flange is

P_w = 4 × 4 × 0.928(3 − 0.75)

= 33.41 kips

> R ... satisfactory

Allowing for ¾ inch corner snips, the capacity of the stiffeners in shear is

P_{st} = 0.4F_y × 2 × 0.375(6 − 0.75)

= 56.70 kips

> R ... satisfactory

2. REINFORCEMENT LENGTH AND WELDING

Continuous fillet welds, on one side of the reinforcement, are required[6] to connect the reinforcement to the web. The shear applied to the weld is given by

$$q_r = VQ/I_x$$

where
$$Q = A_b(\bar{y} - 0.25)$$
$$= 3.218(4.16 - 0.25)$$
$$= 12.58 \text{ inches}^3$$

Hence
$$q_r = 20 \times 12.58/105.81$$
$$= 2.38 \text{ kips per inch}$$

In accordance with UBC Section 2251.Table J2.4, the minimum size of weld required for the ½ inch plate reinforcement is 3/16 inch.

The capacity of the two 3/16 inch fillet welds provided is

$$q_w = 2 \times 3q$$
$$= 2 \times 3 \times 0.928$$
$$= 5.56 \text{ kips per inch}$$
$$> q_r$$

Hence the continuous 3/16 inch fillet welds are satisfactory.

Continuous fillet welds, on both sides of the reinforcement, are required[6] to connect the reinforcement to the web over the development length, beyond the notch. The ½ inch reinforcement must extend into the web of the W14 × 38 a sufficient distance to allow the welding to develop the full capacity of the plate. Thus the development length required is

$$\ell_1 = 0.6F_yA_p/4q$$
$$= 0.6 \times 36 \times 2 \times 0.5 \times 3.0625/(4 \times 3 \times 0.928)$$
$$= 66.15/11.12$$
$$= 5.95 \text{ inches}$$

To develop the shear capacity of the beam web, the development length required is

$$\ell_2 = 0.6F_yA_p/0.4F_yt_w$$
$$= 66.15/(0.4 \times 36 \times 0.31)$$
$$= 14.8 \text{ inches ... governs}$$

The complete detail is shown in Figure 90B-4(ii) ═══════════════════════════

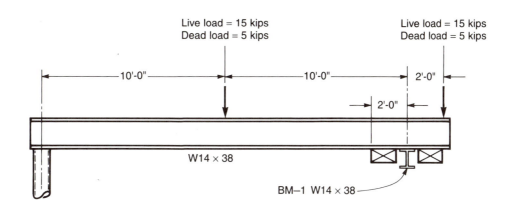

Live load = 15 kips
Dead load = 5 kips

Live load = 15 kips
Dead load = 5 kips

10'-0"

10'-0"

2'-0"

2'-0"

W14 × 38

BM—1 W14 × 38

Detail 1

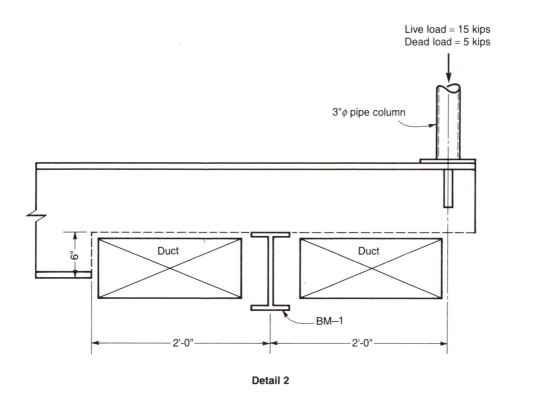

Live load = 15 kips
Dead load = 5 kips

3"ϕ pipe column

6"

Duct

Duct

BM—1

2'-0"

2'-0"

Detail 2

FIGURE 90B—4(i)

253

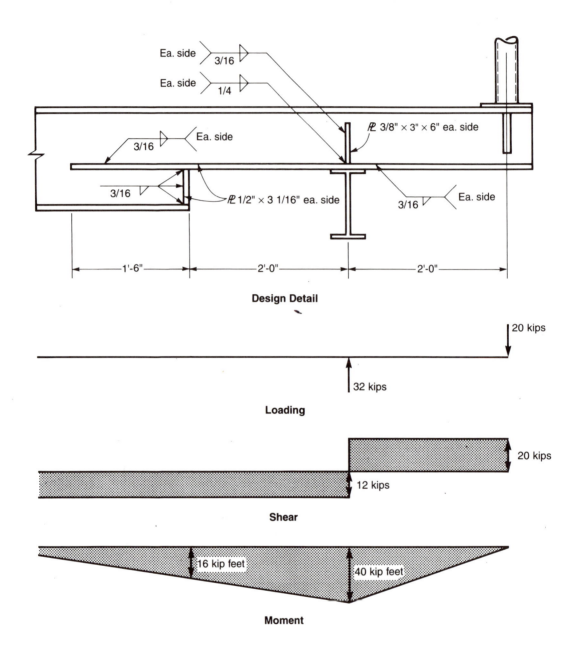

Design Detail

Loading

Shear

Moment

FIGURE 90B–4(ii)

254

STRUCTURAL ENGINEER EXAMINATION – 1987 ===================

PROBLEM B–4 – WT. 5.0 POINTS

GIVEN: The W24 × 68 roof beam, with the loading indicated, must be designed to allow for a mechanical duct, as shown in detail A.

CRITERIA: All steel is Grade A36 with F_y = 36 kips per square inch.
All welding is Grade E70XX electrodes.
Top Flange and Column are laterally braced.
Neglect the dead load of the beam.

REQUIRED: 1.   Check the notched portion of the beam shown.
                  2.   Design the notched portion of beam using a 7 inch wide bottom flange if necessary. A one inch clear space is required around the duct.
                  3.   Design the column connection between the column and the beam.
                  4.   Sketch the revised detail showing the size of all plates, bolts and welds with all the appropriate dimensions.

SOLUTION

1. LOADING

The maximum moment and shear occurring over the length of the notch are

$$M = 25 \times 36$$
$$= 900 \text{ kip inches}$$
$$V = 25 \text{ kips}$$

The allowable bending stress is

$$F_b = 0.6F_y$$
$$= 21.6 \text{ kips per square inch}$$

The required section modulus is

$$S_r = M/F_b$$
$$= 900/21.6$$
$$= 41.7 \text{ inches}^3$$

and to provide this modulus, a bottom flange plate must be welded to the web.

2. DESIGN NOTCH

The relevant properties of the W24 × 68 beam are

$$
\begin{aligned}
\text{Web thickness } t_w &= 0.415 \text{ inches} \\
\text{Flange width } b_f &= 8.965 \text{ inches} \\
\text{Flange thickness } t_f &= 0.585 \text{ inches}
\end{aligned}
$$

The flange area of the beam is

$$
\begin{aligned}
A_f &= b_f \times t_f \\
&= 8.965 \times 0.585 \\
&= 5.25 \text{ square inches}
\end{aligned}
$$

Using a seven inch wide plate to provide a bottom flange of equal area to the top flange requires a plate thickness of

$$
\begin{aligned}
t &= A_f/b \\
&= 5.25/7 \\
&= 0.75 \text{ inches}
\end{aligned}
$$

Using a ¾ inch × 7 inch plate provides a built-up section with the following properties

$$
\begin{aligned}
\text{Overall depth } d &= 10.5 + 0.75 \\
&= 11.25 \text{ inches} \\
\text{Depth between flange centroids } d_c &= d - (t_f + t)/2 \\
&= 11.25 - (0.585 + 0.75)/2 \\
&= 10.58 \text{ inches} \\
\text{Height of centroid } y &= (d_c + t)/2 \\
&= (10.583 + 0.75)/2 \\
&= 5.67 \text{ inches} \\
\text{Inertia } I &= A_f d_c^2/2 + t_w(10.5 - 0.585)^3/12 \\
&= 5.25 \times 10.58^2/2 + 0.415 \times 9.915^3/12 \\
&= 327.54 \text{ inches}^4 \\
\text{Section modulus } S &= I/y \\
&= 327.54/5.67 \\
&= 57.77 \\
&> S_r
\end{aligned}
$$

Hence the built up member is adequate in flexure.

Continuous fillet welds, on each side of the web, are required[6] to connect the reinforcement to the web.

The shear applied to the weld is given by

$$q_r = VQ/I$$

where $Q = A_f d_c/2$

$= 5.25 \times 10.58/2$

$= 27.77 \text{ inches}^3$

Hence $q_r = 25 \times 27.77/327.54$

$= 2.12 \text{ kips per inch}$

In accordance with UBC Section 2251 Table J2.4, the minimum size of weld required for the ¾ inch plate reinforcement is ¼ inch. The capacity of a ¼ inch fillet weld is given by UBC Section 2251.J2.4 as

$$q = 4 \times 0.928$$

$= 3.71 \text{ kips per inch}$

The capacity of the two ¼ inch fillet welds is

$$q_w = 2q$$

$= 2 \times 3.71$

$= 7.42 \text{ kips per inch}$

$> q_r$

Hence the continuous ¼ inch fillet welds are satisfactory.

Continuous fillet welds, on both sides of the reinforcement, are required[6] to connect the reinforcement to the web over the development length, past the notch. The ¾ inch reinforcement must extend into the web of the W24 × 68 a sufficient distance to develop the full tensile capacity of the plate. Thus the development length required, based on the weld strength, is given by

$$\ell_1 = 0.6F_y A_f/4q$$

$= 0.6 \times 36 \times 5.25/(4 \times 3.71)$

$= 7.64 \text{ inches}$

The development length, based on the shear capacity of the beam web, is given by

$$\ell_2 = 0.6F_y A_f/0.4F_y t_w$$

$= 19 \text{ inches ... governs.}$

3. COLUMN CONNECTION

BEARING PLATE

The required thickness of the bearing plate is based on the cantilever bending, shown in Figure 87B-4, and is given by AISC page 2–142 as

$$t \quad = (3f_p n^2 / F_b)^{0.5}$$

where
f_p = bearing pressure on the plate
$\quad$ = $V/(B \times N)$
$\quad$ = $25/(6 \times 5)$
$\quad$ = 0.83 kips per square inch

n = cantilever length
$\quad$ = $B/2 - k$
$\quad$ = $6/2 - (0.75 + 0.25)$
$\quad$ = 2.0 inches

F_b = allowable bending stress given in UBC Section 2251 Equation (F2–1)
$\quad$ = $0.75F_y$
$\quad$ = 27 kips per square inch

and
t = $(3 \times 0.83 \times 2^2/27)^{0.5}$
$\quad$ = 0.61 inches

Use a ⅝ × 6 × 5½ plate.

LOCAL WEB YIELDING

To prevent web yielding, stiffeners must be provided when the end reaction exceeds the value given by UBC Section 2251 Equation (K1–3) as

$$P \quad = 0.66F_y t_w(N + 2.5k)$$
$$= 0.66 \times 36 \times 0.415(5 + 2.5 \times 1.0)$$
$$= 73.95 \text{ kips}$$
$$> V \ldots \text{ stiffeners not required}$$

WEB CRIPPLING

To prevent web crippling, stiffeners must be provided when the end reaction exceeds the value given by UBC Section 2251 Equation (K1–5) as

$$P \quad = 34t_w^2[1.0 + 3(N/d)(t_w/t_f)^{1.5}](F_y t_f/t_w)^{0.5}$$
$$= 34 \times 0.415^2[1.0 + 3 \times (5/11.25)(0.415/0.75)^{1.5}](36 \times 0.75/0.415)^{0.5}$$
$$= 73.15 \text{ kips}$$
$$> V \ldots \text{ stiffeners not required}$$

SIDESWAY WEB BUCKLING

To prevent sidesway web buckling, when the loaded flange is restrained against rotation, stiffeners must be provided if the concentrated load exceeds the value given by UBC Section 2251 Equation (K1–6) as

$$P = R_5(1.0 + 0.4R_6^3)$$

where

$$R_6 = d_c b_f / \ell t_w$$

d_c = web depth clear of fillets

= 10.5 − 1.375 − 0.25

= 8.875 inches

b_f = flange width

= 7.0 inches

ℓ = largest unbraced length of either flange

= 12 × 12

= 144 inches

t_w = web thickness

= 0.415 inches

and

R_6 = 8.875 × 7.0/(144 × 0.415)

= 1.05

< 2.3 ... Equation (K1–6) must be checked

$R_5 = 6800t_w^3/h$

h = clear depth between flanges

= 10.5 − 0.585

= 9.915 inches

and

R_5 = 6800 × 0.415³/9.915

= 49.02

then

P = 49.02(1.0 + 0.4 × 1.05³)

= 71.72 kips

> V ... stiffeners not required

The complete detail is shown in Figure 97B–4.

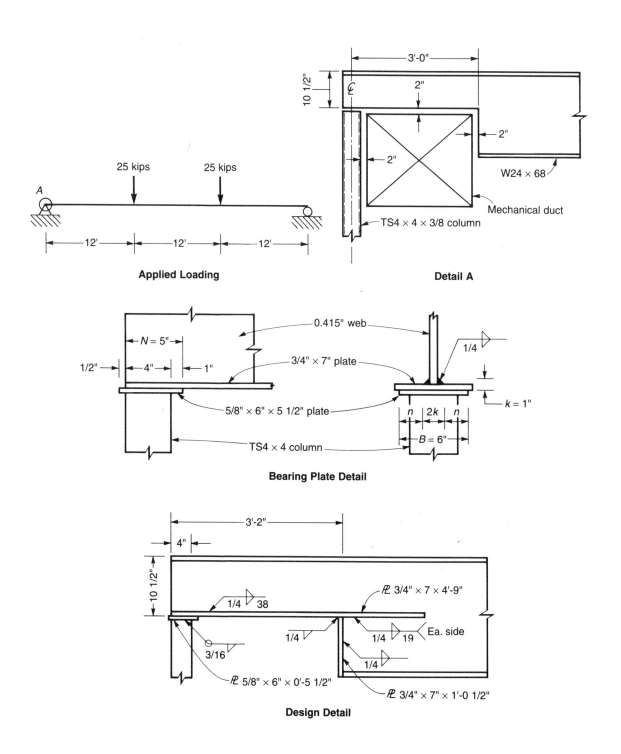

Applied Loading

25 kips   25 kips

A

12'   12'   12'

Detail A

3'-0"

10 1/2"

2"

2"

2"

W24 × 68

Mechanical duct

TS4 × 4 × 3/8 column

Bearing Plate Detail

0.415" web

N = 5"

1/2"   4"   1"

3/4" × 7" plate

5/8" × 6" × 5 1/2" plate

TS4 × 4 column

1/4

k = 1"

n   2k   n

B = 6"

Design Detail

3'-2"

4"

10 1/2"

1/4 ∇ 38

1/4

3/16

ℓ 5/8" × 6" × 0'-5 1/2"

ℓ 3/4" × 7 × 4'-9"

1/4 ∇ 19   Ea. side

1/4

ℓ 3/4" × 7" × 1'-0 1/2"

FIGURE 87B–4

STRUCTURAL ENGINEER EXAMINATION

CHAPTER 2

STRUCTURAL STEEL DESIGN

SECTION 2.6

DIAPHRAGM OPENING

PROBLEMS:

1987 B–5

STRUCTURAL ENGINEER EXAMINATION – 1987

PROBLEM B5 – WT. 3.0 POINTS

GIVEN: The drawings below show the roof framing plan of a one-story building with a metal deck roof and steel roof framing. All exterior walls are masonry shear walls. The building is located in seismic zone 4.

CRITERIA: Roof DL = 20 pounds per square foot
Roof LL = 20 pounds per square foot
Wall Wt = 80 pounds per square foot
1½" metal deck diaphragm
A36 steel, F_u = 58 kips per square inch
All welding is Grade E70XX Electrodes

REQUIRED: Check the adequacy of the connection detail shown in detail A.

SOLUTION

LATERAL FORCE

The seismic force acting on the structure is obtained from UBC Section 1628.2 Formula (28-1) as

$$V = (ZIC/R_w)W$$

where
Z = 0.4 for zone 4 from Table 16-I
I = 1.0 from Table 16-K for a standard occupancy structure
C = 2.75 from Section 1628.2 for an unspecified soil type
R_w = 6 from Table 16-N for a bearing wall system with masonry shear walls
W = structure dead load

Then
V = $(0.4 \times 1 \times 2.75/6)W$
= 0.183W

NORTH–SOUTH SEISMIC

The lateral seismic loads acting at the roof diaphragm level are shown in Figure 4–12 and are given by

Roof, first bay = $0.183 \times 0.02 \times 50$
= 0.183 kips per linear foot
Roof, remaining bays = $0.183 \times 0.02 \times 100$
= 0.366 kips per linear foot
North wall = $0.183 \times 0.08 \times 18.5^2/(2 \times 16)$
= 0.157 kips per linear foot
South wall = 0.157 kips per linear foot

262

The method of analysis employed[3] is to determine the overall diaphragm forces and then to correct for the local effects of the lateral loads at the opening. In accordance with UBC Section 1631.2.9, allowable stresses may not be increased by one-third for seismic loads because of the re-entrant corner in the building. Analyzing the overall diaphragm, the shear force acting on the West wall is given by

$$R_A = 0.68 \times 200/2 - 0.183 \times 40 \times 180/200$$
$$= 61.31 \text{ kips}$$

The resultant loads acting on the diaphragm at the opening are shown in Figure 87B-5. The lateral loading consists of the seismic loads due to the roof plus the North wall, and is given by

$$w = 0.183 + 0.157$$
$$= 0.34 \text{ kips per linear foot}$$

The horizontal force acting on the connection at detail A is given by

$$F_h = 40R_A/50 - w \times 40^2/100$$
$$= 40 \times 61.31/50 - 0.34 \times 40^2/100$$
$$= 43.61$$

and this may be assumed to act at the centroid of the bolt group.
The vertical force acting on the connection is due to the roof dead load and is given by

$$F_v = 0.02 \times 40 \times 5/4$$
$$= 1.0 \text{ kip}$$

and this acts vertically through the bolts' centroid.

PLATE WELD

The flange thickness of the W24 × 62 is approximately 0.5 inches which gives the shear plate size as 3.5 × 11 inches. Assuming a one-inch corner snip is provided to the plate to clear the web fillet on the W24 × 62, the weld profile is as shown in Figure 87B-5. The relevant weld properties for unit size of the weld are

Length of weld L	$= 10.0 + 2.5$
	$= 12.5$ inches
Area A	$= 1 \times L$
	$= 12.5$ inches^2 per inch
Centroid location y	$= 10.0 \times 6.0/12.5$
	$= 4.80$ inches
Centroid location x	$= 2.5 \times 2.25/12.5$
	$= 0.45$ inches
Inertia about x-axis I_x	$= 10.0^3/12 + 10.0 \times 1.20^2 + 2.5 \times 4.80^2$
	$= 155.33$ inches^4 per inch
Inertia about y-axis I_y	$= 2.5^3/12 + 2.5 \times 1.80^2 + 10.0 \times 0.45^2$
	$= 11.43$

$$\text{Polar Inertia, } I_o = I_x + I_y$$
$$= 155.33 + 11.43$$
$$= 166.76$$

The moment acting about the centroid of the weld profile is

$$M_o = F_h(6.5 - y) - F_v(2 - x)$$
$$= 43.61(6.5 - 4.80) - 1(2 - 0.45)$$
$$= 72.59 \text{ kip inch}$$

The co-existent forces acting at the bottom of the weld profile in the x-direction and y-direction are given by

$$f_x = F_h/A + M_o(11.0 - y)/I_o$$
$$= 43.61/12.5 + 72.59(11.0 - 4.80)/166.76$$
$$= 6.19 \text{ kips per inch}$$
$$f_y = F_v/A + M_o x/I_o$$
$$= 1/12.5 + 72.59 \times 0.45/166.76$$
$$= 0.28 \text{ kips per inch}$$

The resultant force at the bottom of the weld profile is

$$f_r = (f_x^2 + f_y^2)^{1/2}$$
$$= (6.19^2 + 0.28^2)^{1/2}$$
$$= 6.19 \text{ kips per inch}$$

The allowable force on the double 3/16-inch E70XX grade fillet weld is obtained from UBC Table J2.5 and, in accordance with UBC Section 1631.2.9 this may not be increased by one-third for seismic loads because of the re-entrant corner. The allowable force is, then,

$$q = 2 \times 3/16 \times 0.707 \times 0.3F_u$$
$$= 2 \times 3/16 \times 0.707 \times 0.3 \times 70$$
$$= 2 \times 3 \times 0.928$$
$$= 5.57 \text{ kips per inch}$$
$$< f_r$$

Hence the weld provided is inadequate

DESIGN FORCE

The resultant force acting on the connection is given by

$$F_r = (F_h^2 + F_v^2)^{1/2}$$
$$= (43.61^2 + 1^2)^{1/2}$$
$$= 43.62 \text{ kips}$$

PLATE CAPACITY

The gross plate area is given by

$$A_g = Bt$$
$$= 11.5 \times 0.375$$
$$= 4.31 \text{ square inches}$$

Based on the gross area, the allowable plate strength is defined in UBC Section 2251.D1 and is given by

$$P_t = 0.6F_yA_g$$
$$= 0.6 \times 36 \times 4.31$$
$$= 93.15 \text{ kips}$$

The maximum allowable value of the effective net area is defined by UBC Section 2251.B3 as

$$A_e = 0.85A_g$$
$$= 0.85 \times 4.31$$
$$= 3.66 \text{ square inches}$$

The actual effective net area is defined by UBC Section 2251.B3 and is given by

$$A_e = t(B - nd_e)$$

where

$$n = 3 = \text{number of bolts of diameter } d_b \text{ in one vertical row}$$
$$d_e = \text{effective width of hole as given by Section 2251.B2 and Table J3.1}$$
$$= d_b + 0.125$$
$$= 0.75 + 0.125$$
$$= 0.875 \text{ inches}$$

Then

$$A_e = 0.375 (11.5 - 3 \times 0.875)$$
$$= 3.33 \text{ square inches}$$
$$< 3.66$$

Hence the applicable net area is

$$A_e = 3.33 \text{ square inches}$$

Based on the effective net area, the allowable plate strength is defined by UBC Section 2251.D1 and is given by

$$P_t = 0.5 F_uA_e$$
$$= 0.5 \times 58 \times 3.33$$
$$= 96.57 \text{ kips}$$
$$> 93.15$$

Hence, the allowable plate capacity, as governed by the gross area is

$$P_t = 93.15 \text{ kips}$$

This exceeds the horizontal design force and is satisfactory.

BOLT SHEAR

The allowable single shear force on a ¾ inch diameter A325-X bolt is given in AISC Table 1-D as 13.3 kips. The allowable force on three bolts, conforming to the minimum edge distance of 1.25 inches given in UBC Table J3.5 is

$$P_s = 3 \times 13.3$$
$$= 39.9 \text{ kips}$$

This is less than the resultant design force and is unsatisfactory.

BOLT BEARING

Bearing on the beam web governs and AISC Table I-F defines the bearing capacity of the three bolts, conforming to minimum edge distance and spacing criteria, as

$$P_b = 52.2nt_w$$
$$= 52.2 \times 3 \times 0.23$$
$$= 36.02 \text{ kips}$$

This is less than the resultant design force and is unsatisfactory.

WEB TEAR-OUT

The net effective area resisting shear is given by AISC page 4-9 as

$$A_v = t_w(2 \times 1.5 - d_h)$$

where t_w = web thickness of the W14 × 22
= 0.23 inches

d_h = nominal diameter of a standard hole in shear as given by UBC Section 2251 Table J3.1
$= d_b + 0.0625$
= 0.8125 inches

and $A_v = 0.23(2 \times 1.5 - 0.8125)$
= 0.50 square inches

The net effective area resisting tension is given by

$$A_t = t_w(6 - 2d_h)$$
$$= 0.23(6 - 2 \times 0.8125)$$
$$= 1.01 \text{ square inches}$$

The resistance to shearing rupture is determined from UBC Section 2251.J4 as

$$P_r = 0.3F_uA_v + 0.5F_uA_t$$
$$= 0.3 \times 58 \times 0.50 + 0.5 \times 58 \times 1.01$$
$$= 38.04 \text{ kips}$$

This is less than the resultant design force and is unsatisfactory.

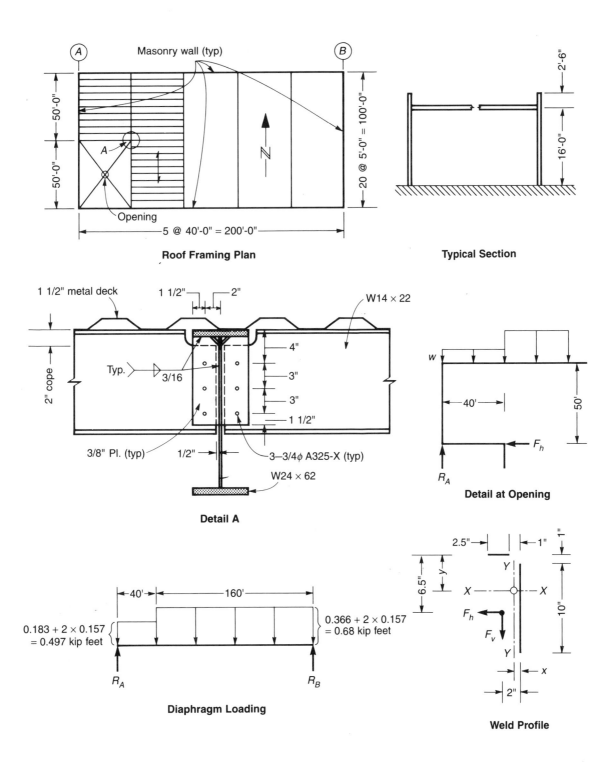

Roof Framing Plan

Typical Section

Detail A

1 1/2" metal deck

1 1/2" — 2"

W14 × 22

4"

3"

3"

1 1/2"

2" cope

Typ. 3/16

3/8" Pl. (typ)   1/2"

3-3/4φ A325-X (typ)

W24 × 62

Detail at Opening

w

40'

50'

R_A F_h

Diaphragm Loading

40'   160'

0.183 + 2 × 0.157
= 0.497 kip feet

0.366 + 2 × 0.157
= 0.68 kip feet

R_A R_B

Weld Profile

2.5"   1"   1"

6.5"   y

X — X

F_h 10"

F_v

x

2"

Masonry wall (typ)

50'-0"

50'-0"

A

Opening

5 @ 40'-0" = 200'-0"

20 @ 5'-0" = 100'-0"

2'-6"

16'-0"

FIGURE 87B-5

267

CHAPTER 2

REFERENCES

1.   William, A. *The analysis of indeterminate structures*. Hart Publishing Company, Inc, New York, 1968, pp 9–41.

2.   Tuma, J.J. *Structural Analysis*. McGraw–Hill, New York, 1969, pp 162–179.

3.   Pask, J.W. *Manual on connections for beam and column construction*. British Constructional Steelwork Association Ltd, London, 1982, pp 92–104.

4.   Becker, R. Naeim, F., Teal, E.J. *Seismic design practice for steel buildings*. Steel Committee of California, El Monte, 1988, p. 26.

5.   Naeim, F. Editor. *Seismic design handbook*. Van Nostrand Reinhold, New York, 1989, pp 254–255.

6.   Darwin, D. *Steel and composite beams with web openings*. American Institute of Steel Construction, Chicago, 1990, p. 15.

7.   Applied Technology Council, *Guidelines for the design of horizontal wood diaphragms*. Berkeley, CA, 1981, pp. 13–14.

3

STRUCTURAL CONCRETE DESIGN

SECTION 3.1

REINFORCED CONCRETE BEAMS

PROBLEMS:

STRUCTURAL ENGINEER EXAMINATION – 1991 ════════════════

PROBLEM C–2A – WT. 8.0 POINTS

GIVEN: The loading diagram and resulting forces acting on a cast–in–place continuous concrete beam are shown in Figure 91C-2. The beam cross section is shown. The longitudinal reinforcement required, for bending only, is 1.38 square inches, top and bottom.

CRITERIA: Materials:
- Concrete: f'_c = 4000 psi (Normal weight concrete)
- Reinforcement: f_y = 60,000 psi

Assumptions:
- The forces shown are due to gravity loads only
- Longitudinal reinforcement consists of No.6 bars
- Stirrups consist of No.3 bars
- Clear cover to stirrups is 1.50 inches
- Beam effective depth is 21.50 inches

REQUIRED:
1. Determine the combined transverse shear and torsion reinforcement requirements for the given beam.
2. Determine the combined flexural and longitudinal torsion reinforcement requirements for the given beam.

SOLUTION

1. SHEAR AND TORSION REINFORCEMENT

The critical section for shear and torsion in accordance with UBC Sections 1911.1.3.1 and 1911.6.4, is located a distance from the support equal to the effective depth of the section. At this location, the factored applied shear force and torsional moment are

$$V_u = 32 \text{ kips}$$
$$T_u = 30 \text{ kip feet}$$

In accordance with UBC Section 1911.6.1.1, the flange width to be included in the section properties is given by

$$y = 3h_f$$
$$= 3 \times 6$$
$$= 18 \text{ inches}$$

270

The torsional section property is, then

$$\Sigma x^2 y = (x^2 y)_{flange} + (x^2 y)_{beam}$$
$$= 6^2 \times 18 + 16^2 \times 24$$
$$= 6792 \text{ inches}^3$$

Torsional effects may be neglected, as indicated by UBC Section 1911.6.1, when the applied factored torsional moment is

	T	$< 0.5\phi(f'_c)^{0.5}\Sigma x^2 y$
where	ϕ	= strength reduction factor
		= 0.85 from UBC Section 1909.3.2.3
Then	T	$= 0.5 \times 0.85 \times (4000)^{0.5} \times 6792/12,000$
		= 15.21 kip feet
		$< T_u$

Hence torsional effects must be considered, the shear capacity of the concrete section is governed by UBC Formula (11–5) and the minimum area of closed stirrups is governed by UBC Formula (11–16). The torsional cracking moment, as given by UBC Section 1911.6.3, is

	T	$= 4\phi(f'_c)^{0.5}\Sigma x^2 y/3$
		$= 4 \times 0.85 \times (4000)^{0.5} \times 6792/36,000$
		= 40.56 kip feet
		$> T_u$

Hence no redistribution of the torsion occurs and the torsional reinforcement must be designed for the full torsional moment

$$T_u = 30 \text{ kip feet}$$

The design torsional strength of the concrete section is given by UBC Formula (11–22) as

	ϕT_c	$= 0.8\phi(f'_c)^{0.5}\Sigma x^2 y/[1.0 + (0.4V_u/C_t T_u)^2]^{0.5}$
where	C_t	$= b_w d/\Sigma x^2 y$
		$= 16 \times 21.5/6,792$
		= 0.0506
Hence	ϕT_c	$= 0.8 \times 0.85 \times (4000)^{0.5} \times 6792/[1.0 + (0.4 \times 32/0.0506 \times 360)^2]^{0.5}$
		= 239,040 pound inches
		= 19.92 kip feet

The design torsional strength required from the torsional reinforcement is obtained from UBC Formula (11–21) as

	ϕT_s	$= T_u - \phi T_c$
		= 12(30 – 19.92)
		= 120.96 kip inches
		$< 4\phi T_c$

Hence the requirement of UBC Section 1911.6.9.4 is satisfied

The area of closed stirrups required for torsional effects is given by UBC Formula (11–23) as

$$A_t/s = \phi T_s/(\phi f_y \alpha_t x_1 y_1)$$

where

$$
\begin{aligned}
x_1 &= b_w - d_b - 2(\text{clear cover}) \\
&= 16 - 0.375 - 2 \times 1.5 \\
&= 12.625 \text{ inches} \\
y_1 &= h - d_b - 2(\text{clear cover}) \\
&= 24 - 0.375 - 2 \times 1.5 \\
&= 20.625 \\
\alpha_t &= 0.66 + 0.33 y_1/x_1 \\
&= 0.66 + 0.33 \times 20.625/12.625 \\
&= 1.20 \\
&< 1.50
\end{aligned}
$$

Then

$$
\begin{aligned}
A_t/s &= 12 \times 120.96/(0.85 \times 60 \times 1.2 \times 12.625 \times 20.625) \\
&= 0.0911 \text{ square inches/foot/leg}
\end{aligned}
$$

The design shear strength of the concrete section is given by UBC Formula (11–5) as

$$
\begin{aligned}
\phi V_c &= 2\phi(f_c')^{0.5} b_w d/[1.0 + (2.5 C_t T_u/V_u)^2]^{0.5} \\
&= 2 \times 0.85 \times (4000)^{0.5} \times 16 \times 21.5/[1.0 + (2.5 \times 0.0506 \times 360/32)^2]^{0.5} \\
&\quad \times 1/1000 \\
&= 21.26 \text{ kips}
\end{aligned}
$$

The design shear strength required from the shear reinforcement is obtained from UBC Formula (11–2) as

$$
\begin{aligned}
\phi V_s &= V_u - \phi V_c \\
&= 32 - 21.26 \\
&= 10.74 \text{ kips}
\end{aligned}
$$

The area of shear reinforcement required is given by UBC Formula (11–17) as

$$
\begin{aligned}
A_v/s &= \phi V_s/\phi d f_y \\
&= 12 \times 10.74/(0.85 \times 21.5 \times 60) \\
&= 0.1175 \text{ square inches/foot}
\end{aligned}
$$

The total area of shear and torsional stirrups required, in accordance with UBC Section 1911.6.7.1, is given by

$$
\begin{aligned}
A/s &= A_v/s + 2A_t/s \\
&= 0.1175 + 2 \times 0.0911 \\
&= 0.30 \text{ square inches/foot}
\end{aligned}
$$

Closed stirrups consisting of No.3 bars at a spacing of nine inches provide an area of

$$A/s = 0.30 \text{ which is satisfactory}$$

The minimum area of torsional plus shear reinforcement, which may be provided, is given by UBC Formula (11–16) as

$$A/s = 50b_w/f_y$$
$$= 12 \times 50 \times 16/6,000$$
$$= 0.160$$
$$< 0.30 \text{ square inches/foot}$$

The value of the expression

$$4\phi(f'_c)^{0.5}b_wd = 4 \times 0.85 \times (4000)^{0.5} \times 16 \times 21.5/1000$$
$$= 74 \text{ kips}$$
$$> \phi V_s$$

Hence in accordance with UBC Section 1911.5.4 the maximum spacing of shear reinforcement is

$$s = d/2$$
$$= 21.5/2$$
$$= 10.8 \text{ inches}$$
$$> 9$$

The maximum spacing of torsional reinforcement is defined in UBC Section 1911.6.8 as

$$s = (x_1 + y_1)/4$$
$$= (12.625 + 20.625)/4$$
$$= 8.3 \text{ inches}$$
$$< 9$$

Hence closed stirrups consisting of No.3 bars must be provided at a spacing of eight inches.

2. LONGITUDINAL REINFORCEMENT

The longitudinal torsional reinforcement required is given by UBC Formula (11–24) as

$$A_\ell = 2A_t(x_1 + y_1)/s$$
$$= 2 \times 0.0911(12.625 + 20.625)/12$$
$$= 0.505 \text{ square inches}$$

Since $\quad 50b_w/f_y \; < \; 2A_t/s$

the longitudinal torsional reinforcement shall not be less than the value given by UBC Formula (11–25) as

$$\begin{aligned}
A_\ell \quad &= [400xs/f_y \times T_u/(T_u + V_u/3C_t) - 2A_t](x_1 + y_1)/s \\
&= [400 \times 16 \times 12/60{,}000 \times 360/(360 + 32/3 \times 0.0506) - 2 \times \\
&\quad 0.0911](12.625 + 20.625)/12 \\
&= 1.73 \text{ square inches}
\end{aligned}$$

and this value governs

In accordance with UBC Section 1911.6.8, one bar shall be placed in each corner of the closed stirrup and the remaining bars distributed around the perimeter at a maximum spacing of twelve inches. Using six longitudinal torsional bars, consisting of four corner bars and two side bars, gives a total area of reinforcement required at the top and bottom of the section of

$$\begin{aligned}
A_t \quad &= A_{st} + 2A_\ell/6 \\
&= 1.38 + 2 \times 1.73/6 \\
&= 1.96 \text{ square inches}
\end{aligned}$$

Five No.6 bars provide an area of

$$\begin{aligned}
A \quad &= 2.20 \text{ square inches} \\
&> A_t
\end{aligned}$$

The area of reinforcement required at the sides of the beam is

$$\begin{aligned}
A_s \quad &= A_\ell/6 \\
&= 1.73/6 \\
&= 0.29 \text{ square inches}
\end{aligned}$$

One No.6 bar provides an area of

$$\begin{aligned}
A \quad &= 0.44 \text{ square inches} \\
&> A_s
\end{aligned}$$

The total steel area required is

$$\begin{aligned}
\Sigma A \quad &= 2A_{st} + A_\ell \\
&= 2 \times 1.38 + 1.73 \\
&= 4.49 \text{ square inches}
\end{aligned}$$

The twelve No.6 bars proposed provide an area of

$$\begin{aligned}
A \quad &= 5.28 \text{ square inches} \\
&> \Sigma A
\end{aligned}$$

The required reinforcement is detailed in Figure 91C–2.

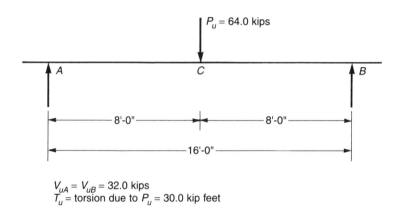

$V_{uA} = V_{uB} = 32.0$ kips
T_u = torsion due to P_u = 30.0 kip feet

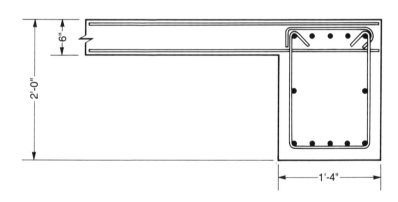

FIGURE 91C-2

275

STRUCTURAL ENGINEER EXAMINATION – 1988 ════════════════

PROBLEM C–1 – WT. 4.0 POINTS

GIVEN:

The architect wants to add a plaster ceiling to the roof of a building just completed. A section through the roof framing indicating a typical joist section is shown.

CRITERIA:

Design Loads:

Assume the total proposed service dead loads (including joist weight and plaster ceiling) is equivalent to 680 plf and the service live load is 160 plf for each joist.

Materials:

- Concrete: $f'_c = 3000$ psi (normal weight with $E_c = 3.12 \times 10^6$ psi)
- Reinforcement: $f_y = 60,000$ psi

Assumptions:

- The fifty feet span joists are simply supported and adequately supported laterally.
- All dead loads shall be considered as sustained loads for the expected 40 year life of the structure.
- Neglect the compression reinforcement in all computations.

REQUIRED:

What would be your recommendations to the architect regarding the existing concrete joists in terms of structural capacity and serviceability. Provide complete computations to support your conclusions.

SOLUTION

APPLIED LOADS

The dead load moment is given by

$$M_D = w_D \ell^2/8$$
$$= 0.68 \times 50^2/8$$
$$= 212.5 \text{ kip feet}$$

The live load moment is given by

$$M_L = M_D \times 160/680$$
$$= 50 \text{ kip feet}$$

The maximum factored moment is obtained from UBC Formula (9-1) as

$$M_u = 1.4M_D + 1.7M_L$$
$$= 1.4 \times 212.5 + 1.7 \times 50$$
$$= 382.50 \text{ kip feet}$$

The factored shear force at the support is given by

$$V_u' = \ell(1.4w_D + 1.7w_L)/2$$
$$= 30.6 \text{ kips}$$

The critical section for shear, in accordance with UBC Section 1911.1.3.1, is located a distance from the support equal to the effective depth of the section. The factored shear force at this location is

$$V_u = V_u' (1.0 - 2d/\ell)$$
$$= 30.6 [1.0 - 2 \times 27/(50 \times 12)]$$
$$= 27.85 \text{ kips}$$

DESIGN MOMENT STRENGTH

The design moment strength is obtained by using calculator program A.1.2 in Appendix A.

The reinforcement ratio is

$$\rho = A_s/bd$$
$$= 0.0093$$

The design moment strength is

$$\phi M_n = 0.9A_s f_y d(1 - 0.59\rho f_y/f_c')/12$$
$$= 433 \text{ kip feet}$$
$$> M_u$$

In accordance with UBC Section 1910.3.3, the maximum allowable reinforcement ratio is

ρ_{max} = $0.75 \times 0.85 \times 87\beta_1 f'_c/f_y(87 + f_y)$

where β_1 = 0.85 as defined in UBC Section 1910.2.7

Hence ρ_{max} = 0.016

$> \rho$

In accordance with UBC Section 1910.5.1, the minimum allowable reinforcement ratio is given by

ρ_{min} = $200/f_y$

= 0.0033

$< \rho$

Hence the joist is satisfactory in flexure.

DESIGN SHEAR STRENGTH

The design shear strength provided by the concrete is obtained from UBC Section 1911.3.1.1 and UBC Formula (11–3) and by utilizing calculator program A.1.2 as

ϕV_c = $0.85 \times 2bd(f'_c)^{0.5}$

= 40.22 kips

$> V_u$

$< 2V_u$

Hence in accordance with UBC Section 1911.5.5, minimum shear reinforcement is required and this is given by UBC Formula (11–14) as

A_v = $50bs/f_y$

= 0.16 square inches

for a stirrup spacing s of twelve inches.

The shear reinforcement provided is

A'_v = 2×0.11

= 0.22 square inches

$> A_v$

Hence the shear reinforcement provided is adequate

The maximum allowable stirrup spacing is given by UBC Section 1911.5.4 as

s_{max} = $d/2$

= 13.5 inches

> 12

Hence the stirrup spacing is satisfactory and the joint is satisfactory in shear.

CRACK CONTROL

In order to provide satisfactory crack control for interior exposure, UBC Section 1910.6.4 limits the numerical value of the factor z to 175. The factor z is given by UBC Formula (10–4) as

$$z \quad = f_s(d_cA)^{0.33}$$

where
$$f_s \quad = \text{reinforcement stress at service load}$$
$$= 0.6f_y$$
$$d_c \quad = \text{concrete cover to center of reinforcement}$$
$$= h - d$$
$$A \quad = \text{effective tension area of reinforcement}$$
$$= 2d_cb/n$$

Thus
$$z \quad = 0.6 \times 60[(30 - 27) \times 2 \times 3 \times 16/4]^{0.33}$$
$$= 150$$
$$< 175$$

Hence flexural cracking is adequately controlled.

DEFLECTION CONTROL

The minimum thickness stipulated in UBC Table 19-C-1 to provide adequate stiffness to limit deflection of the joist is

$$h \quad = \ell/16$$
$$= 50 \times 12/16$$
$$= 37.5 \text{ inches}$$
$$> 30 \text{ inches}$$

Hence in accordance with UBC Section 1909.5.2, computation of deflections is required.

Assuming that the given value of the effective moment of inertia for the computation of deflection is applicable at the service load level, the immediate deflection due to the applied live load is

$$\delta_L \quad = 22.5w_LL^4/EI$$
$$= 22.5 \times 0.16 \times 50^4/(3120 \times 16{,}500)$$
$$= 0.44 \text{ inches}$$

To assess the effects of long–term deflection, the multiplier for long–term deflection is obtained from UBC Section 1909.5.2.5 as

$$\lambda \quad = \xi/(1.0 + 50\rho')$$
$$= 2.0$$

since compression reinforcement may be neglected.

The applicable load for assessing long-term effects is given in UBC Table 19-I as the service dead load w_D.

Assuming that the given value of the effective moment of inertia is applicable at this load level, the corresponding deflection is given by

$$\begin{aligned}
\delta &= 22.5\lambda w_D L^4/EI \\
&= 22.5 \times 2 \times 0.68 \times 50^4/(3120 \times 16{,}500) \\
&= 3.72 \text{ inches}
\end{aligned}$$

The total deflection due to sustained load plus live load is

$$\begin{aligned}
\delta &= \delta_L + \delta_D \\
&= 0.44 + 3.72 \\
&= 4.16 \text{ inches}
\end{aligned}$$

and
$$\begin{aligned}
\delta/L &= 4.16/600 \\
&= 1/144
\end{aligned}$$

The plaster ceiling is likely to be damaged by large deflections and the maximum allowable deflection/span ratio for long-term loading is given in UBC Table 19-I as

$$\begin{aligned}
\delta/L &= 1/480 \\
&< 1/144
\end{aligned}$$

Hence the long-term deflection of the joist is unsatisfactory and is likely to cause damage to the plaster ceiling.

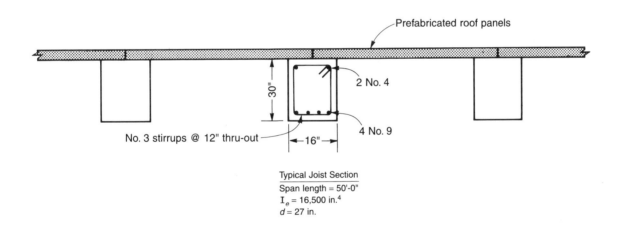

Prefabricated roof panels

30"

2 No. 4

No. 3 stirrups @ 12" thru-out

16"

4 No. 9

Typical Joist Section
Span length = 50'-0"
I_e = 16,500 in.^{4}
d = 27 in.

FIGURE 88C-1

STRUCTURAL ENGINEER EXAMINATION – 1987 ═══════════

PROBLEM C–3 – WT. 5.0 POINTS

GIVEN: A three-span concrete beam as shown in detail A

CRITERIA:
- Concrete beam width b=20 inches
- Concrete beam total depth h=27 inches
- Concrete beam reinforcing steel depth d=24 inches
- Concrete: $f'_c = 3000$ psi
- Reinforcement: $f_y = 60,000$ psi
- Do not calculate alternative span (skip) loadings
- Show all calculations

REQUIRED:
1. Determine the maximum design negative moment.
2. Determine the maximum design positive moment.
3. Sketch to an appropriate scale the shear diagram showing all maximum values.
4. Determine the critical design shear.
5. Determine the required spacing at the critical design shear location for No.4 stirrups with four vertical legs as shown in detail C.
6. The longitudinal reinforcing shown in detail B has been proposed to you by a junior engineer. Is it adequate? Refer to detail C for the beam cross section.

SOLUTION

APPLIED LOADS

The factored load is obtained from UBC Section 1909.2 Formula (9-1) as

$$w_u = 1.4w_D + 1.7w_L$$

For spans 12 and 34 the factored load is

$$\begin{aligned} w_{12} &= 1.4 \times 3 + 1.7 \times 10 \\ &= 21.2 \text{ kips per foot} \end{aligned}$$

For span 23 the factored load is

$$\begin{aligned} w_{23} &= 1.4 \times 3 + 1.7 \times 2 \\ &= 7.6 \text{ kips per foot} \end{aligned}$$

The factored moment and shear acting on the beam may be obtained by the moment distribution procedure[1,2]. Advantage may be taken of the hinged supports and the symmetry of the structure and the loading to modify the stiffness factors, as shown in the Table so that distribution is required in only half the structure and no carry-over is required. The initial fixed-end moments at joint 2 are given by

$$\begin{aligned} M_{21}^F &= w_{12}\ell_{12}^2/8 \\ &= 21.2 \times 20^2/8 \\ &= 1060 \text{ kip feet} \end{aligned}$$

$$\begin{aligned} M_{23}^F &= -w_{23}\ell_{23}^2/12 \\ &= -7.6 \times 10^2/12 \\ &= -63 \text{ kip feet} \end{aligned}$$

The convention adopted is that clockwise moments acting from the joint on a member are positive and the distribution, for the left half of the structure, is shown in the following Table

Joint	1	2	2
Member	12	21	23
Relative EI/L	1/20	1/20	1/10
Modified stiffness	4/20	3/20	2/10
Distribution factor	1	3/7	4/7
Fixed end moments	0	1060	-63
Distribution	0	-427	-570
Final moment, kip ft.	0	633	-633

1. MAXIMUM NEGATIVE MOMENT

The maximum negative moment occurs at joints 2 and 3, as shown in Figure 87C-3, and is

$$M_{21} = 633 \text{ kip feet}$$

2. MAXIMUM POSITIVE MOMENT

The shear acting on the beam at joint 1 is given by

$$
\begin{aligned}
V_{12} &= w_{12}\ell_{12}/2 - M_{21}/l_{12} \\
&= 21.2 \times 20/2 - 633/20 \\
&= 180 \text{ kips}
\end{aligned}
$$

The shear reduces to zero at a distance from joint 1 given by

$$
\begin{aligned}
x &= V_{12}/w_{12} \\
&= 180/21.2 \\
&= 8.5 \text{ feet}
\end{aligned}
$$

Then the maximum positive moment occurs in spans 12 and 34 and is given by

$$
\begin{aligned}
M_p &= V_{12}x/2 \\
&= 180 \times 8.5/2 \\
&= 765 \text{ kip feet}
\end{aligned}
$$

3. SHEAR DIAGRAM

The shear at the left-hand-side of joint 2 is given by

$$
\begin{aligned}
V_{21} &= V_{12} - w_{12}\ell_{12} \\
&= 180 - 21.2 \times 20 \\
&= -244 \text{ kips}
\end{aligned}
$$

The shear at the right-hand-side of joint 2 is given by

$$
\begin{aligned}
V_{23} &= w_{23} \times \ell_{23}/2 \\
&= 7.6 \times 10/2 \\
&= 38 \text{ kips}
\end{aligned}
$$

The shear diagram is shown in Figure 87C-3.

4. CRITICAL SHEAR

The critical section for shear, in accordance with UBC Section 1911.1.3.1, is located a distance from the support equal to the effective depth of the section. The factored shear force at this location is

$$
\begin{aligned}
V_u &= V_{21} - w_{12}d \\
&= -244 + 21.2 \times 2 \\
&= -202 \text{ kips}
\end{aligned}
$$

5. STIRRUP SPACING

The design shear strength provided by the concrete is obtained from UBC Section 1911.3.1.1 and Formula (11–3) and by utilizing calculator program A.1.2 as

$$\phi V_c = 0.85 \times 2b_w d(f'_c)^{0.5}$$
$$= 44.7 \text{ kips}$$

The design shear strength required from shear reinforcement is given by UBC Formula (11–2) as

$$\phi V_s = V_u - \phi V_c$$
$$= 202 - 44.7$$
$$= 157.3 \text{ kips}$$
$$> 2 \times \phi V_c$$
$$< 4 \times \phi V_c$$

Hence the concrete section is adequate and the maximum stirrup spacing is given by UBC Section 1911.5.4 as

$$s = d/4$$
$$= 24/4$$
$$= 6 \text{ inches}$$

The area of shear reinforcement required is given by UBC Formula (11–17) as

$$A_v/s = \phi V_s/\phi d f_y$$
$$= 157.3 \times 12/(0.85 \times 24 \times 60)$$
$$= 1.54 \text{ inches}^2/\text{foot}$$

Shear reinforcement consisting of four vertical legs of No.4 stirrups at six inch spacing provides a reinforcement area of

$$A_v/s = 4 \times 0.39$$
$$= 1.56 \text{ inches}^2/\text{foot}$$
$$> 1.54$$

The minimum area of shear reinforcement, which may be provided, is given by UBC Formula (11–14) as

$$A_v/s = 50b_w/f_y$$
$$= 0.20$$
$$< 1.54$$

Hence the shear reinforcement indicated is satisfactory at a spacing of six inches.

6. FLEXURAL REINFORCEMENT

For span 12, the effective slab width is defined by UBC Section 1908.10.2 as

$$b = b_w + 16h_f = 20 + 16 \times 8 = 148 \text{ inches}$$

Assuming that the equivalent rectangular stress block is within the slab thickness, the design moment strength is obtained by using calculator program A.1.2 in Appendix A.

The reinforcement ratio is

$$
\begin{aligned}
\rho &= A_s/bd \\
&= 6 \times 1.56/(148 \times 24) \\
&= 0.0026
\end{aligned}
$$

The depth of the rectangular stress block is given by

$$
\begin{aligned}
a &= \rho d f_y/(0.85 f_c') \\
&= 1.49 \text{ inches} \\
&< 8.00
\end{aligned}
$$

Hence the stress block is within the slab thickness.

The design moment strength is

$$
\begin{aligned}
\phi M_n &= 0.9 A_s f_y d (1-0.59\rho f_y/f_c')/12 \\
&= 979 \text{ kip feet} \\
&= M_p
\end{aligned}
$$

In accordance with UBC Section 1910.3.3, the maximum allowable reinforcement ratio is given by

$$\rho_{max} = 0.75 \times 0.85 \times 87\beta_1 f_c'/f_y(87 + f_y)$$

where β_1 = 0.85 as defined in UBC Section 1910.2.7

Hence ρ_{max} = 0.016

$$> \rho$$

In accordance with UBC Section 2610(f)1, the minimum allowable reinforcement ratio is given by

$$
\begin{aligned}
\rho_{min} &= 200/f_y \\
&= 0.0033 \\
&< \rho
\end{aligned}
$$

In order to provide satisfactory crack control, UBC Section 1910.6.4 limits the numerical value of the factor z to 175 for interior exposure and 145 for exterior exposure. The factor z is given by UBC Formula (10-4) as

$$z = f_s(d_c A)^{0.33}$$
$$f_s = \text{reinforcement stress at service load}$$
$$= 0.6 f_y$$
$$d_c = \text{concrete cover to center of reinforcement}$$
$$= h - d$$
$$A = \text{effective tension area of reinforcement}$$
$$= 2 d_c b / n$$

Thus
$$z = 0.6 \times 60[(27 - 24) \times 2 \times 3 \times 20/6]^{0.33}$$
$$= 141$$
$$< 175$$
$$< 145$$

Hence flexural cracking is adequately controlled and the reinforcement provided in spans 12 and 34 is satisfactory.

In span 23, the moment at mid span is given by

$$M = M_{23} - V_{23} \times \ell_{23}/4$$
$$= 633 - 38 \times 10/4$$
$$= 538 \text{ kip feet}$$

This is a negative moment and, since no negative reinforcement is provided in span 23, this span is inadequate.

At joint 2, the design moment strength is obtained by using calculator program A.1.2.

The reinforcement ratio is

$$\rho = A_s / b_w d$$
$$= 4 \times 1.27/(20 \times 24)$$
$$= 0.0106$$

The design moment strength is

$$\phi M_n = 480 \text{ kip feet}$$
$$< M_{21}$$

Hence the reinforcement provided at joints 2 and 3 is inadequate.

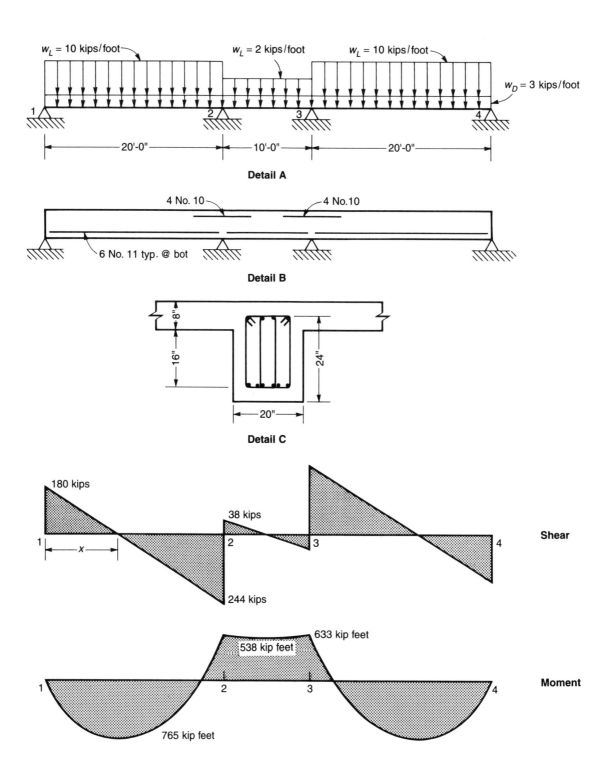

FIGURE 87C-3

288

STRUCTURAL ENGINEER REGISTRATION

CHAPTER 3

STRUCTURAL CONCRETE DESIGN

SECTION 3.2

REINFORCED CONCRETE COLUMNS

PROBLEMS:

1986 C–5

STRUCTURAL ENGINEER EXAMINATION – 1986 ═══════════════

PROBLEM C5 – WT. 6.0 POINTS

GIVEN: You have just completed designing the reinforced concrete tied columns shown in Figure 86C–5 when it was requested that the rainwater leader be placed inside the columns as shown. Three columns are affected, with factored loads and moments as follows:

$$\text{Col. No.1: } P_u = 60 \text{ kips ; } M_u = 30 \text{ kip feet}$$
$$\text{Col. No.2: } P_u = 150 \text{ kips; } M_u = 17 \text{ kip feet}$$
$$\text{Col. No.3: } P_u = 166 \text{ kips; } M_u = 14 \text{ kip feet}$$

CRITERIA:
- Concrete: $f'_c = 3500$ psi
- Reinforcement: $f_y = 60$ ksi
- $E_s = 29,000$ ksi
- $C_m = 1.0$
- $P_c = \pi^2 EI/(kL_u)^2$
  $= 325$ kips

REQUIRED: Are the columns still adequate? Show all calculations and diagrams supporting your answer. Consider only the direction of bending moments as indicated.

SOLUTION

AXIAL LOAD ONLY

The hypothetical design axial load strength at zero eccentricity is given by[3]

$$\phi P_o = \phi[0.85 f_c'(A_g - A_s) + A_s f_y]$$

where
$$\phi = \text{strength reduction factor}$$
$$= 0.7 \text{ from UBC Section 1909.3.2}$$
$$f_c' = \text{cylinder strength}$$
$$= 3.5 \text{ kips per square inch}$$
$$A_g = \text{gross area of the section}$$
$$= 12 \times 12 - 3 \times 7.5$$
$$= 121.5 \text{ square inches}$$
$$A_s = \text{reinforcement area}$$
$$= 2.4 \text{ square inches}$$
Thus
$$\phi P_o = 0.7[0.85 \times 3.5(121.5 - 2.4) + 2.4 \times 60]$$
$$= 348.83 \text{ kips}$$

and this is plotted on the interaction diagram in Figure 86C-5.

For a tied column, the actual design axial load strength at zero eccentricity is limited by UBC Section 1910.3.5 Formula (10–2), to account for accidental eccentricities, to the value

$$\phi P_{n(max)} = 0.8 \phi P_o$$
$$= 0.8 \times 348.83 = 279.06 \text{ kips}$$

and this is plotted on the interaction diagram in Figure 86C-5.

BALANCED DESIGN

For the balanced strain condition, under combined flexure and axial load, the maximum strain in the concrete and in the tension reinforcement must simultaneously reach the values specified in UBC Section 1910.3.2 as

$$\varepsilon_c = 0.003$$
$$\varepsilon_s = f_y/E_s = 60/29,000$$

From Figure 86C-5, the depth to the neutral axis and the stress in the compression reinforcement may be derived as

$$c = d\varepsilon_c/(\varepsilon_c + \varepsilon_s)$$
$$= 87d/(87 + f_y)$$
$$= 87 \times 9.5/(87 + 60) = 5.62 \text{ inches}$$
and
$$f_s' = E_s \varepsilon_c(c - d')/c$$
$$= 87(1.0 - d'/c)$$
$$= 87(1.0 - 2.5/5.62)$$
$$= 48.30 \text{ kips per square inch}$$
$$< f_y$$

In accordance with UBC Section 1910.2.7 the depth of the equivalent rectangular stress block is given by

$$a = c\beta_1$$

where β_1 = 0.85 as defined in UBC Section 1910.2.7

Hence a = 5.62 × 0.85

$\qquad$ = 4.78 inches

Then the forces in the concrete, compression reinforcement, and tensile reinforcement are given by

$$
\begin{aligned}
C_c &= 0.85f_c'(ab - A_s/2) \\
&= 0.85 \times 3.5(4.78 \times 9 - 1.2) \\
&= 124.41 \text{ kips} \\
C_s &= f_s'A_s/2 \\
&= 48.30 \times 1.2 \\
&= 57.96 \\
T &= f_yA_s/2 \\
&= 60 \times 1.2 \\
&= 72 \text{ kips}
\end{aligned}
$$

The nominal axial load capacity at the balanced strain condition is

$$
\begin{aligned}
P_b &= C_c + C_s - T \\
&= 124.41 + 57.96 - 72 \\
&= 110.37 \text{ kips}
\end{aligned}
$$

The design axial load capacity at the balanced strain condition is

$$
\begin{aligned}
\phi P_b &= 0.7 \times 110.37 \\
&= 77.26 \text{ kips}
\end{aligned}
$$

The nominal moment capacity at the balanced strain condition is obtained by summing moments about the mid–depth of the section and is given by

$$
\begin{aligned}
M_b &= C_c(h/2 - a/2) + C_s(h/2 - d') + T(h/2 - d_s) \\
&= 124.41(12/2 + 4.78/2) + 57.96(12/2 - 2.5) + 72(12/2 - 2.5) \\
&= 903.98 \text{ kip inches}
\end{aligned}
$$

The design moment capacity at the balanced strain condition is

$$
\begin{aligned}
\phi M_b &= 0.7 \times 903.98/12 \\
&= 52.73 \text{ kip feet}
\end{aligned}
$$

and the point corresponding to the balanced strain condition is plotted on the interaction diagram.

FLEXURE ONLY

Neglecting compression reinforcement, the design moment strength of the section, without applied axial load, is obtained by using calculator program A.1.2 in Appendix A and is given by

$$\phi M_n = 0.9 A_s f_y d(1 - 0.59 \rho f_y / f_c')/12$$
$$= 44.02 \text{ kip feet}$$

The ratio of the distance between the compression and tension reinforcement to the overall depth of the section is given by

$$(h - d' - d_s)/h = (12 - 2.5 - 2.5)/12$$
$$= 0.583$$
$$< 0.7$$

Also the value of

$$0.1 f_c' A_g = 0.1 \times 3.5 \times 121.5$$
$$= 42.53 \text{ kips}$$
$$< \phi P_b$$

Hence, in accordance with UBC Section 1909.3.2, the strength reduction factor ϕ decreases from a value of 0.9, at zero axial load, to a value of 0.7, at an axial load of $0.1 f_c' A_g$. Using a value of 0.7 for the strength reduction factor gives a value for the design moment strength of

$$\phi M_n' = \phi M_n \times 7/9$$
$$= 44.02 \times 7/9$$
$$= 34.24$$

The values of ϕM_n, $\phi M_n'$ and $0.1 f_c' A_g$ are plotted on the interaction diagram.

APPLIED LOADING

For column 1, assuming negligible sway moments and allowing for slenderness effects, the magnified factored moment is obtained from UBC Section 1910.11.5 Formula (10–6) as

$$M_c = \delta_b M_{2b} + \delta_s M_{2s}$$
$$= \delta_b M_u$$

where
$$\delta_b = C_m/(1.0 - P_u/\phi P_c)$$
$$= 1.0/[1.0 - 60/(0.7 \times 325)]$$
$$= 1.36$$

Hence
$$M_c = 1.36 \times 30$$
$$= 40.75 \text{ kip feet}$$

From the interaction diagram, the allowable factored moment at an axial load of 60 kips is

$$M_c = 49.0 \text{ kip feet}$$
$$> 40.75$$

Hence column 1 is satisfactory.

For column 2, the magnification factor is

$$\delta_b = 1.0/(1.0 - 150/227.5)$$
$$= 2.94$$

The magnified moment is

$$M_c = \delta_b M_u$$
$$= 2.94 \times 17$$
$$= 49.90 \text{ kip feet}$$

from the interaction diagram, the allowable factored moment at an axial load of 150 kips is

$$M_c = 39.0 \text{ kip feet}$$
$$< 49.90$$

Hence column 2 is unsatisfactory.

For column 3, the magnification factor is

$$\delta_b = 1.0/(1.0 - 166/227.5)$$
$$= 3.70$$

The magnified moment is

$$M_c = \delta_b M_u$$
$$= 3.70 \times 14$$
$$= 51.79 \text{ kip feet}$$

From the interaction diagram, the allowable factored moment at an axial load of 166 kip is

$$M_c = 35.0 \text{ kip feet}$$
$$< 51.79$$

Hence column 3 is unsatisfactory.

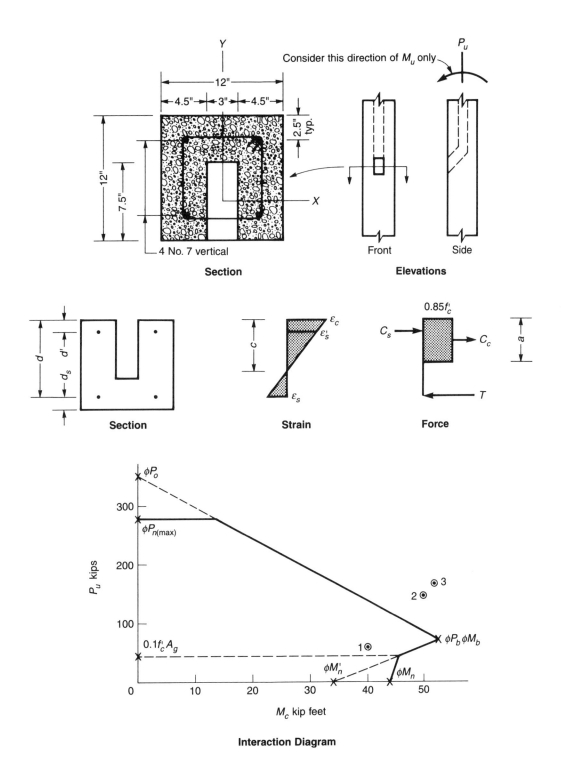

FIGURE 86C-5

STRUCTURAL ENGINEER REGISTRATION

CHAPTER 3
STRUCTURAL CONCRETE DESIGN

SECTION 3.3

REINFORCED CONCRETE SLABS

PROBLEMS:

1989 C–1
1989 C–3

STRUCTURAL ENGINEER EXAMINATION – 1989 ══════════════

PROBLEM C-1 – WT. 5.0 POINTS

GIVEN: The fourth level of a five-level concrete parking structure is currently under design. See the fourth-floor plan for details.

- Lateral load F = 370 kips (Including walls)

CRITERIA: Seismic zone 4
Strength design

Materials:
- Concrete: f'_c = 3000 psi
- Reinforcement: f_y = 60,000 psi
- Normal weight concrete

Assumptions:
- Consider north-south forces only

REQUIRED:
1. Design the reinforcement required for the maximum chord force at line "Y" of the main diaphragm.
2. Determine if boundary members are required for the main diaphragm
3. Design "collector" (drag) reinforcement for the shear wall at the west elevation considering diaphragm torsion. Relative rigidities "R" are shown.
4. Design and detail the shear transfer reinforcement from the slab to the west elevation shear wall and show collector reinforcement only on section "A".

SOLUTION

CHORD REINFORCEMENT

Due to the north–south seismic force, the factored moment at the mid span of the diaphragm, in accordance with UBC Section 1921.2.7, is given by

$$M_u = 1.4 \times F\ell/8$$
$$= 1.4 \times 370 \times 192/8 = 12{,}432 \text{ kip feet}$$

Assuming that the centroid of the chord reinforcement is located six inches from the opening in the diaphragm, the force in the chord reinforcement is given by

$$T = M_u/d$$
$$= 12{,}432/(64 - 0.5) = 195.8 \text{ kips}$$

Using the strength reduction factor for axial tension given in UBC Section 1909.3.1 the required chord reinforcement area is

$$A_s = T/\phi f_y$$
$$= 195.8/(0.9 \times 60) = 3.63 \text{ square inches}$$

Providing six No.7 bars gives a reinforcement area of

$$A_s = 3.60 \text{ square inches} \dots \text{satisfactory.}$$

BOUNDARY MEMBERS

In accordance with UBC Section 1921.6.2.3, boundary members must be provided with special transverse reinforcement when the maximum extreme fibre stress is

$$f_{cu} > 0.2 f'_c$$
$$> 0.2 \times 3000$$
$$> 600 \text{ pounds per square inch}$$

The relevant section modulus of the diaphragm is given by

$$S = bh^2/6$$
$$= 5 \times 64^2/(6 \times 12)$$
$$= 284 \text{ feet}^3$$

The maximum extreme fibre stress is given by

$$f_{cu} = M_u/S$$
$$= 12{,}432/284$$
$$= 43.7 \text{ kips per square foot}$$
$$= 303 \text{ pounds per square inch}$$
$$< 600 \text{ pounds per square inch}$$

Hence special transverse reinforcement is not required.

DRAG REINFORCEMENT

Since the roof diaphragm is rigid, in determining the distribution of lateral forces to the shear walls, torsional effects must be considered. In locating the center of rigidity for lateral force in the north-south direction, the south wall, which has no stiffness in this direction, is omitted. The center of rigidity is, then, located a distance from the east edge of the diaphragm given by

$$x_r = R_1\ell/(R_1 + R_3)$$
$$= 3.5 \times 192/(3.5 + 5)$$
$$= 79 \text{ feet}$$

By inspection, for lateral force in the east-west direction, the center of rigidity lies on grid line W. The distance of each wall from the center of rigidity is

$$r_1 = \ell - x_r$$
$$= 192 - 79$$
$$= 113 \text{ feet}$$
$$r_2 = 0$$
$$r_3 = x_r$$
$$= 79 \text{ feet}$$

The polar moment of inertia of the walls is

$$\Sigma r^2 R = r_1^2 R_1 + r_2^2 R_2 + r_3^2 R_3$$
$$= 113^2 \times 3.5 + 0 + 79^2 \times 5$$
$$= 75,896 \text{ square feet}$$

The sum of the shear wall rigidities for lateral force in the north-south direction is

$$\Sigma R_{NS} = R_1 + R_3$$
$$= 3.5 + 5$$
$$= 8.5$$

For lateral force in the north-south direction, the eccentricity is

$$e_x = \bar{x} - x_r$$
$$= 96 - 79$$
$$= 17 \text{ feet}$$

To comply with UBC Section 1628.5, accidental eccentricity must be considered and this amounts to five percent of the diaphragm dimension perpendicular to the direction of the seismic force. Hence the accidental eccentricity is given by

$$e_a = \pm 0.05 \times 192$$
$$= \pm 9.6 \text{ feet}$$

The net eccentricity is, then

$$e = e_x + e_a$$
$$= 17 \pm 9.6$$
$$= 26.6 \text{ or } 7.4 \text{ and the critical value is}$$
$$e = 26.6 \text{ feet}$$

The torsional moment acting about the center of rigidity is

$$T_u = 1.4Fe$$
$$= 1.4 \times 370 \times 26.6$$
$$= 13,778 \text{ kip feet}$$

The total shear in the diaphragm is the sum of the in-plane shear force and the torsional shear force. For the west edge of the diaphragm, these values are given by:

The in-plane shear force

$$F_S = 1.4FR_1/\Sigma R_{NS}$$
$$= 1.4 \times 370 \times 3.5/8.5$$
$$= 213.3 \text{ kips}$$

The torsional shear force

$$F_T = T_u r_1 R_1/\Sigma r^2 R$$
$$= 13,778 \times 3.5 \times 113/75,896$$
$$= 71.8 \text{ kips}$$

The total shear force at the west edge of the diaphragm is

$$F_F = F_S + F_T$$
$$= 213.3 + 71.8$$
$$= 285.1 \text{ kips}$$

The unit diaphragm shear along the west edge of the diaphragm is

$$q = F_F/B$$
$$= 285.1/96$$
$$= 2.97 \text{ kips per linear foot}$$

The resultant drag force at the west wall is

$$F_1 = 64q$$
$$= 64 \times 2.97$$
$$= 190 \text{ kips}$$

The required drag force reinforcement area is given by

$$A_s = F_1/\phi f_y$$
$$= 190/(0.9 \times 60)$$
$$= 3.5 \text{ square inches}$$

Providing six No.7 bars gives a reinforcement area of

$$A_s = 3.6 \text{ square inches}$$

Assuming the shear wall is seven inches thick, these bars may be located as shown on section "A" in Figure 89C-1

4. SHEAR TRANSFER

The unit shear between the slab and the shear wall is

$$q' = F_F/\ell_w$$
$$= 285.1/32$$
$$= 8.91 \text{ kips per linear foot}$$

In accordance with UBC Section 1911.7.5, shear-friction may be utilized to transfer shear from the diaphragm to the shear wall provided the shear does not exceed the value

$$q_{max} = 0.2\phi f'_c A_c$$
$$= 0.2 \times 0.85 \times 3 \times 7 \times 12$$
$$= 42.8 \text{ kips per linear foot}$$
$$> q'$$

Hence this requirement is satisfied.

For normal weight concrete, assuming the construction joint is intentionally roughened, the coefficient of friction at the interface is obtained from UBC Section 1911.7.4.3 as

$$\mu = 1.0\lambda$$
$$= 1.0$$

The required area of shear-friction reinforcement is given by UBC Formula (11-26) as

$$A_{vf} = q'/\phi f_y \mu$$
$$= 8.91/(0.85 \times 60 \times 1.0)$$
$$= 0.175 \text{ square inches per linear foot}$$

Providing No.4 bars at twelve inches on center gives a reinforcement area of

$$A_{vf} = 0.20 \text{ square inches per linear foot}$$

To comply with UBC Section 1911.7.8 anchorage must be provided to develop the full tensile strength of the bars on both sides of the shear plane. In accordance with UBC Section 1912.2, a minimum development length of twelve inches is required on either side of the slab. ═══

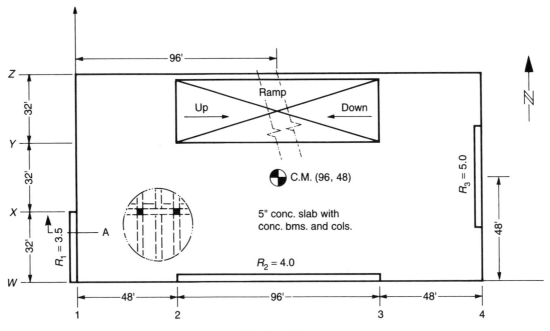

4th Floor Plan

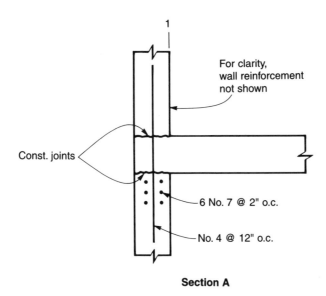

Section A

FIGURE 89C-1

302

STRUCTURAL ENGINEER EXAMINATION – 1989

PROBLEM C-3 – WT. 7.0 POINTS

GIVEN:
A concrete flat plate floor plan is shown.
Service loads are as follows:
- Dead load = 135 psf
- Reduced live load = 30 psf

CRITERIA:
Materials:
- Concrete: $f'_c = 4000$ psi
- Reinforcement: $f_y = 60,000$ psi
- Normal weight concrete.

Assumptions:
- Columns are of equal height above and below the floor.
- Use the direct design method.

REQUIRED:
1. Determine the minimum slab thickness required by the code when actual deflections are not calculated.

   For the remaining required items, assume a slab thickness of 8½ inches is adopted and an effective depth for positive and negative moments of 7 inches is provided.

   For an interior bay:
2. Compute the shear in the slab at the critical section for Column D-2. Compare this with the allowable shear stress.
3. Compute the positive and negative moments of the column strip and of the middle strip for the north-south direction of Grid D.
4. Compute the flexural reinforcing in the column strip on Grid D. Show this reinforcing on section "X". Use only No.5 bars. Detail the bar development lengths.
5. If the slab were to be post-tensioned instead of conventionally reinforced, list two positive and two negative concerns to be considered. Calculations are not required.

<u>SOLUTION</u>

APPLIED LOADS

The factored load is obtained from UBC Section 1909.2 Formula (9–1) as

$$
\begin{aligned}
w_u \quad &= 1.4w_D + 1.7w_L \\
&= 1.4 \times 135 + 1.7 \times 30 \\
&= 240 \text{ pounds per square foot}
\end{aligned}
$$

1. SLAB THICKNESS

For a flat plate floor without edge beams, the minimum slab thickness required, when deflections are not calculated, is given by UBC Table 19–C–2 as

$$
\begin{aligned}
h \quad &= \ell_n/30 \dots \text{ in an exterior panel} \\
&= (\ell_1 - c_1)/30 \\
&= (28 \times 12 - 24)/30 \\
&= 10.4 \text{ inches}
\end{aligned}
$$

2. SLAB SHEAR

The critical section for punching shear in a flat plate, as defined by UBC Section 1911.12.1 is located on the perimeter of a rectangle with sides of length

$$
\begin{aligned}
b_1 \quad &= c_1 + d \\
&= 24 + 7 \\
&= 31 \text{ inches} \\
b_2 \quad &= c_2 + d \\
&= 28 + 7 \\
&= 35 \text{ inches}
\end{aligned}
$$

The length of the critical perimeter is

$$
\begin{aligned}
b_o \quad &= 2(b_1 + b_2) \\
&= 2(31 + 35) \\
&= 132 \text{ inches}
\end{aligned}
$$

Since the openings in the slab are located within a column strip, the critical perimeter must be reduced as defined in UBC Section 1911.12.5 and as illustrated[4]. The ineffective length produced by the rectangular opening is given by

$$x_1 = 16 \times 17.5/44$$
$$= 6.36 \text{ inches}$$

The half angle subtended by the twelve inch diameter sleeve at the center of the column is given by

$$\theta = (\sin^{-1}6)/(12 + 14)$$
$$= 13.34°$$

The ineffective length produced by the sleeve is

$$x_2 = 2 \times 17.5\tan\theta$$
$$= 8.30 \text{ inches}$$

The effective perimeter is then

$$b_e = b_o - x_1 - x_2$$
$$= 132 - 6.36 - 8.30$$
$$= 117.34$$

The ratio of the long side to the short side of the column is

$$\beta = c_2/c_1$$
$$= 28/24$$
$$= 1.17$$
$$< 2$$

Hence the shear capacity of the flat plate for two-way action is given by UBC Formula (11-38) as

$$\phi V_c = 4\phi b_e d(f'_c)^{0.5}$$

where ϕ = strength reduction factor = 0.85 from UBC Section 1909.3

Thus
$$\phi V_c = 4 \times 0.85 \times 117.34 \times 7(4000)^{0.5}/1000$$
$$= 176.63 \text{ kips}$$

The factored applied shear at the critical perimeter is

$$V_u = w_u(\ell_1 \times \ell_2 - b_1 \times b_2)$$
$$= 0.24(28 \times 21.5 - 31 \times 35/144)$$
$$= 142.67 \text{ kips}$$
$$< \phi V_c$$

Hence the shear capacity of the slab is satisfactory.

3. COLUMN AND MIDDLE STRIP DESIGN MOMENTS

The clear span in the north–south direction is

$$\ell_n = \ell_1 - c_1$$
$$= 28 - 2$$
$$= 26 \text{ feet}$$
$$> 0.65\ell_1$$

Then, in accordance with UBC Section 1913.6.2.5, the value to be adopted for the clear span is

$$\ell_n = 26 \text{ feet}$$

For an interior bay, the widths of a column strip and of a middle strip in the north–south direction are defined by UBC Section 1913.2 as

$$w_c = \ell_2/2$$
$$= 21.5/2$$
$$= 10.75 \text{ feet}$$

and

$$w_m = \ell_2 - w_c$$
$$= 21.5 - 10.75$$
$$= 10.75 \text{ feet}$$

The total factored static moment is given by

$$M_o = w_u \ell_2 \ell_n^2 / 8$$
$$= 0.24 \times 21.5 \times 26^2 / 8$$
$$= 436 \text{ kip feet}$$

This moment is distributed in accordance with UBC Section 1913.6.3 to give positive and negative moment across the full width between columns and then distributed to the column and middle strips in accordance with UBC Section 1913.6.4 and UBC Section 1913.6.6 as indicated in the following Table.

Strip	Section	Line 1	Span 12	Line 2	Span 23
Full Width	Distribution Coeff	0.26	0.52	0.70	0.35
	Moment	113	227	305	153
Column Strip	Distribution Coeff	1.00	0.60	0.75	0.60
	Moment	113	136	229	92
Middle Strip	Distribution Coeff	0.00	0.40	0.25	0.40
	Moment	0	91	76	61

4. COLUMN STRIP REINFORCEMENT

The minimum reinforcement for temperature and shrinkage stresses is defined for Grade 60 reinforcement in UBC Section 1907.12 as

$$A_{s(min)} = 0.0018bh$$
$$= 0.0018 \times 12 \times 8.5$$
$$= 0.184 \text{ square inches per foot}$$

Maximum bar spacing is defined in UBC Section 1913.4.2 as

$$s_{max} = 2h$$
$$= 2 \times 8.5$$
$$= 17 \text{ inches}$$

The maximum reinforcement in accordance with UBC Section 1910.3.3 is

$$A_{s(max)} = 0.6375bd\beta_1 f'_c \times 87,000/f_y(87,000 + f_y)$$
$$= 1.76 \text{ square inches per foot}$$

The required reinforcement area is obtained by using calculator program A.1.1 in Appendix A and is given by

$$A_s = 0.85bdf'_c[1.0 - (1.0 - 2K/0.765f'_c)^{0.5}]/f_y$$

where $K = 12M/bd^2$

The reinforcement area and the required spacing of No.5 bars are indicated in the following Table.

[handwritten: 0.7 × 436 × 75% in column strip = 229 k' over 10.75 col. strip = 21.3 k'/ft]

Section	Line 1	Span 12	Line 2	Span 23
Moment, kip feet/foot	10.55	12.65	21.29	8.52
K	0.215	0.258	0.435	0.174
A_s required	0.35	0.42	0.73	0.28
Spacing No.5 bars, ins	10	8	5	12
A_s provided	0.37	0.46	0.74	0.31

Additional reinforcement is required over the exterior column to resist that fraction of the unbalanced moment transferred by flexure to the column. These additional bars are placed within an effective slab width given by UBC Section 1913.3.3.2 as

$$b_c = c_2 + 3h$$
$$= 28 + 3 \times 8.5$$
$$= 53.5 \text{ inches}$$

The required fraction of the unbalanced moment at column D1 is given by UBC Formula (13–1) as

$$\gamma_f = 1.0/[1.0 + 0.67(b_1/b_2)^{0.5}]$$
$$= 1.0/\{1.0 + 0.67[(24 + 7/2)/35]^{0.5}\}$$
$$= 0.63$$

The moment to be transferred is, then

$$\gamma_f M_u = 0.63 \times 113$$
$$= 71 \text{ kip feet}$$

Using calculator program A.1.1, the additional reinforcement required is

$$A_s = 2.39 \text{ square inches}$$

This requires a total of eight No.5 bars, at 7.5 inch centers over the 53.5 inch width.

The reinforcement layout and development length are determined from UBC Figure 19–1 and are shown in Figure 89C–3(ii).

5. POST-TENSIONED SLAB

The advantages of a post-tensioned flat slab include:

- The elimination of cracking.
- The reduction in deflections.
- The reduction in construction depth.
- The elimination of reinforcement congestion.

The disadvantages of a post-tensioned slab include:

- The elastic shortening produced during the stressing operation which causes lateral forces to be applied to the columns.
- Long-term creep and shrinkage producing additional lateral forces.
- Possible safety problems during demolition.

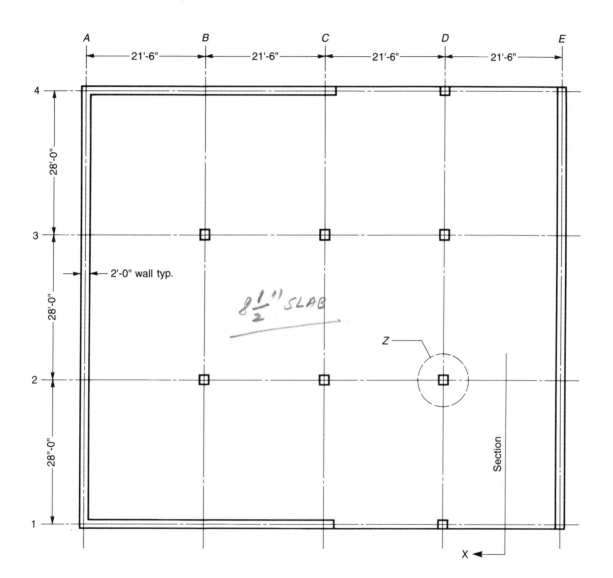

Floor Plan

FIGURE 89C-3(i)

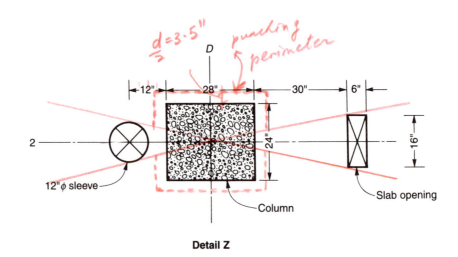

Detail Z

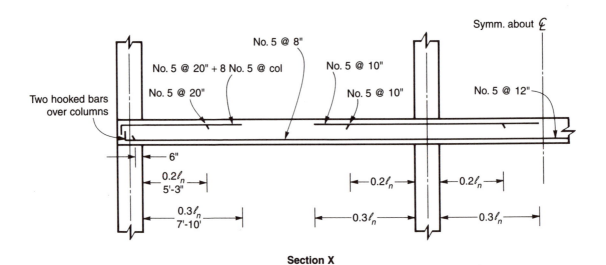

Section X

FIGURE 89C–3(ii)

STRUCTURAL ENGINEER REGISTRATION

CHAPTER 3

STRUCTURAL CONCRETE DESIGN

SECTION 3.4

REINFORCED CONCRETE WALLS

PROBLEMS:

1990 C–2
1989 C–2
1987 C–4

STRUCTURAL ENGINEER EXAMINATION – 1990

PROBLEM C–2 – WT. 7.0 POINTS

GIVEN: A precast concrete wall as shown.

CRITERIA: Seismic zone 4 with $I_p=1.0$
Use the slender wall design criteria.

Materials:
- Concrete: $f'_c = 3000$ psi
- Reinforcement: $f_y = 60,000$ psi
- Normal weight concrete: $E = 3.16 \times 10^6$ psi

Assumptions:
- Wind does not govern.
- Ignore the effect of the parapet.
- The wall is pinned at the top and bottom.
- The vertical reinforcing is at the center line of the wall.

REQUIRED:
1. Determine the ultimate moment at midheight. Include the p–delta moment and assume that $A_{se} = 0.24$ in^2/ft. for calculating Δ_n. The value of "c" used in calculating I_{cr} shall be 0.84 inches.
2. Determine the adequacy of the No.4 verticals at 10 inches on center and show that the first three items of limitations in the UBC Section 1914.8.2 for slender wall design have been met.

$.24$ in^2/ft

SOLUTION

BASIC THEORY

The P–delta moment in a slender wall is determined[5] as shown by the free body diagram in Figure 90C–2. The horizontal reaction at the top of the panel is given by

$$H_T = w\ell_c/2 + 4P_2\Delta/3\ell_c - P_1e/\ell_c$$

The bending moment at the midheight of the panel is given by

$$M = w\ell_c^2/8 + P_1e/2 + (P_1 + P_2)\Delta$$

At the ultimate load, UBC Section 1914.8.3 specifies that the deflection used to determine the P–delta effects is

$$\Delta = \Delta_n$$
where Δ_n = the maximum potential deflection at the nominal moment

312

FACTORED LOADS

Since the concrete panel is located in seismic zone 4, in accordance with UBC Section 1921.2.7, the required strength is

$$U = 1.4(D + E)$$

where $\quad D$ = dead load

and $\quad E$ = seismic load

Considering a one foot width of panel, the factored roof dead load is given by

$$P_{u1} = 1.4 \times 200$$
$$= 280 \text{ pounds}$$

The eccentricity of the roof load about the panel center line is

$$e = 1.75 + 6.25/2$$
$$= 4.875 \text{ inches}$$

The factored self weight of the panel is

$$q_u = 1.4 \times 150 \times 6.25/12$$
$$= 109.375 \text{ pounds per foot}$$

The factored self weight of the top half of the panel, neglecting the parapet, is

$$P_{u2} = q_u \ell_c/2$$
$$= 109.375 \times 22/2$$
$$= 1203 \text{ pounds}$$

The total factored axial load at midheight of the wall, neglecting the parapet, is

$$P_u = P_{u1} + P_{u2}$$
$$= 280 + 1203$$
$$= 1483 \text{ pounds}$$

The factored seismic load on the panel is given by UBC Section 1630 Formula (30–1) as

$$w_u = ZI_pC_pq_u$$

where $\quad Z$ = 0.4 as given

$\quad I_p$ = 1.0 as given

$\quad C_p$ = 0.75 from UBC Table 16–O, item 1.1.b

$\quad q_u$ = 109.375 as calculated

Then $\quad w_u = 0.4 \times 1.0 \times 0.75 \times 109.375$

$$= 32.81 \text{ pounds per foot}$$

0.3 W

MOMENT STRENGTH

The depth of the equivalent rectangular stress block is given by UBC Section 1910.2.7 as

	a	$= \beta_1 c$
where	β_1	$= 0.85$ as defined in UBC Section 1910.2.7
	c	$= 0.84$ the given depth to the neutral axis
Then	a	$= 0.85 \times 0.84 = 0.714$ inches

The nominal moment strength is given by[5]

	M_n	$= A_{se}f_y(d - a/2)$
where	A_{se}	$= 0.24$ the given effective reinforcement area
Then	M_n	$= 0.24 \times 60(6.25/2 - 0.714/2)$
		$= 39.86$ kip inches

In accordance with UBC Section 1909.3.2, the strength reduction factor for combined axial load and flexure is given by

$$\phi = 0.9 - 2P_u/f'_cA_g$$
$$= 0.9 - 2 \times 1483/(3000 \times 12 \times 6.25) = 0.887$$

Hence the design moment strength is given by

$$\phi M_n = 0.887 \times 39.86$$
$$= 35.36 \text{ kip inches}$$

1. ULTIMATE MOMENT

The modular ratio is

$$n = E_s/E_c$$
$$= 29,000/3160 = 9.18$$

From UBC Section 1914.8.4, the moment of inertia of the cracked section is

$$I_{cr} = nA_{se}(d - c)^2 + bc^3/3$$
$$= 9.18 \times 0.24(6.25/2 - 0.84)^2 + 12 \times 0.84^3/3$$
$$= 13.87 \text{ inches}^4$$

The deflection occurring at the midheight of the wall under the effect of the nominal moment is

$$\Delta_n = 5M_n\ell_c^2/48E_cI_{cr}$$
$$= 5 \times 39.86 \times 22^2 \times 144/(48 \times 3160 \times 13.87)$$
$$= 6.60 \text{ inches}$$

Then the applied ultimate moment at the midheight of the panel is given by[5]

$$M_u = w_u\ell_c^2/8 + P_{u1}e/2 + (P_{u1} + P_{u2})\Delta_n$$
$$= 32.81 \times 22^2 \times 12/8000 + 0.280 \times 4.875/2 + 1.483 \times 6.6$$
$$= 34.29 \text{ kip inches}$$

2. SECTION CAPACITY

The effective reinforcement area is given by UBC Section 1914.8.4 as

$$A_{se} = (P_u + A_sf_y)/f_y$$
$$= (1.483 + 0.24 \times 60)/60$$
$$= 0.265 \text{ square inches}$$

The depth of the equivalent rectangular stress block is

$$a = A_{se}f_y/0.85f'_cb$$
$$= 0.265 \times 60/(0.85 \times 3 \times 12)$$
$$= 0.52 \text{ inches}$$

The design moment strength is, then

$$\phi M_n = \phi A_{se}f_y(d - a/2)$$
$$= 0.887 \times 0.265 \times 60(6.25/2 - 0.52/2)$$
$$= 40.41 \text{ kip inches}$$
$$> M_u \text{ ... satisfactory}$$

2(i). SERVICE LOAD STRESS LIMITATION

The total service axial load at midheight of the wall is

$$P = P_u/1.4$$
$$= 1483/1.4 = 1059 \text{ pounds}$$

The service axial load stress at midheight of the wall is

$$f_a = P/A_g$$
$$= 1059/(12 \times 6.25)$$
$$= 14.12 \text{ pounds per square inch}$$
$$< 0.04f'_c \text{ ... satisfactory}$$

2(ii). REINFORCEMENT RATIO LIMITATION

The reinforcement ratio producing balanced strain conditions is given by UBC Section 1908.4.3 Formula (8–1) as

$$\rho_b = 87 \times 0.85\beta_1f'_c/(87 + f_y)f_y$$
$$= 87 \times 0.85 \times 0.85 \times 3/60(87 + 60)$$
$$= 0.0214$$

The actual reinforcement ratio provided is

$$\rho = A_s/bd$$
$$= 0.24/(12 \times 3.125)$$
$$= 0.0064$$
$$< 0.6\rho_b \text{ ... satisfactory}$$

2(iii). CRACKING MOMENT LIMITATION

The moment of inertia of the gross concrete section about its centroidal axis, neglecting reinforcement is

$$
\begin{aligned}
I_g \quad &= bh^3/12 \\
&= 12 \times 6.25^3/12 \\
&= 244 \text{ inches}^4
\end{aligned}
$$

The section modulus of the gross section is given by

$$
\begin{aligned}
S_g \quad &= 2I_g/h \\
&= 2 \times 244/6.25 \\
&= 78 \text{ inches}^3
\end{aligned}
$$

The cracking moment is given by UBC Section 1914.0 as

$$
\begin{aligned}
M_{cr} \quad &= 5S_g(f_c')^{0.5} \\
&= 5 \times 78 \times (3000)^{0.5}/1000 \\
&= 21.36 \text{ kip inches} \\
&< \phi M_n \dots \text{satisfactory}
\end{aligned}
$$

2(iv). SERVICE LOAD DEFLECTION LIMITATION

The allowable deflection occurring at the midheight of the panel, under the effects of the service loading, is given by UBC Section 1914.8.4 as

$$
\begin{aligned}
\Delta_{s(allow)} \quad &= \ell_c/150 \\
&= 22 \times 12/150 \\
&= 1.76 \text{ inches}
\end{aligned}
$$

Then the applied service load moment at the midheight of the panel is given by[5]

$$
\begin{aligned}
M_s \quad &\approx w\ell_c^2/8 + P_1e/2 + (P_1 + P_2)\Delta_{s(allow)} \\
&= 23.44 \times 22^2 \times 12/8000 + 0.20 \times 4.875/2 + 1.06 \times 1.76 \\
&= 19.37 \text{ kip inches} \\
&< M_{cr}
\end{aligned}
$$

Then the deflection occurring at the midheight of the wall under the effects of the service loading is given by UBC Section 1914.84 as

$$
\begin{aligned}
\Delta_s \quad &= 5M_s\ell_c^2/48E_cI_g \\
&= 5 \times 19.37 \times 22^2 \times 144/(48 \times 3160 \times 244) \\
&= 0.18 \text{ inches} \\
&< \Delta_{s(allow)} \dots \text{satisfactory}
\end{aligned}
$$

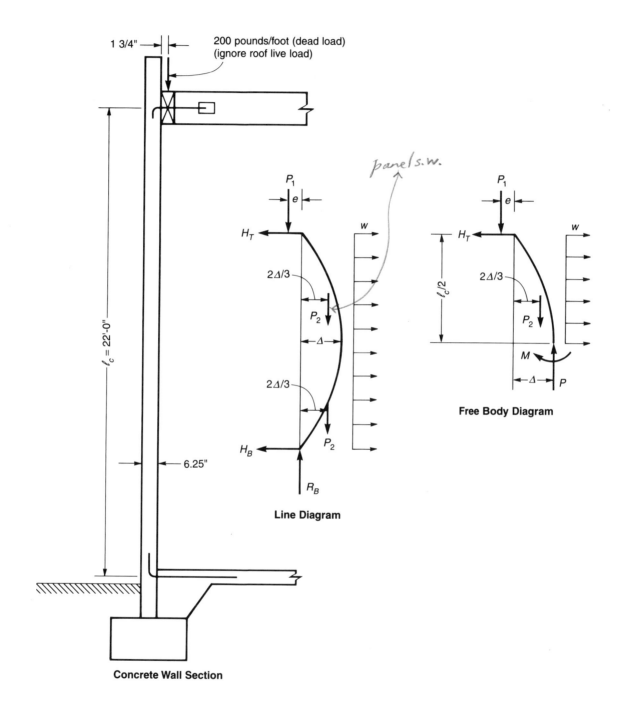

Concrete Wall Section

1 3/4"

200 pounds/foot (dead load)
(ignore roof live load)

$\ell_c = 22'-0''$

6.25"

P_1

e

H_T

w

$2\Delta/3$

P_2

Δ

$2\Delta/3$

P_2

H_B

R_B

panel s.w.

Line Diagram

P_1

e

H_T

w

$\ell_c/2$

$2\Delta/3$

P_2

M

Δ

P

Free Body Diagram

FIGURE 90C-2

STRUCTURAL ENGINEER EXAMINATION – 1989

PROBLEM C–2 – WT. 6.0 POINTS

GIVEN: The five-story concrete shearwall indicated is to be designed to resist the lateral loads shown and the following service loads.

- Dead load = 90 kips per level
- Live load = 30 kips per level
- Wall load = 150 kips total

CRITERIA: Seismic zone 3

Materials:
- Concrete: f'_c = 4000 psi
- Reinforcement: f_y = 60,000 psi
- Normal weight concrete

Assumptions:
- The footing is to be roughened for shear transfer.
- Use only No.11 bars for boundary element reinforcement.
- Use only No.4 bars for tie and stirrup reinforcement.
- Use only No.5 bars for wall reinforcement.

REQUIRED: At the ground level only, determine the following:
1. The required shear reinforcement.
2. The necessary longitudinal and transverse boundary reinforcement required. Show a sketch of the cross section.
3. The method of transferring the shear at the footing.
4. What is the maximum nominal shear force in kips for which the wall can be designed?

<u>SOLUTION</u>

FACTORED LOADS

For seismic zone 3 the required strength, in accordance with UBC Section 1921.2.7, is given by

	U	$= 1.4(D + L + E)$
or	U	$= 0.9D \pm 1.4E$
where	D	= dead load
		$= 90 \times 5 + 150$
		= 600 kips
	L	= live load
		$= 30 \times 5$
		= 150 kips
	E	= seismic load

The alternative factored gravity loads are

$$P_{u1} = 1.4(D + L)$$
$$= 1.4(600 + 150)$$
$$= 1050 \text{ kips}$$
$$P_{u2} = 0.9D$$
$$= 0.9 \times 600$$
$$= 540 \text{ kips}$$

The factored moment at the base is

$$M_u = 1.4 \times 12(75 \times 5 + 60 \times 4 + 45 \times 3 + 35 \times 2 + 20)$$
$$= 14,112 \text{ kip feet}$$

The factored shear at the base is

$$V_u = 1.4(75 + 60 + 45 + 35 + 20)$$
$$= 329 \text{ kips}$$

SECTION PROPERTIES

The web area of the shear wall is

$$A_{cv} = b_w \ell_w$$
$$= 20 \times 12 \times 10$$
$$= 2400 \text{ square inches}$$

1. SHEAR REINFORCEMENT

The design shear force limit given in UBC Section 1921.6.2.1 is

$$2A_{cv}(f'_c)^{0.5} = 2 \times 2400(4000)^{0.5}/1000$$
$$= 303.6 \text{ kips}$$
$$< V_u$$

Hence the required minimum reinforcement ratios along the longitudinal and transverse axes are

$$\rho_v = \rho_n = 0.0025$$

In accordance with UBC Section 1921.6.2.2 two curtains of shear reinforcement are required in the wall.

The wall height to length ratio is

$$h_w/\ell_w = 5 \times 12/20$$
$$= 3$$
$$> 2$$

Hence from UBC Section 1921.6.4 the shear capacity of the wall is given by UBC Formula (21–6) as

$$\phi V_n = V_u$$
$$= 329 \text{ kips}$$
$$= \phi A_{cv}[2(f'_c)^{0.5} + \rho_n f_y]$$

where ϕ = 0.6 from UBC Section 1909.3.4

and ρ_n = ratio of transverse shear reinforcement to web area

Then 329 = $0.6 \times 2400[2(4000)^{0.5}/1000 + 60\rho_n]$
$$= 182 + 86,400\rho_n$$

and ρ_n = $(329 - 182)/86,400$
$$= 0.0017$$
$$< 0.0025$$

Hence the reinforcement required, per foot of wall, is

$$A_s = 10 \times 12 \times 0.0025$$
$$= 0.30 \text{ square inches per foot}$$

Two curtains of No.5 bars at twenty–four inches spacing provides

$$A_s = 0.31 \text{ square inches per foot}$$
$$> 0.30$$

However UBC Section 1921.6.2.1 stipulates a maximum spacing of eighteen inches and the required reinforcement is No.5 bars at eighteen inch spacing, longitudinally and transversely, in each face.

2. BOUNDARY ZONE REQUIREMENTS

CONFINEMENT REINFORCEMENT

An exemption from the provision of boundary zone confinement reinforcement is given by UBC Section 1921.6.5.4 provided that

$$P_{u1} \leq 0.10 A_g f'_c$$
$$= 0.10 \times 21.33 \times 144 \times 4$$
$$= 1229 \text{ kips}$$
$$> 1051 \text{ kips}$$

and
$$V_u \leq 3\ell_w h(f'_c)^{0.5}$$
$$= 3 \times 240 \times 10(4000)^{0.5}/1000$$
$$= 455 \text{ kips}$$
$$> 329$$

Both conditions are satisfied and boundary zone confinement reinforcement need not be provided.

FLEXURAL REINFORCEMENT

To resist the factored moment at the base, provide ten No.11 reinforcing bars in each end section as shown in Figure 89C-2. To confirm that the flexural capacity of the wall is adequate, it is necessary to determine that the design forces M_u and P_{u1} lie within an interaction diagram for the wall. As a conservative approximation, the contribution of the shear reinforcement to the flexural capacity is neglected. In addition, since the axial force is less than $0.10 A_g f'_c$, it is sufficient to establish that ϕM_n for the wall is not less than M_u without consideration of the axial force.

Assume the depth to the neutral axis is given by

$$c = 13 \text{ inches}$$

The depth of the equivalent rectangular concrete stress block is given by Section 1910.2.7.1 as

$$a = c\beta_1$$
where β_1 = compression zone factor
$$= 0.85 \text{ as defined in UBC Section 1910.2.7.1}$$
then $a = 13 \times 0.85$
$$= 11.05 \text{ inches}$$

The strain distribution across the section and the forces developed are shown in Figure 89C-2. The maximum strain in the concrete specified in UBC Section 1910.3.2 is

$$\varepsilon_c = 0.003$$

The yield stress in the tension reinforcement is reached at a strain of

$$\varepsilon_s = f_y/E_s$$
$$= 60/29,000$$
$$= 0.00207$$

The bars on line 1 are in compression at the yield stress and the compressive force is

$$C_1 = 60A_s$$
$$= 60 \times 4 \times 1.56$$
$$= 374 \text{ kips}$$

The bars on line 2 are in compression and the compressive force is

$$C_2 = 0.003A_sE_s(c - 12)/c$$
$$= 0.003 \times 2 \times 1.56 \times 29,000(13 - 12)/13$$
$$= 21 \text{ kips}$$

The force in the concrete stress block is

$$C_c = 0.85f'_c(ah - A'_s]$$
$$= 0.85 \times 4(11.05 \times 24 - 6 \times 1.56)$$
$$= 870 \text{ kips}$$

The total compressive force on the section is

$$\Sigma C = C_1 + C_2 + C_c$$
$$= 374 + 21 + 870$$
$$= 1265 \text{ kips}$$

The bars on line 3 are in tension and the tensile force is

$$T_3 = 0.003A_sE_s(21.3 - c)/c$$
$$= 0.003 \times 4 \times 1.56 \times 29,000(21.3 - 13)/13$$
$$= 347 \text{ kips}$$

The bars in the left–hand end section are in tension at the yield stress and the tensile force is

$$T_4 = 60A_s$$
$$= 60 \times 10 \times 1.56$$
$$= 936 \text{ kips}$$

The sum of the tensile forces on the section is

$$\Sigma T = T_3 + T_4$$
$$= 347 + 936$$
$$= 1283 \text{ kips}$$
$$\approx \Sigma C$$

Hence the assumed depth to the neutral axis is satisfactory.

The nominal moment capacity of the section, without axial load, is obtained by summing moments about the centroid of the bars in the left hand end section and is given by

$$M_n = C_c(228 - 11.05/2) + C_1(228 - 2.7) + C_2(228 - 12) - T_3(228 - 21.3)$$
$$= 210,626 \text{ kip inches}$$
$$= 17,552 \text{ kip feet}$$

In accordance with UBC Section 1909.3.2 the design moment capacity without axial load is

$$\phi M_n = 0.9M_n$$
$$= 15,797 \text{ kip feet}$$

The values of the design forces P_u and M_u are plotted on the interaction diagram and lie within the valid region for an under-reinforced section. Hence the proposed reinforcement is satisfactory.

LATERAL TIE REINFORCEMENT

Transverse reinforcement, consisting of lateral ties, must be provided throughout the height of the end sections to provide confinement and to prevent buckling of the longitudinal bars. The minimum size required is No.4 and the tie spacing requirements, in accordance with UBC Sections 1907.10 and 1907.11, are:

$$S_{max} \leq 48 \times \text{tie bar diameters}$$
$$= 48 \times 0.5$$
$$= 24 \text{ inches}$$

or
$$\leq \text{least dimension of column}$$
$$= 24 \text{ inches}$$

or
$$\leq 16 \times \text{longitudinal bar diameters}$$
$$= 16 \times 1.41$$
$$= 22.6 \text{ inches ... governs}$$

Provide No.4 ties at a spacing of 22 inches.

As specified in UBC Section 1907.10.5.3, every corner bar and alternate bar shall have lateral support provided by the corner of a tie or crosstie with no unsupported bar more than 6 inches from a supported bar. The arrangement shown in Figure 89C-2 is suitable.

3. SHEAR TRANSFER

In accordance with UBC Section 1911.7.5, shear-friction may be utilized for shear transfer at the footing provided the shear does not exceed the value

$$\phi V_n = 0.2\phi f_c' A_{cv}$$

where ϕ = 0.85 from UBC Section 1909.3

Then
$$\phi V_n = 0.2 \times 0.85 \times 4 \times 2400$$
$$= 1632 \text{ kips}$$
$$< 800 A_{cv}/1000$$
$$> V_u$$

Hence this requirement is satisfied.

For normal weight concrete, assuming the construction joint is intentionally roughened to an amplitude of 0.25 inch, the coefficient of friction at the interface is obtained from UBC Section 1911.7.4 as

$$\mu = 1.0 \times \lambda$$
$$= 1.0$$

The required area of shear-friction reinforcement, per linear foot of wall, is given by UBC Formula (11-26) as

$$A_{vf} = V_u/\phi f_y \mu \ell_w$$
$$= 329/(0.85 \times 60 \times 1.0 \times 20)$$
$$= 0.323 \text{ square inches per linear foot}$$

Providing No.5 bars at eleven inches on center gives a reinforcement area of

$$A_{vf} = 0.34 \text{ square inches per linear foot}$$

To comply with UBC Section 1912.2.2 a minimum development length of twelve inches is required on either side of the interface.

4. MAXIMUM SHEAR FORCE

For a single shear wall, the maximum nominal shear force is given by UBC Section 1921.6.4.6 as

$$V_n = 8 A_{cv} (f_c')^{0.5}$$
$$= 8 \times 2400 \times (4000)^{0.5}/1000$$
$$= 1214 \text{ kips}$$

The maximum design shear force is given by

$$\phi V_n = 0.6 \times 1214$$
$$= 728 \text{ kips}$$
$$> V_u$$

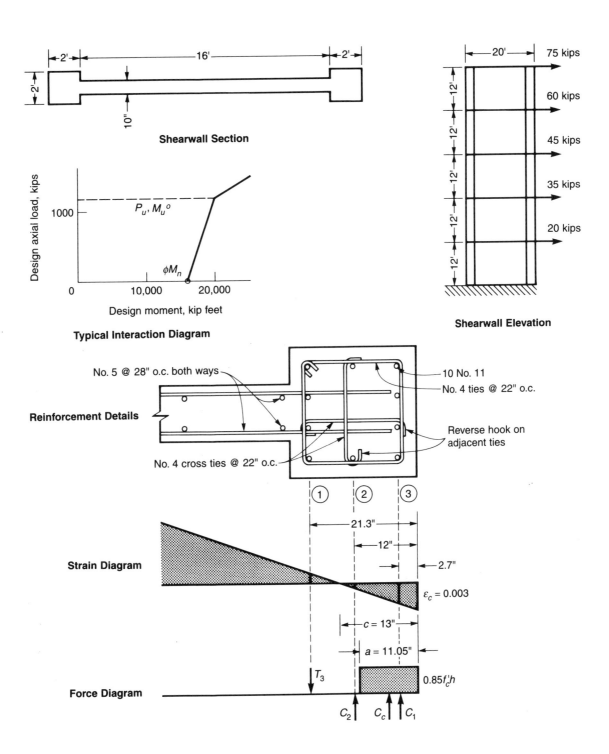

Shearwall Section

Typical Interaction Diagram

Shearwall Elevation

Reinforcement Details

Strain Diagram

Force Diagram

FIGURE 89C-2

325

STRUCTURAL ENGINEER EXAMINATION – 1987 ══════════════

PROBLEM C–4 – WT. 6.0 POINTS

<u>GIVEN:</u>	The concrete retaining wall as shown.

<u>CRITERIA:</u>
- Lateral soil pressure: equivalent fluid pressure of 30 lb/ft³
- Weight of soil: 100 lb/ft.³
- Weight of concrete: 150 lb/ft.³
- Weight of surcharge: 200 lb/ft.²
- Passive pressure: 450 lb/ft²/ft.
- Coefficient of friction: 0.3
- Concrete: $f_c' = 3000$ psi
- Reinforcement: $f_y = 60,000$ psi

<u>REQUIRED:</u>
- a) Overturning moment
- b) Resisting moment
- c) Factor of safety against overturning
- d) Determine the soil pressure at the toe of the footing
- e) Determine the soil pressure at the heel of the footing
- f) Determine the spacing of No.9 footing bars (ignore effects of shear key)
- g) Determine the spacing of No.6 footing bars
- h) Determine the size, spacing, and location of temperature reinforcing in both the wall and the footing.
- i) The soils engineer has determined that the effects of friction may be combined with the passive soil pressure. Determine the required depth of the shear key D. Reinforcing for the key is not required as part of this problem.
- j) Is the vertical reinforcing adequate and, if not, recommend two alternative methods of correction.

SOLUTION

SERVICE LOADS

H_b = triangular lateral soil pressure from backfill
      = $ph^2/2$
      = $30 \times 19.5^2/2$ = 5704 pounds

H_s = uniform surcharge pressure
      = whp/γ_b
      = $200 \times 19.5 \times 30/100$ = 1170 pounds

W_s = weight of surcharge
      = $6.5w$
      = 6.5×200 = 1300 pounds

W_b = weight of backfill
      = $6.5 \times 18\gamma_b$
      = $6.5 \times 18 \times 100$ = 11,700 pounds

W_f = weight of footing
      = $150 \times 10 \times 1.5$ = 2250 pounds

W_w = weight of wall
      = $150 \times 18 \times 1$ = 2700 pounds

FACTORED LOADS

The applicable load factors are obtained from UBC Section 1909.2.4 and the factored loads are

γH_b = 1.7×5704 = 9697 pounds
γH_s = 1.7×1170 = 1989 pounds
γW_s = 1.7×1300 = 2210 pounds
γW_b = $1.4 \times 11,700$ = 16,380 pounds
γW_f = 1.4×2250 = 3150 pounds
γW_w = 1.4×2700 = 3780 pounds

a. OVERTURNING MOMENT

The overturning moment about the bottom corner of the toe for service and factored loading, is shown in the Table below

Part	H	γH	y	Hy	γHy
Backfill pressure	5704	9697	6.50	37,076	63,030
Surcharge pressure	1170	1989	9.75	11,408	19,393
Total	6874	–	–	48,484	82,423

b. RESISTING MOMENT

The resisting moment about the bottom corner of the toe, for service and factored loading, is shown in the Table below

Part	W	γW	x	Wx	γWx
Surcharge	1,300	2,210	6.75	8,775	14,918
Backfill	11,700	16,380	6.75	78,975	110,565
Footing	2,250	3,150	5.00	11,250	15,750
Wall	2,700	3,780	3.00	8,100	11,340
Total	17,950	25,520	–	107,100	152,573

c. OVERTURNING FACTOR OF SAFETY

The factor of safety against overturning for service loads is given by

$$\Sigma Wx/\Sigma Hy = 107,100/48,484$$
$$= 2.2$$

d. SOIL PRESSURE AT TOE

The eccentricity of the applied loads about the toe is

$$e = (\Sigma Wx - \Sigma Hy)/\Sigma W$$
$$= (107,100 - 48,484)/17,950$$
$$= 3.27 \text{ feet}$$
$$< 1/3$$

Hence the pressure distribution is as shown in Figure 87C–4 and the pressure at the toe is given by

$$q = 2\Sigma W/3e$$
$$= 2 \times 17,950/(3 \times 3.27)$$
$$= 3665 \text{ pounds per square foot}$$

e. SOIL PRESSURE AT HEEL

The pressure distribution terminates at a distance from the toe of

$$3e = 3 \times 3.27$$
$$= 9.81 \text{ feet}$$

Hence there is zero pressure at the heel.

FACTORED SOIL PRESSURE

The eccentricity of the factored loads about the toe is

$$
\begin{aligned}
e \quad &= (\Sigma\gamma Wx - \Sigma\gamma Hy)/\Sigma\gamma W \\
&= (152{,}573 - 82{,}423)/25{,}520 \\
&= 2.75 \text{ feet} \\
&< \ell/3
\end{aligned}
$$

Hence the pressure distribution, as shown in Figure 87C-4, terminates at a distance from the toe of

$$
\begin{aligned}
3e \quad &= 3 \times 2.75 \\
&= 8.25 \text{ feet}
\end{aligned}
$$

The pressure at the toe is given by

$$
\begin{aligned}
q \quad &= 2\Sigma\gamma W/3e \\
&= 2 \times 25{,}520/(3 \times 2.75) \\
&= 6186 \text{ pounds per square feet}
\end{aligned}
$$

f. HEEL REINFORCEMENT

The critical section for moment at the heel is at the rear face of the wall and the factored moment is

$$
\begin{aligned}
M_u \quad &= [3.25(\gamma W_s + \gamma W_b + 6.5\gamma W_f/10) - 4.75^2 \times 3562/6]/1000 \\
&= 53.68 \text{ kip feet}
\end{aligned}
$$

The effective depth of the heel reinforcement is

$$
\begin{aligned}
d \quad &= 18 - 3 - 1.125/2 \\
&= 14.4 \text{ inches}
\end{aligned}
$$

The required tension reinforcement area is obtained by using calculator program A.1.1 in Appendix A and is given by

$$\rho \quad = 0.85f'_c[1.0 - (1.0 - 2K/0.765f'_c)^{0.5}]/f_y$$

where
$$
\begin{aligned}
K \quad &= 12M_u/bd^2 \\
&= 12 \times 53.68/(12 \times 14.4^2) \\
&= 0.259
\end{aligned}
$$

Then $\quad \rho \quad = 0.0051$

In accordance with UBC Section 1910.3.3, the maximum allowable reinforcement ratio is given by

$$\rho_{max} \quad = 0.75 \times 0.85 \times 87\beta_1 f'_c/f_y(87 + f_y)$$

where $\quad \beta_1 \quad = 0.85$ as defined in UBC Section 1910.2.7

Hence $\quad \rho_{max} \quad = 0.016$

$$> \rho \ \dots \text{ satisfactory}$$

In accordance with UBC Section 1910.3.3, the minimum allowable reinforcement ratio is given by

$$\rho_{min} \quad = 0.0018$$

$$< \rho \ \dots \text{ satisfactory}$$

Thus the required reinforcement area is given by

$$A_s = \rho bd$$
$$= 0.0051 \times 12 \times 14.4$$
$$= 0.88 \text{ square inches per foot}$$

This may be provided by No.9 bars at a spacing of thirteen inches giving an area of

$$A_s = 0.92 \text{ square inches per foot}$$

g. TOE REINFORCEMENT

The critical section for moment at the toe is at the front face of the wall and the factored moment is

$$M_u = [2.5^2(4312 + 2 \times 6186)/6 - 2.5^2\gamma W_f/20]/1000$$
$$= 16.39 \text{ kip feet}$$

The effective depth of the toe reinforcement is

$$d = 18 - 3 - 0.75/2$$
$$= 14.6 \text{ inches}$$

Then using calculator program A.1.1

$$K = 0.0769$$
$$\rho = 0.00145$$
$$< \rho_{max}$$
$$< \rho_{min}$$

Thus the required reinforcement area is given by

$$A_s = \rho_{min}bd$$
$$= 0.0018 \times 12 \times 14.6$$
$$= 0.315 \text{ square inches per foot}$$

This may be provided by No.6 bars at a spacing of sixteen inches giving an area of

$$A_s = 0.33 \text{ square inches per foot}$$

h. TEMPERATURE REINFORCEMENT

For the wall, which is cast–insitu and exceeds ten inches in thickness, UBC Section 1914.3 requires two layers of horizontal reinforcement each with an area of

$$A_h = 0.0020 \times 12 \times 12/2$$
$$= 0.144 \text{ square inches per foot}$$

This may be provided by No.4 bars at a spacing of sixteen inches giving an area of

$$A_h = 0.15 \text{ square inches per foot}$$

in each face of the wall.

In the outer face of the wall, vertical reinforcement is required with a cover of two inches and with an area of

$$A_v = 0.0012 \times 12 \times 12/3$$
$$= 0.058 \text{ square inches per foot}$$

The maximum permitted spacing of the reinforcement, in accordance with UBC Section 1914.3.5, is eighteen inches and using No.3 bars at this spacing gives an area of

$$A_v = 0.07 \text{ square inches per foot}$$

For the footing, UBC Section 1907.12 requires a minimum area of longitudinal reinforcement, in both the top and bottom faces, of

$$A_f = 0.0018 \times 18 \times 12/2$$
$$= 0.194 \text{ square inches per foot}$$

This may be provided by No.4 bars at a spacing of twelve inches giving an area of

$$A_f = 0.20 \text{ square inches per foot}$$

i. SHEAR KEY

The horizontal applied force on the retaining wall is

$$H = H_b + H_s$$
$$= 5704 + 1170$$
$$= 6874 \text{ pounds}$$

The frictional resistance available to counteract this force is

$$H_f = \mu \Sigma W$$
$$= 0.3 \times 17,950$$
$$= 5385 \text{ pounds}$$

Adopting a factor of safety against sliding of 1.5 requires a passive resistance of

$$H_p = 1.5H - H_f$$
$$= 1.5 \times 6874 - 5385$$
$$= 4926 \text{ pounds}$$

This must be provided by the passive pressure developed on the front face of the shear key and the footing. The total required depth of the shear key plus footing is given by

$$H_p = p_p h^2/2$$
and $$h^2 = 4926 \times 2/450$$
$$h = 4.68 \text{ feet}$$

Hence the depth of shear key required is

$$D = h - 1.5$$
$$= 4.68 - 1.5$$
$$= 3.18 \text{ feet}$$

j. WALL REINFORCEMENT

The critical section for moment in the wall is at the base of the wall and the factored moment is

$$
\begin{aligned}
M_u &= 1.7(py^3/6 + y^2wp/2\gamma_b) \\
&= 1.7[30 \times 18^3/6 + 18^2 \times 200 \times 30/(2 \times 100)]/1000 \\
&= 66.10 \text{ kip feet}
\end{aligned}
$$

The effective depth of the wall reinforcement is

$$
\begin{aligned}
d &= 12 - 3 - 0.875/2 \\
&= 8.6 \text{ inches}
\end{aligned}
$$

Then using calculator program A.1.1,

$$
\begin{aligned}
K &= 0.894 \\
\rho &= 0.0225 \\
\rho_{max} &= 0.0160 \\
&< \rho
\end{aligned}
$$

Hence the reinforcement ratio required exceeds the maximum allowable. The reinforcement ratio provided is

$$
\begin{aligned}
\rho &= 2.4/(12 \times 8.6) \\
&= 0.0233 \\
&> 0.0225
\end{aligned}
$$

The distance between bars is

$$
\begin{aligned}
s - d_b &= 3 - 0.875 \\
&= 2.125 \text{ inches} \\
&> 1.0
\end{aligned}
$$

Hence the requirements of UBC Section 1907.6.1 are satisfied. The required development length is given by UBC Section 1912.2 as

$$
\begin{aligned}
\ell_d &= 0.04A_bf_y/(f'_c)^{0.5} \\
&= 0.04 \times 0.6 \times 60,000/(3000)^{0.5} \\
&= 26 \text{ inches}
\end{aligned}
$$

The anchorage length provided is

$$
\begin{aligned}
\ell_a &= 4.68 \times 12 - 3 \\
&= 53 \text{ inches} \\
&> \ell_d
\end{aligned}
$$

In order to correct for the excessive reinforcement ratio, either the wall width must be increased or compression reinforcement must be provided in the front face of the wall.. ==========

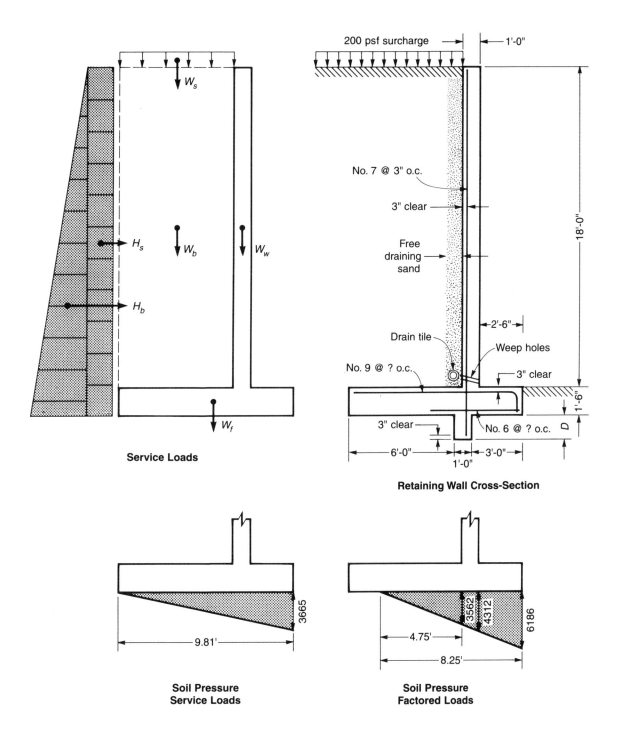

Service Loads

Retaining Wall Cross-Section

**Soil Pressure
Service Loads**

**Soil Pressure
Factored Loads**

FIGURE 87C–4

333

STRUCTURAL ENGINEER REGISTRATION

CHAPTER 3

STRUCTURAL CONCRETE DESIGN

SECTION 3.5

MOMENT-RESISTING FRAMES

PROBLEMS:

1990 C-4
1988 C-5

STRUCTURAL ENGINEER EXAMINATION – 1990 =====================

PROBLEM C–4 – WT. 6.0 POINTS

GIVEN: A three–story concrete special moment–resisting frame as shown. The summary of factored moments and factored axial load for column AB and factored moment for beam BC are tabulated below. Column AB has an ultimate moment capacity as shown in the illustration.

Load Case	COLUMN AB			BEAM BC
	Axial Load	Moment End B	Moment End A	Moment End B
1.4(D + L + E)	288 kips	258 kip ft	266 kip ft	–578 kip ft

Note: Negative moment means producing tension on the top face of the beam.

CRITERIA: Seismic zone 4

Materials:
- Concrete: $f'_c = 4000$ psi
- Reinforcement: $f_y = 60,000$ psi
- Use No.4 for column tie bars
- Use No.9 for other bars

Assumptions: The column is classified as slender

REQUIRED:
1. Check the adequacy of the flexural strength of column AB at joint B.
2. Determine the column transverse reinforcement requirements for shear.
3. Determine the column transverse reinforcement requirements for confinement.
4. Draw reinforcing details for column AB showing vertical bars, tie bars, cross ties, spacing, and lap splices.

SOLUTION

1. COLUMN MOMENT CAPACITY

The factored axial compressive force in the column is given as

$$P_u \quad = 288 \text{ kips}$$

This exceeds the limiting value specified in UBC Sections 1921.4.1 and 1921.4.2 as

$$\begin{aligned} 0.1 A_g f_c' \quad &= 0.1 \times 24 \times 24 \times 4 \\ &= 230 \text{ kips} \end{aligned}$$

Then member AB qualifies as a column and the moment capacity of the columns relative to that of the beam must be determined in accordance with UBC Formula (21-1).

The design moment strength of the beam is obtained by using calculator program A.1.2 in Appendix A. The reinforcement ratio is

$$\begin{aligned} \rho \quad &= A_s/bd \\ &= 5/(28 \times 27.44) \\ &= 0.0065 \\ &< 0.025 \\ &> 200/f_y \end{aligned}$$

Hence the requirements of UBC Section 1921.3.2 are satisfied.

The design moment strength, ignoring compression reinforcement, is

$$\begin{aligned} \phi M_n \quad &= 0.9 A_s f_y d(1.0 - 0.59\rho f_y/f_c')/12 \\ &= 582 \text{ kip feet} \\ &= \Sigma M_g \end{aligned}$$

The design moment strength of the column is given as 485 kip feet. Then, the sum of the design moment strengths of the columns framing into the joint is

$$\begin{aligned} \Sigma M_c \quad &= 2 \times 485 \\ &= 970 \\ &> 1.2 \Sigma M_g \end{aligned}$$

Hence UBC Formula (21-1) is satisfied and hoop reinforcement is not required over the full height of the column.

STRONG COL-WEAK BEAM

2. COLUMN SHEAR

In accordance with UBC Sections 1921.4.5 and 1921.4.4.7 the design shear force for column AB may be calculated from the probable moment strength at the top and bottom of the column. The column probable moment strength is determined[6] by assuming a strength reduction factor of unity and a tensile reinforcement stress of $1.25f_y$. The factored axial compressive force in the column is

$$P_u = 288 \text{ kips}$$
$$> 0.1A_g f'_c$$

The strength reduction factor for this condition of axial compression with flexure is given by UBC Section 1909.3.2 as

$$\phi = 0.7$$

The probable moment strength may be determined from the column design moment strength as

$$M_{pr} \approx \phi M_n \times 1.25/0.7 \quad \checkmark$$
$$= 485 \times 1.25/0.7$$
$$= 866 \text{ kip feet}$$

The clear height of column AB is

$$H_n = 12 - 2.5$$
$$= 9.5 \text{ feet}$$

The design shear force is then

$$V_e = 2M_{pr}/H_n$$
$$= 2 \times 866/9.5$$
$$= 182 \text{ kips}$$

However, in accordance with UBC Section 1921.4.5.1, the maximum design shear force in the column need not exceed that determined from the probable flexural strength of girder BC which frames into joint B. The probable girder strength, assuming a strength reduction factor of unity and a tensile reinforcement stress of $1.25f_y$, may be derived[6] as

$$M_{pg} = A_s f_y d(1.25 - 0.92\rho f_y/f'_c)/12$$
$$= 5 \times 60 \times 27.44(1.25 - 0.92 \times 0.0065 \times 60/4)/12$$
$$= 796 \text{ kip feet}$$

This moment may be distributed equally to the upper and lower column at joint B and, for a sway displacement, the same value is carried over to end A, as shown in Figure 90C-4(ii), giving

$$M_{AB} = M_{BA}$$
$$= M_{pg}/2$$
$$= 796/2$$
$$= 398 \text{ kip feet}$$

337

Then the maximum probable design shear force acting on the column is

$$V_e = (M_{AB} + M_{BA})/H_n$$
$$= 2 \times 398/(12 - 2.5)$$
$$= 83.79 \text{ kips}$$
$$< 182 \text{ kips}$$

Then $V_e = 83.79$ kips governs

The compressive stress value given by

$$A_g f'_c/20 = 24^2 \times 4/20$$
$$= 115 \text{ kips}$$
$$< P_u \left(= 288 k\right)$$

Hence, since the total design shear is due to earthquake forces, in accordance with UBC Section 1921.4.5.2 the design shear strength provided by the concrete may be utilized and this is given by UBC Formula (11–3) as

$$\phi V_c = 0.85 \times 2bd(f'_c)^{0.5}$$
$$= 0.85 \times 2 \times 24 \times(24 - 1.5 - 0.5 - 0.56)(4000)^{0.5}/1000$$
$$= 55.32 \text{ kips}$$
$$< V_e$$

The design shear strength required from shear reinforcement is given by UBC Formula (11–2) as

$$\phi V_s = V_e - \phi V_c$$
$$= 83.79 - 55.32$$
$$= 28.47 \text{ kips}$$
$$< 4 \times \phi V_c$$

Hence the proposed shear reinforcement, in accordance with UBC Section 1911.5.6.8, is satisfactory. The maximum hoop spacing, in accordance with UBC Section 1921.4.4.6, may not exceed the lesser value given by

$$s = 6d_b$$
$$= 6 \times 1.128$$
$$= 6.8 \text{ inches}$$

or $s = 6$ inches

Hence a spacing of six inches governs.

The minimum size of crosstie required for a No.9 longitudinal bar is specified, by UBC Section 1907.10.5, as a No.3 bar and at least one crosstie is required to satisfy the lateral support requirements of UBC Section 1921.4.4.3. The minimum area of shear reinforcement, which may be provided at a spacing of six inches, is given by UBC Formula (11–14) as

$$A_{v(min)} = 50b_w s/f_y$$
$$= 50 \times 24 \times 6/60,000$$
$$= 0.12 \text{ square inches}$$

The area of shear reinforcement which is required, at a spacing of six inches, to provide a shear strength of ϕV_s, is specified by UBC Formula (11–17) as

$$A_v = \phi V_s s/\phi d f_y$$
$$= 28.47 \times 6/(0.85 \times 21.44 \times 60)$$
$$= 0.16 \text{ square inches}$$
$$> 0.12 \text{ square inches}$$

Providing one No.4 hoop and one No.4 crosstie gives an area of

$$A_v = 3 \times 0.2$$
$$= 0.6 \text{ square inches}$$
$$> 0.16 \text{ square inches}$$

3. CONFINEMENT REINFORCEMENT

To conform with UBC Section 1921.4.3 and SEAOC Section 3D.3, a tension splice must be provided within the center half of the clear column height. The lap length required is specified by UBC Section 1912.15.1 as being equal to the tensile development length. The basic development length is given by UBC Section 1912.2.2 as

$$\ell_{db} = 0.04 A_b f_y/(f_c')^{0.5}$$
$$= 0.04 \times 1.0 \times 60{,}000/(4000)^{0.5}$$
$$= 38 \text{ inches}$$

Hoop reinforcement, at a maximum spacing of four inches, is provided over the splice length in accordance with SEAOC Section 3D.3. Hence, in accordance with UBC Section 1912.2.3.5 a modification factor of 0.75 is applicable. The modified development length is given by

$$\ell_d = 0.75\ell_{db}$$
$$= 0.75 \times 38$$
$$= 28.5 \text{ inches}$$

The minimum development length is specified by UBC Section 1912.2.3.6 as

$$\ell_{d(min)} = 0.03 d_b f_y/(f_c')^{0.5}$$
$$= 0.03 \times 1.128 \times 60{,}000/(4000)^{0.5}$$
$$= 32 \text{ inches}$$

The minimum lap length recommended in SEAOC Commentary Section 3D.3 is

$$\ell_{(min)} = 30 d_b$$
$$= 30 \times 1.128$$
$$= 34 \text{ inches}$$

Hence the required lap length is thirty four inches.

As shown in Figure 90C–4(ii), the point of contraflexure of the column, for the given loading, lies within the center half of the column clear height. Hence in accordance with UBC Section 1921.4.4.4 confinement reinforcement at a maximum spacing of four inches is required only for a distance from each joint face given by the greater of

$$\ell_o \quad = h$$
$$= 24 \text{ inches}$$
or $\quad \ell_o \quad = H_n/6$
$$= 9.5 \times 12/6$$
$$= 19 \text{ inches}$$
or $\quad \ell_o \quad = 18 \text{ inches}$

Hence the length of 24 inches governs.

Using No.4 hoop reinforcement bars, and providing one and a half inches clear cover to the bars, gives a core dimension, measured center-to-center of the confining bars, of

$$h_c \quad = 24 - 3 - 0.5$$
$$= 20.5 \text{ inches}$$

The dimension measured out-to-out of the confining bars is, in both directions,

$$h_c' \quad = h_c + 0.5$$
$$= 20.5 + 0.5$$
$$= 21 \text{ inches}$$

The area, calculated out-to-out of the confining bars, is

$$A_{ch} \quad = (h_c')^2$$
$$= 21^2$$
$$= 441 \text{ square inches}$$

The required area of confinement reinforcement is given by the greater value obtained from UBC Formulas (21-3) and (21-4)

Then $\quad A_{sh} \quad = 0.3sh_c(A_g/A_{ch} - 1.0)f_c'/f_y$
$$= 0.3 \times 4 \times 20.5(576/441 - 1.0)4/60$$
$$= 0.50 \text{ square inches}$$
or $\quad A_{sh} \quad = 0.09sh_c f_c'/f_y$
$$= 0.09 \times 4 \times 20.5 \times 4/60$$
$$= 0.49 \text{ square inches}$$

Hence three No.4 bars are required providing an area of confinement reinforcing of

$$A_{sh} \quad = 0.60 \text{ square inches}$$

The center-to-center spacing between the column longitudinal bars is

$$s_\ell \quad = (h_c - 1.128 - 0.5)/3$$
$$= 18.9/3$$
$$= 6.3 \text{ inches}$$

The maximum center-to-center spacing between the crossties and the hoop reinforcement is

$$s' \quad = 18.9 - s_\ell$$
$$= 18.9 - 6.3$$
$$= 12.6 \text{ inches}$$
$$< 14 \text{ inches}$$

Hence the requirements of UBC Section 1921.4.4.3 are satisfied.

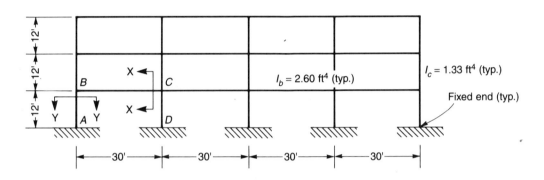

Frame Elevation

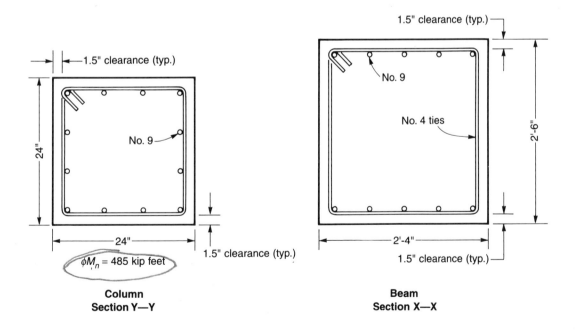

**Column
Section Y—Y**

**Beam
Section X—X**

FIGURE 90C–4(i)

341

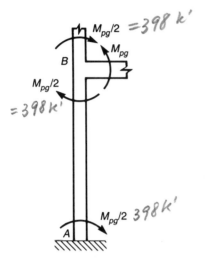

Distribution of Probable Girder Strength

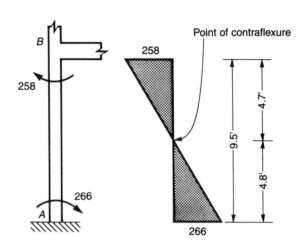

Factored Moments in Column AB

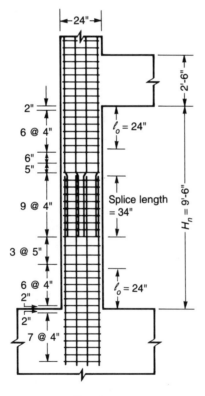

Hoop Spacing

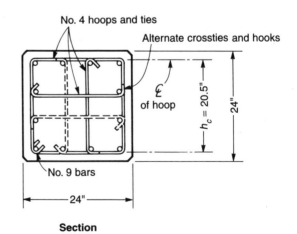

Section

FIGURE 90C–4(ii)

STRUCTURAL ENGINEER EXAMINATION – 1988

PROBLEM C–5 – WT. 6.0 POINTS

GIVEN: The column detail shown, for a special moment–resisting frame, has been submitted to you for your review.

CRITERIA: Materials:
- Concrete: $f'_c = 4000$ psi
- Reinforcement: $f_y = 60,000$ psi

Assumptions:
- Column vertical reinforcing steel is adequate.
- Joint reinforcing is adequate.
- M_C = Flexural strength of the column at the center of the joint.
- M_G = Flexural strength of the girder at the center of the joint.

REQUIRED: 1.   List three reasons why the detail does not meet code requirements.
2.   Sketch the details required to meet the code requirements and show all calculations. Provide appropriate notes on the details to show the following:
   a.   Column vertical reinforcing steel lap length and location.
   b.   Column tie reinforcing steel size and spacing.
   c.   Construction joint(s) and location(s).
   d.   Concrete cover requirements.

SOLUTION

1. CODE CONTRAVENTIONS

In accordance with UBC Section 1921.4.3, lap splices are permitted only within the center half of the clear column height. A tension splice is specified which requires a lap length of thirty eight inches for the No. 10 bars. Hoop reinforcement, at four inch spacing, is required over the length of the splice, in accordance with UBC Section 1921.4.4.2.

In accordance with UBC Section 1921.4.4.4, hoop requirement, at four inches spacing, is required in the column for a length of thirty inches, from each joint face. In addition, crossties are required on the hoop reinforcement at a maximum spacing of fourteen inches perpendicular to the longitudinal axis of the column, to conform to UBC Section 1921.4.4.3.

2a. LAP SPLICES

To conform with UBC Section 1921.4.3, a tension lap splice must be provided within the center half of the clear column height. The lap length required is specified by UBC Section 1912.15 as being equal to the tensile development length which is given by UBC Section 1912.2 as being the greater of

$$\ell_d = 0.03 d_b f_y/(f'_c)^{0.5}$$
$$= 0.03 \times 1.27 \times 60{,}000/(4000)^{0.5}$$
$$= 36 \text{ inches}$$

or $\quad \ell_d \quad = 0.75\ell_{db} \ldots$ for hoop reinforcement at four inches spacing
$$= 0.75 \times 0.04 A_b f_y/(f'_c)^{0.5}$$
$$= 0.75 \times 0.04 \times 1.27 \times 60{,}000/(4000)^{0.5}$$
$$= 36 \text{ inches}$$

The minimum lap length recommended by SEOC Commentary Section 3D.3 is

$$\ell_{(min)} = 30 d_b$$
$$= 30 \times 1.27$$
$$= 38 \text{ inches}$$

Hence the required lap length is thirty eight inches.

2b. TRANSVERSE REINFORCEMENT

The limiting compressive stress value given by

$$A_g f'_c/10 = 30^2 \times 4/10$$
$$= 360 \text{ kips}$$
$$< P_u$$

Hence the UBC Section 1921.4 provisions relating to members subjected to combined bending and axial load are applicable. The sum of the design strengths of the girders framing into the joint is

$$\Sigma M_G = 2 \times 550$$
$$= 1100 \text{ kip feet}$$

The sum of the design moment strengths of the columns framing into the joint, allowing for the factored axial force of 400 kips, is

$$\Sigma M_C = 2 \times 720$$
$$= 1440$$
$$> 6\Sigma M_G/5$$

Then UBC Formula (21–1) is satisfied and, in addition, the point of contraflexure may be assumed to be within the middle half of the column. Hence, in accordance with UBC Sections 1921.4.2.3 and 1921.4.4.1.6, hoop reinforcement is not required over the full height of the column. Hoop reinforcement, at a maximum spacing of four inches, is required only over the length of the splice and for a distance from each joint face given by UBC Section 1921.4.4.4 as the greater of

$$\ell_o = \ell_u/6$$
$$= 13 \times 12/6$$
$$= 26 \text{ inches}$$
or $\quad \ell_o \quad = 18 \text{ inches}$
or $\quad \ell_o \quad = h$
$$= 30 \text{ inches} \ldots \text{governs}$$

Using No.5 hoop reinforcement bars, and providing one and a half inches clear cover to the bars, gives a core dimension, measured center-to-center of the confinement reinforcement, of

$$h_c \quad = 30 - 3 - 0.625$$
$$= 26.375 \text{ inches}$$

The dimension measured out-to-out of the confining bars is, in both directions,

$$h_c' \quad = h_c + 0.625$$
$$= 27.0 \text{ inches}$$

The area, calculated out-to-out of the confining bars, is

$$A_{ch} \quad = (h_c')^2$$
$$= 27^2$$
$$= 729 \text{ square inches}$$

The required area of confinement reinforcement is given by the greater value obtained from UBC Formulae (21-3) and (21-4)

Then $\quad A_{sh} \quad = 0.3sh_c(A_g/A_{ch} - 1.0)f_c'/f_y$
$\qquad\qquad\quad = 0.3 \times 4 \times 26.375(900/729 - 1.0) \times 4/60$
$\qquad\qquad\quad = 0.495 \text{ square inches}$

or $\quad A_{sh} \quad = 0.12sh_c f_c'/f_y$
$\qquad\qquad\quad = 0.12 \times 4 \times 26.375 \times 4/60$
$\qquad\qquad\quad = 0.844 \text{ square inches ... governs}$

Hence three No.5 bars are required providing an area of confinement reinforcement of

$$A_{sh} \quad = 0.93 \text{ square inches}$$

The center-to-center spacing between the crosstie and the hoop reinforcement is

$$s' \quad = h_c/2 + d_b/2 + d_{hoop}/2$$
$$= (26.375 + 1.27 + 0.625)/2$$
$$= 14.14 \text{ inches}$$
$$\approx 14 \text{ inches}$$

Hence the requirements of UBC Section 1921.4.4.3 are satisfied.

Since the column is in a state of double curvature, the maximum design shear force which can be developed in the column may be determined from the probable flexural strength of the girders framing into the joint[6]. Hence, in accordance with UBC Section 1921.4.5.1 the probable girder strength is derived from the design strength as

$$M_{PG} \quad = M_G \times 1.25/\phi$$
$$= 550 \times 1.25/0.9$$
$$= 764 \text{ kip feet}$$

At the joint, the sum of the maximum moments which may be developed in the column is

$$\Sigma M_C = \Sigma M_{PG}$$

and

$$M_C = \Sigma M_{PG}/2$$
$$= 764 \text{ kip feet}$$

Then the maximum probable design shear force acting on the column is

$$V_e = 2M_C/H_n$$
$$= 2 \times 764/13$$
$$= 117.5 \text{ kips}$$

In accordance with UBC Sections 1921.4.5 and 1921.4.4.7 the design shear force for the column may also be determined from the probable moment strength at the top and bottom of the column. The column probable moment strength is determined[6] by assuming a strength reduction factor of unity and a tensile reinforcement stress of $1.25f_y$. The factored axial compressive force in the column is

$$P_u = 400 \text{ kips}$$
$$> 0.1A_g f'_c$$

The strength reduction factor for this condition of axial compression with flexure is given by UBC Section 1909.3.2 as

$$\phi = 0.7$$

The probable moment strength may be determined from the column design moment strength as

$$M_{PC} = M_C \times 1.25/\phi$$
$$= 720 \times 1.25/0.7$$
$$= 1286 \text{ kip feet}$$

The design shear force is then

$$V_e = 2M_{PC}/H_n$$
$$= 2 \times 1286/13$$
$$= 198 \text{ kips}$$
$$> 117.5 \text{ kips}$$

Then

$$V_e = 117.5 \text{ kips governs}$$

The limiting compressive stress value given by

$$A_g f'_c/20 = 30^2 \times 4/20$$
$$= 180 \text{ kips}$$
$$< P_u$$

Hence, in accordance with UBC Section 1921.4.5.2, the design shear strength provided by the concrete may be utilized and this is given by UBC Formula (11–3) as

$$\phi V_c = 0.85 \times 2bd(f'_c)^{0.5}$$
$$= 0.85 \times 2 \times 30 \times 27.24 \times (4000)^{0.5}/1000$$
$$= 87.86 \text{ kips}$$
$$< V_e$$

The design shear strength required from shear reinforcement is given by UBC Formula (11-2) as

$$\phi V_s = V_e - \phi V_c$$
$$= 117.5 - 87.86$$
$$= 29.64 \text{ kips}$$
$$< 2 \times \phi V_c \text{ ... satisfactory}$$

The maximum hoop spacing, in accordance with UBC Section 1921.4.4.6, may not exceed the lesser value given by

$$s = 6d_b$$
$$= 6 \times 1.27$$
$$= 7.62 \text{ inches}$$

or

$$s = 6 \text{ inches}$$

Hence a spacing of 6 inches governs. The minimum size of tie required for a No.10 longitudinal bar is specified, by UBC Section 1907.10.5, as a No.3 bar and crossties are required to provide support to each central bar. The minimum area of shear reinforcement, which may be provided at a spacing of six inches, is given by UBC Formula (11-14) as

$$A_v = 50b_w s/f_y$$
$$= 50 \times 30 \times 6/60,000$$
$$= 0.15 \text{ square inches}$$

The area of shear reinforcement, which is required at a spacing of six inches to provide a shear strength of ϕV_s, is specified by UBC Formula (11-17) as

$$A_v = \phi V_s s/\phi df_y$$
$$= 29.64 \times 6/(0.85 \times 27.24 \times 60)$$
$$= 0.13 \text{ square inches}$$
$$< 0.15 \text{ square inches}$$

Providing a No.5 hoop and crosstie at six inches spacing gives an area of

$$A_v = 0.93 \text{ square inches}$$
$$> 0.15$$

2c. CONSTRUCTION JOINT

The most practical location of the construction joint is at the lower end of the splice as shown.

2d. COVER

In accordance with UBC Section 1907.7, a clear cover of one and a half inches is required to the hoop reinforcement for an interior column not exposed to the weather. ══════

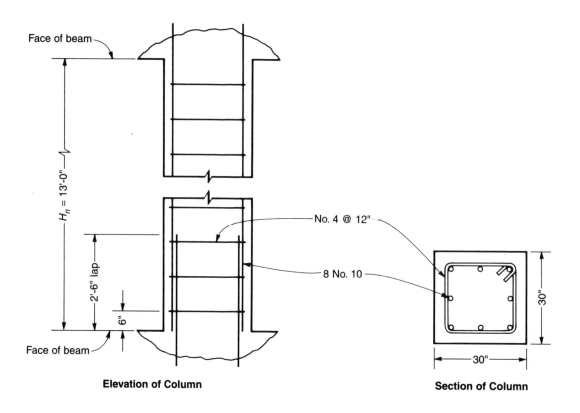

Elevation of Column

Section of Column

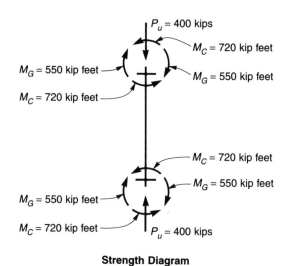

Strength Diagram

FIGURE 88C–5(i)

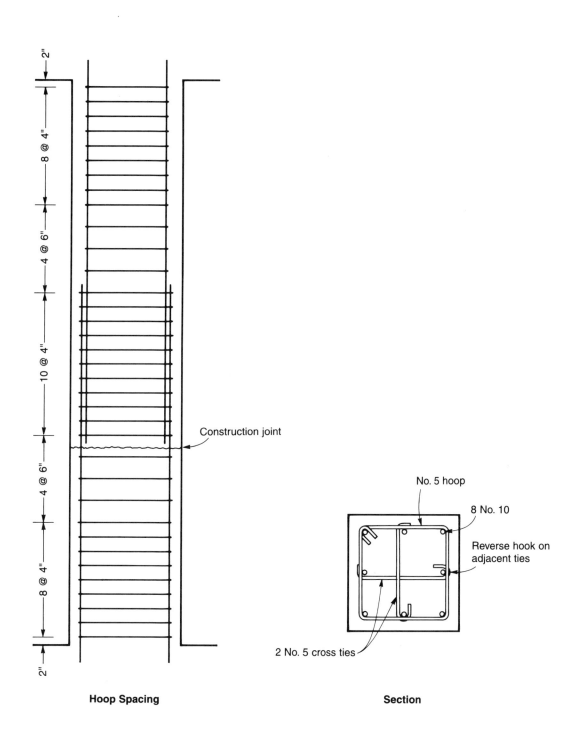

Hoop Spacing

Section

FIGURE 88C–5(ii)

STRUCTURAL ENGINEER REGISTRATION

CHAPTER 3

STRUCTURAL CONCRETE DESIGN

SECTION 3.6

REINFORCED CONCRETE FOUNDATIONS

PROBLEMS:

STRUCTURAL ENGINEER EXAMINATION – 1991 ===============

PROBLEM C–3 – WT. 5.0 POINTS

GIVEN: The concrete column and footing, as shown

CRITERIA: Materials:
- Concrete: $f'_c = 4000$ psi
- Reinforcement: $f_y = 60,000$ psi

Assumptions:
- Assume the column lap splices are adequate.
- Neglect any other loads or weights except for the ones shown.
- The allowable soil bearing pressure is not exceeded.
- The average effective depth is 20 inches in each direction.

REQUIRED:
1. Check the adequacy of the footing for shear.
2. Check the adequacy of the footing for flexure.
3. Ignoring vertical reinforcement, check the adequacy of the column size shown to transfer the 800 kip load to the footing.

SOLUTION:

LOADING

The net factored loading on the footing is

$$
\begin{aligned}
q_u &= P_u/A_f \\
&= 800/(15 \times 10) \\
&= 5.33 \text{ kips per square foot}
\end{aligned}
$$

1. SHEAR

FLEXURAL SHEAR

The critical section for flexural shear in a footing, as defined by UBC Sections 1915.5.2 and 1911.1.3.1, is located a distance from the long side of the column equal to the effective depth. The distance of the critical section from the short edge of the footing is, then

$$
\begin{aligned}
x &= \ell_2/2 - c_2/2 - d \\
&= 15/2 - 18/24 - 20/12 \\
&= 5.083 \text{ feet}
\end{aligned}
$$

The shear capacity of the footing for beam action is given by UBC Formula (11–3) as

$$
\begin{aligned}
\phi V_c &= 2\phi bd(f'_c)^{0.5} \\
&= 2 \times 0.85 \times 10 \times 12 \times 20(4000)^{0.5}/1000 \\
&= 258.04 \text{ kips}
\end{aligned}
$$

The factored applied shear at the critical section is

$$V_u = q_u bx$$
$$= 5.33 \times 10 \times 5.083$$
$$= 270.92 \text{ kips}$$
$$> \phi V_c$$

Hence beam action shear is unsatisfactory according to the simplified Formula and it is necessary to check with the more detailed UBC Formula (11–6). Then, the shear capacity is given by

$$\phi V_c = \phi[1.9(f'_c)^{0.5} + 2500\rho V_u d/M_u]bd$$

where ρ = reinforcement ratio
$$= A_s/\ell_1 d$$
$$= 29.64/(10 \times 12 \times 20)$$
$$= 0.0124$$

and M_u = factored moment occurring with V_u
$$= q_u bx^2/2$$
$$= 5.33 \times 10 \times 5.083^2/2$$
$$= 688.55 \text{ kip feet}$$

Hence $V_u d/M_u$
$$= 270.92 \times 20/(12 \times 688.55)$$
$$= 0.66$$
$$< 1.0 \dots \text{satisfactory}$$

and ϕV_c
$$= 0.85(1.9(4000)^{0.5} + 2500 \times 0.0124 \times 0.66)10 \times 12 \times 20/1000$$
$$= 286.6 \text{ kip feet}$$
$$> V_u \dots \text{satisfactory}$$
$$< 3.5 \times 258.04/2 \dots \text{satisfactory}$$

Hence the flexural shear capacity is satisfactory.

TWO-WAY SHEAR

The critical section for punching shear in a footing, as defined by UBC Sections 1915.5.2 and 1911.12.1.2, is located on the perimeter of a rectangle with sides of length

$$b_1 = c_1 + d$$
$$= 24 + 20$$
$$= 44 \text{ inches}$$

and b_2
$$= c_2 + d$$
$$= 18 + 20$$
$$= 38 \text{ inches}$$

The length of the critical perimeter is

$$b_o = 2(b_1 + b_2)$$
$$= 2(44 + 38)$$
$$= 164 \text{ inches}$$

The ratio of the long side to the short side of the column is

$$\beta_c = c_1/c_2$$
$$= 24/18$$
$$= 1.33$$
$$< 2$$

Hence the punching shear capacity of the footing is given by UBC Formula (11–38) as

$$\phi V_c = 4\phi b_o d(f'_c)^{0.5}$$
$$= 4 \times 0.85 \times 164 \times 20(4000)^{0.5}/1000$$
$$= 705.31 \text{ kips}$$

The factored applied shear at the critical perimeter is

$$V_u = q_u(\ell_1\ell_2 - b_1b_2)$$
$$= 5.33(10 \times 15 - 44 \times 38/144)$$
$$= 737.61 \text{ kips}$$
$$> \phi V_c$$

Hence, the punching shear capacity is inadequate.

2. FLEXURE

The critical section for flexure in a footing, as defined by UBC Section 1915.4.2, is located along the long side of the column. The factored applied moment at this section, for the longitudinal direction, is

$$M_u = q_u\ell_1(\ell_2/2 - c_2/2)^2/2$$
$$= 5.33 \times 10(15/2 - 18/24)^2/2$$
$$= 1214 \text{ kip feet}$$

The required reinforcement ratio is derived from UBC Section 1910.2 and is

$$\rho_r = 0.85f'_c[1.0 - (1.0 - 2M_u/0.765bd^2f'_c)^{0.5}]/f_y$$
$$= 0.0059$$
$$< \rho \text{ ...satisfactory}$$

The maximum allowable reinforcement ratio is given by UBC Section 1910.3 as

$$\rho_{max} = 0.75 \times 0.85 \times 87\beta_1 f'_c/f_y(87 + f_y)$$

where β_1 = 0.85 as defined in UBC Section 1910.2.7.3

and ρ_{max} = 0.0214

$$> \rho \text{ ...satisfactory}$$

The minimum reinforcement area in a footing slab for Grade 60 bars is given by UBC Section 1907.12.2.1 as

$$A_{s(min)} = 0.18 \text{ percent of the gross area}$$
$$= 0.0018 \times \ell_1 \times h$$
$$= 0.0018 \times 120 \times 24$$
$$= 5.18 \text{ square inches}$$
$$< A_s \text{ ... satisfactory}$$

UBC Sections 1912.2.2 and 1912.2.3 are applicable, with clear cover exceeding $2d_b$ and clear spacing exceeding $3d_b$, and the development length of the No.11 bars is

$$\ell_d' = 0.04A_bf_y/(f_c')^{0.5}$$
$$= 0.04 \times 1.56 \times 60,000/(4000)^{0.5}$$
$$= 59.2$$

and
$$\ell_{d(min)} = 0.03d_bf_y/(f_c')^{0.5}$$
$$= 0.03 \times 1.41 \times 60,000/(4000)^{0.5}$$
$$= 40.1 \text{ inches}$$
$$< \ell_d' \text{ ... satisfies UBC Section 1912.2.3.6}$$

The development length of the No.11 bars, allowing for the excess reinforcement in accordance with UBC Section 1912.2.5, is

$$\ell_d = (\rho_r/\rho) \times \ell_d'$$
$$= (0.0059/0.0124) \times 59.2$$
$$= 28 \text{ inches}$$

Allowing for an end cover of three inches, the available anchorage length for the bars is

$$\ell_{da} = (\ell_2 - c_2)/2 - c_e$$
$$= (180 - 18)/2 - 3$$
$$= 78 \text{ inches}$$
$$> \ell_d \text{ ... satisfactory}$$

Hence, the flexural reinforcement is adequate in the longitudinal direction. The factored applied moment at the critical section for the transverse direction is

$$M_u = q_u\ell_2(\ell_1/2 - c_1/2)^2/2$$
$$= 5.33 \times 15(10/2 - 2/2)^2/2$$
$$= 640 \text{ kip feet}$$

The required reinforcement ratio is derived from UBC Section 1910.2 and is

$$\rho_r = 0.85f_c'[1.0 - (1.0 - 2M_u/0.765bd^2f_c')^{0.5}]/f_y$$
$$= 0.00201$$

The required reinforcement area is

$$A_r = \rho_r \times \ell_2 \times d$$
$$= 0.00201 \times 180 \times 20$$
$$= 7.24 \text{ square inches}$$
$$< A_s \text{ ... satisfactory}$$

The ratio of the long side to the short side of the footing is

$$\beta = \ell_2/\ell_1$$
$$= 15/10$$
$$= 1.5$$

From UBC Formula (15-1) the area of reinforcement required in a central band width equal to the length of the short side of the footing is

$$A_b \quad = A_r \times 2/(\beta + 1)$$
$$= 7.24 \times 2/(1.5 + 1)$$
$$= 5.79 \text{ square inches}$$

The number of No.11 bars provided in the central band width of ten feet is

$$n \quad = 19 \times 10/14.5$$
$$= 13$$

and the area of reinforcement provided over this band width is

$$A_b' \quad = 20.28 \text{ square inches}$$
$$> 5.79 \text{ ... satisfactory}$$

The available anchorage length exceeds the required development length and the flexural reinforcement is adequate in the transverse direction.

3. LOAD TRANSFER

Neglecting vertical reinforcement, the column load must be transferred to the footing by bearing on the concrete. The bearing capacity of the column concrete at the interface is given by UBC Section 1910.15.1 as

$$\phi P_n \quad = 0.85\phi f_c' A_1$$
where $\quad \phi \quad$ = strength reduction factor
$$= 0.7 \text{ from UBC Section 1909.3}$$
and $\quad A_1 \quad$ = loaded area
$$= 24 \times 18$$
$$= 432 \text{ square inches}$$
Hence $\quad \phi P_n \quad = 0.85 \times 0.7 \times 4 \times 432$
$$= 1028 \text{ kips}$$
$$> 800 \text{ kips ... satisfactory}$$

The area of the base of the pyramid, with side slopes of 1:2, formed within the footing by the loaded area is

$$A_2 \quad = (c_1 + 4d)(c_2 + 4d)$$
$$= (24 + 4 \times 20)(18 + 4 \times 20)$$
$$= 10,192 \text{ square inches}$$
Hence $\quad (A_2/A_1)^{0.5} \quad = (10,192/432)^{0.5}$
$$= 4.9$$
$$> 2$$

Then the bearing capacity of the footing concrete is given by UBC Section 1910.15.1 as

$$\phi P_n \quad = 2 \times 0.85\phi f_c' A_1$$
$$= 2 \times 1028$$
$$= 2056 \text{ kips}$$
$$> 800 \text{ kips ... satisfactory}$$

Hence the concrete alone is adequate to transfer the column load to the footing. ======

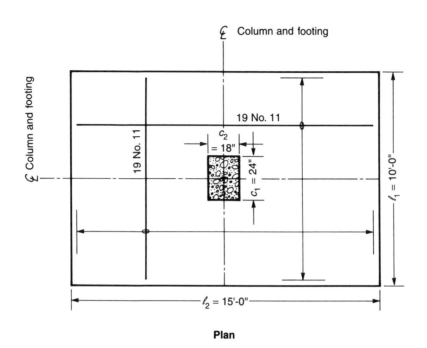

Plan

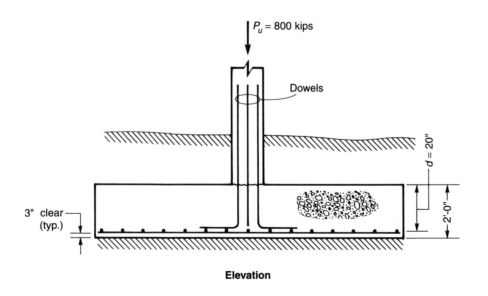

Elevation

FIGURE 91C–3

356

STRUCTURAL ENGINEER EXAMINATION – 1988

PROBLEM C-2 – WT. 4.0 POINTS

GIVEN: The continuous foundation shown in elevation in Figure 88C-2 supports a wood frame stucco wall. The geotechnical engineer for the project recommends designing the footing for ½" differential settlement over 25 feet.

CRITERIA:

Materials:
- Concrete: $f_c' = 2500$ psi
- Reinforcement: $f_y = 60,000$ psi

Assumptions:
- Footing EI = 9.65×10^9 lb in²
- Full fixity exists beyond the design points.
- $W_{DL} = 500$ lb/ft. (including footing and soil).
- $W_{LL} = 1000$ lb/ft.

REQUIRED: 1. Design the footing for ½" differential settlement over 25 feet.
2. Sketch the reinforcing steel required in the section.
3. List one comment that you would make to the owner regarding the differential settlement.

SOLUTION

MEMBER FORCES

The applied dead and live loads produce no moment or shear in the footing since they are balanced by the soil pressure. The bending moment produced by the half inch differential settlement over a length of twenty-five feet produces the bending moments and shears shown in Figure 88C-2. The bending moment is given by

$$
\begin{aligned}
M &= 6EI\delta/\ell^2 \\
&= 6 \times 9.65 \times 10^9 \times 0.5/(25^2 \times 144 \times 12,000) \\
&= 26.81 \text{ kip feet}
\end{aligned}
$$

The shear produced by the settlement is given by

$$V = 2M/\ell$$
$$= 2.14 \text{ kips}$$

The factored forces are given by UBC Formula (9–6) as

$$M_u = 1.4M$$
$$= 37.5 \text{ kip feet}$$
$$V_u = 1.4V$$
$$= 3.0 \text{ kips}$$

1. DESIGN FOR FLEXURE AND SHEAR

The critical case for flexure is with sagging moment producing tensile stresses in the bottom of the footing giving an effective width of

$$b = 8"$$

Assuming three inch cover to No.3 links and using No.6 main bars gives an effective depth of

$$d = 30 - 3 - 0.75$$
$$= 26.25 \text{ inches}$$

Neglecting compression reinforcement, the required tension reinforcement area is obtained by using calculator program A.1.1 in Appendix A and is given by

$$\rho = 0.85f_c'[1.0 - (1.0 - 2K/0.765f_c')^{0.5}]/f_y$$

where $K = 12M_u/bd^2$
$$= 0.0816$$

Then $\rho = 0.0015$

In accordance with UBC Section 1910.3.3, the maximum allowable reinforcement ratio is given by

$$\rho_{max} = 0.75 \times 0.85 \times 87\beta_1f_c'/f_y(87 + f_y)$$

where $\beta_1 = 0.85$ as defined in UBC Section 1910.2.7.3

Hence $\rho_{max} = 0.0134$
$$> \rho \text{ ... satisfactory}$$

In accordance with UBC Section 1910.5.1, the minimum allowable reinforcement ratio is given by

$$\rho_{min} = 200/f_y$$
$$= 0.0033$$
$$> \rho$$

Hence in accordance with UBC Section 1910.5.2 the required reinforcement area is given by

$$A_s = 1.33\rho bd$$
$$= 1.33 \times 0.0015 \times 8 \times 26.25$$
$$= 0.42 \text{ square inches}$$

This may be provided with one No.6 bar, at both top and bottom of the section, giving an area of

$$A_s = 0.44 \text{ square inches}$$
$$> 0.42 \text{ square inches}$$

The design shear strength, provided by the concrete, is given by UBC Formula (11–3) as

$$\phi V_c = 0.85 \times 2bd(f_c')^{0.5}$$
$$= 17.85 \text{ kips}$$
$$> 2V_u$$

Hence, in accordance with UBC Section 1911.5.5.1, no stirrups are necessary. However, in order to locate the main reinforcement in position, a link consisting of a single arm will be provided at twelve inches on center. Providing the minimum tie reinforcement at the construction joint, specified by UBC Section 1917.6.1 and Formula (11–14), requires an area of

$$A_v = 50bs/f_y$$
$$= 50 \times 8 \times 12/60,000$$
$$= 0.08 \text{ square inches}$$

This may be provided with one No.3 link at twelve inches on center, giving an area of

$$A_v = 0.11 \text{ square inches}$$

2. REINFORCEMENT DETAILS

The reinforcement details are given in Figure 88C–2

3. RECOMMENDATION TO OWNER

It may be anticipated that settlement after the first twelve months will be negligible. The owner should be advised to monitor the cracking of the stucco to confirm this and, if confirmed, repair any existing cracks.

359

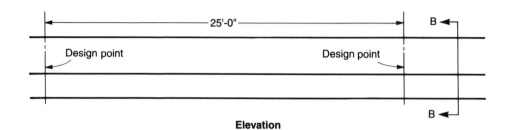

Elevation

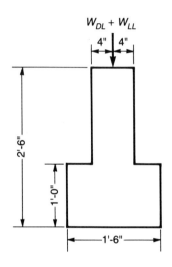

Section B—B

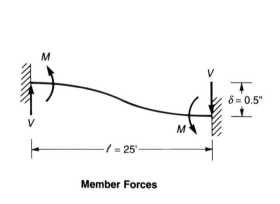

Member Forces

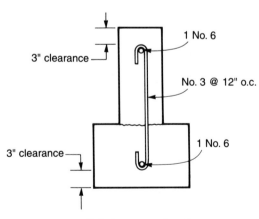

Reinforcement Details

FIGURE 88C–2

STRUCTURAL ENGINEER EXAMINATION – 1988 ═══════════════════

PROBLEM C-3 – WT. 5.0 POINTS

GIVEN: The two-column footing with the service loads shown.

CRITERIA: Materials:
- Concrete: $f'_c = 4000$ psi
- Reinforcement: $f_y = 60,000$ psi
- The allowable soil bearing value $= 3000$ psf

                     Assumptions:
- The footing width is 16 feet.
- Neglect the weight of the concrete footing.
- All the given loads are nonfactored.
- Neglect load cases.
- Use No.6 bars for all reinforcement.
- Assume the footing is rigid.

REQUIRED:
1. Determine if the depth of the footing is adequate.
2. Design the footing flexural reinforcement.
3. Complete the section showing all footing reinforcements.

SOLUTION

LOADING

The factored load on column J is obtained from UBC Formula (9-1) as

$$W_J = 1.4 \times 500 + 1.7 \times 250$$
$$= 1125 \text{ kips}$$

The factored load on column K is

$$W_K = 1.4 \times 350 + 1.7 \times 150$$
$$= 745 \text{ kips}$$

The centroid of the factored loads is located a distance from column K given by

$$x_o = 20W_J/(W_J + W_K)$$
$$= 20 \times 1125/(1125 + 745)$$
$$= 12 \text{ feet}$$

Hence the centroid of the factored loads coincides with the centroid of the footing giving a uniform factored soil pressure under the footing of

$$q = (W_J + W_K)/\ell$$
$$= (1125 + 745)/26$$
$$= 71.92 \text{ kips per linear foot}$$

The shear force and bending moment diagrams for the footing are shown in Figure 88C-3. The point of zero shear is located a distance from column K given by

$$x = 673.08/q$$
$$= 673.08/71.92$$
$$= 9.36 \text{ feet}$$

The maximum negative moment occurs at this point and is given by

$$M_u = xW_K - q(x + 1.0)^2/2$$
$$= 9.36 \times 745 - 71.92(9.36 + 1.0)^2/2$$
$$= 3114 \text{ kip feet}$$

1. FOOTING DEPTH

The footing depth may be checked by determining the capacity of the footing in two-way shear and flexural shear. Assuming three inches clear cover with No.6 bars in both directions gives an average effective depth in each direction of

$$d = h - c_c - d_b$$
$$= 36 - 3 - 0.75$$
$$= 32.25 \text{ inches}$$

TWO-WAY SHEAR

The critical section for punching shear, as defined by UBC Sections 1915.5.2 and 1911.12.1.2, is located a distance equal to half the effective depth from the face of the column. For column K the critical perimeter forms three sides of a rectangle with a length of

$$b_o = (c_1 + d) + 2(c_2 + d/2)$$
$$= (24 + 32.25) + 2(24 + 32.25/2)$$
$$= 136.5 \text{ inches}$$

The ratio of the long side of the column to the short side is

$$\beta_c = 24/24$$
$$= 1.0$$
$$< 2$$

Hence the punching shear capacity of the footing is given by UBC Formula (11-38) as

$$\phi V_c = 4\phi b_o d(f_c')^{0.5}$$

where $\phi = 0.85$ from UBC Section 1909.3.2

and

$$\phi V_c = 4 \times 0.85 \times 136.5 \times 32.25(4000)^{0.5}$$
$$= 947 \text{ kips}$$
$$> W_K$$

At column J the critical perimeter is

$$b_o \quad = 4(c_1 + d)$$
$$= 4(24 + 32.25)$$
$$= 225 \text{ inches}$$

The punching shear capacity at column J is

$$\phi V_c \quad = 947 \times 225/136.5$$
$$= 1561 \text{ kips}$$
$$> W_J$$

Hence the footing depth is adequate for punching shear.

FLEXURAL SHEAR

The critical section for flexural shear in a footing, as defined by UBC Sections 1915.5.2 and 1911.1.3.1, is located a distance equal to the effective depth from the face of the column. The factored applied shear at this critical section is

$$V_u \quad = 765.4 - q(d + c_2/2)$$
$$= 765.4 - 71.92(32.25 + 12)/12$$
$$= 500 \text{ kips}$$

The shear capacity of the footing for beam action is given by UBC Formula (11–3) as

$$\phi V_c \quad = 2\phi bd(f'_c)^{0.5}$$
$$= 2 \times 0.85 \times 16 \times 12 \times 32.25(4000)^{0.5}/1000$$
$$= 666 \text{ kips}$$
$$> V_u$$

Hence the footing depth is adequate for flexural shear.

2. FLEXURAL REINFORCEMENT

Neglecting compression reinforcement, the required reinforcement ratio in the top of the footing in the longitudinal direction is derived from UBC Section 1910.2 and is

$$\rho \quad = 0.85f'_c[1.0 - (1.0 - 2M_u/0.765bd^2f'_c)^{0.5}]/f_y$$

where $\quad M_u \quad = 3114$ kip feet

and $\quad \rho \quad = 0.0036$

In accordance with UBC Section 1910.3 the maximum allowable reinforcement ratio is given by

$$\rho_{max} \quad = 0.75 \times 0.85 \times 87\beta_1 f'_c/f_y(87 + f_y)$$

where $\quad \beta_1 \quad = 0.85$ as defined in UBC Section 1910.2.7.3

and $\quad \rho_{max} \quad = 0.0214$

$$> \rho \text{ ... satisfactory}$$

In accordance with UBC Section 1907.12.2.1 the minimum allowable reinforcement ratio, expressed as a percentage of the gross concrete area, is given by

$$\rho_{min} = 0.18 \text{ percent}$$
$$< \rho \text{ ... satisfactory}$$

Thus the required reinforcement area is given by

$$A_s = \rho bd$$
$$= 0.0036 \times 192 \times 32.25$$
$$= 22.16 \text{ square inches}$$

This may be provided by fifty No.6 bars at a spacing of 3.8 inches giving an area of

$$A_s = 22.0 \text{ square inches}$$

BOTTOM REINFORCEMENT

The critical section for positive moment, as defined by UBC Section 1915.4.2, is located along the face of column J. The factored applied moment at this section is given by

$$M_u = q \times 4^2/2$$
$$= 71.92 \times 4^2/2$$
$$= 575 \text{ kip feet}$$

Neglecting compression reinforcement, the required reinforcement ratio in the bottom of the footing is derived from UBC Section 1910.2 and is

$$\rho = 0.85f_c'[1.0 - (1.0 - 2M_u/0.765bd^2f_c')^{0.5}]/f_y$$

where $M_u = 575$ kip feet

and $\rho = 0.00064$

$$< \rho_{max} \text{ ... satisfactory}$$
$$< \rho_{min}$$

Hence, in accordance with UBC Section 1910.5.2 the required reinforcement area is given by

$$A_s = 1.33\rho bd \text{ ... since the top reinforcement provides the temperature steel}$$
$$= 1.33 \times 0.00064 \times 192 \times 32.25$$
$$= 5.31 \text{ square inches}$$

This may be provided by twelve No.6 bars giving an area of

$$A_s = 5.28 \text{ square inches}$$

The bar spacing is

$$s = 186/11$$
$$= 16 \text{ inches}$$
$$< 3h$$
$$< 18 \text{ inches}$$

Hence the requirements of UBC Section 1907.6.5 are satisfied.

BOTTOM TRANSVERSE REINFORCEMENT

The factored applied moment at the critical section for the transverse direction is

$$\begin{aligned}
M_u &= (W_J + W_K)/b \times 7^2/2 \\
&= 1870 \times 7^2/(16 \times 2) \\
&= 2863 \text{ kip feet}
\end{aligned}$$

The required reinforcement ratio in the bottom of the footing in the transverse direction is derived from UBC Section 1910.2 and is

$$\begin{aligned}
\rho &= 0.85f_c'[1.0 - (1.0 - 2M_u/0.765bd^2f_c')^{0.5}]/f_y \\
&= 0.0020 \\
&< \rho_{max} \ldots \text{ satisfactory} \\
&> \rho_{min} \ldots \text{ satisfactory}
\end{aligned}$$

and $\quad A_s \quad = 20.08$ square inches

The ratio of the long side to the short side of the footing is

$$\begin{aligned}
\beta &= \ell/b \\
&= 26/16 \\
&= 1.63
\end{aligned}$$

From UBC Formula (15–1) the area of reinforcement required in a band width equal to the length of the short side of the footing is given by

$$\begin{aligned}
A_b &= 2A_s/(\beta + 1) \\
&= 2 \times 20.08/(1.63 + 1) \\
&= 15.27 \text{ square inches}
\end{aligned}$$

This is equivalent to an area per linear foot of band width of

$$\begin{aligned}
A_t &= A_b/b \\
&= 15.27/16 \\
&= 0.954 \text{ square inches per linear foot}
\end{aligned}$$

Because of the proximity of the two band widths from column J and column K, it is convenient to provide this reinforcement over the whole length of the footing. Hence use No.6 bars at a spacing of 5.5 inches to give an area of

$$A_t \quad = 0.96 \text{ square inches per linear foot, a total of fifty–five bars.}$$

TOP TRANSVERSE REINFORCEMENT

Provide No.6 bars at a spacing of eighteen inches in order to locate the top longitudinal bars. Complete reinforcement details are shown in Figure 88C-3. ========

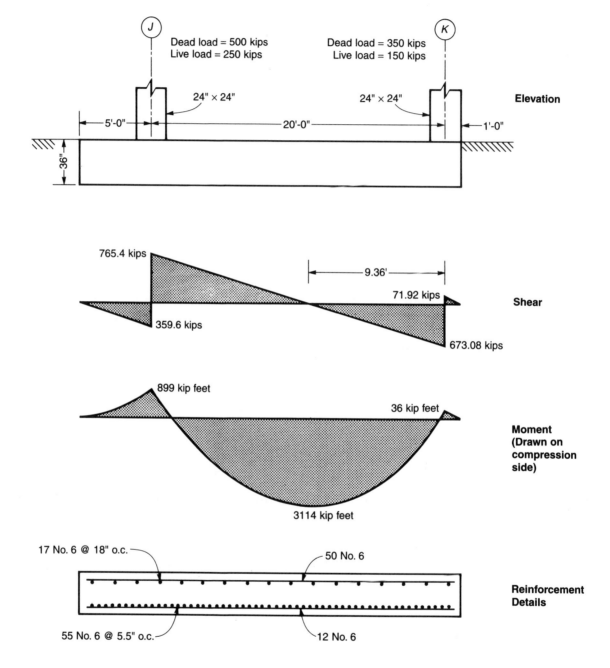

FIGURE 88C–3

STRUCTURAL ENGINEER REGISTRATION

CHAPTER 3

STRUCTURAL CONCRETE DESIGN

SECTION 3.7

REINFORCED CONCRETE ELEMENTS

PROBLEMS:

1991 C–1
1990 C–1

STRUCTURAL ENGINEER EXAMINATION – 1991

PROBLEM C–1A – WT. 8.0 POINTS

GIVEN:　　　　　The bracket attached to a concrete wall as shown.

CRITERIA:　　　Code: UBC strength design method with $\phi = 0.65$

Materials:
- Concrete: f'_c = 4000 psi, normal weight
- Bolts:　　　　= ¾" diameter, Grade A325. Head diam. = 1.31"

Assumptions:
- The steel bracket is infinitely rigid.
- The center of the rectangular compression block is at the center of bottom bolts.
- The anchor bolts are embedded in the tension zone of the concrete.
- The concrete wall is adequate to carry the given loads.
- Special inspection is provided during construction.

REQUIRED:　　　1.　Determine the maximum factored tension and shear forces on the bolts.
　　　　　　　　2.　Determine if the connection is adequate for the given loads.

SOLUTION:

1. FACTORED LOADS

Assuming that the location of the bracket is in seismic zone 4, in accordance with UBC Section 1921.2.7, the required strength is

$$U = 1.4(D + E)$$

where
$$D = \text{dead load} = 23 \text{ kips}$$
$$E = \text{seismic load} = 5 \text{ kips}$$

An additional load factor of 2 is required by Section 1925.2 when the anchors are embedded in the tension zone of the concrete and special inspection is provided. Hence, the factored shear force acting on each bolt is given by

$$V_u = 2 \times 1.4P_D/4$$
$$= 2 \times 1.4 \times 23/4$$
$$= 16.10 \text{ kips}$$

Similarly, by taking moments about the bottom bolts, the factored tensile force in each top bolt is

$$P_u = 2 \times 1.4(4P_D + 10P_T)/(2 \times 12)$$
$$= 2 \times 1.4(4 \times 23 + 10 \times 5)/24$$
$$= 16.57 \text{ kips}$$

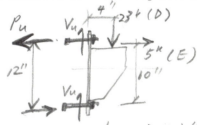

Load Comb.: $2 \times 1.4 \times (D+E)$

M (Lower bolt): $P_u = 2 \times 1.4 \left(\dfrac{23 \times 4'' + 5^k \times 10''}{12} \right)$

$= 33.13$ resisted by 2 bolts

Each bolt take 16.57^k

2. ANCHOR CAPACITY

The bolt strength in tension is given by

$$P_{ss} = 0.9A_b f'_s$$

where
$$A_b = \text{bolt area} = 0.442 \text{ square inches from AISC Table I–A.}$$
$$f'_s = \text{bolt tensile strength} = 120 \text{ kips per square inch from AISC Table I–C.}$$
Thus
$$P_{ss} = 0.9 \times 0.442 \times 120$$
$$= 47.74 \text{ kips}$$
$$> P_u$$

$V_u = \dfrac{2 \times 1.4 \times 23^k}{2} = 32.2$

Each bolt takes 16.1^k.

The bolt strength in shear is given by

$$V_{ss} = 0.75A_b f'_s$$
$$= 0.75 \times 0.442 \times 120$$
$$= 39.78 \text{ kips}$$
$$> V_u$$

Hence the bolt strength is satisfactory in both tension and shear.

The ratio of anchor spacing to embedment length is

$$s/d_m = 10/5$$
$$= 2$$

Hence the failure surface in the concrete forms a truncated cone, radiating from the bolt head, with a surface area of

$$A_s = 1.414\pi(d_h + d_m)d_m$$

where d_h = bolt head diameter = 1.31 inches, as given

Thus $A_s = 1.414\pi(1.31 + 5) \times 5$
$$= 140.17 \text{ square inches}$$

The ratio of the edge distance to the embedment length is

$$d_e/d_m = (8.5 + 12)/5$$
$$= 4.1$$
$$> 1$$

Hence the concrete capacity in tension for normal weight concrete is given by

$$\phi P_c = \phi\lambda(2.8A_s)(f_c')^{0.5}$$
$$= 0.65 \times 1 \times (2.8 \times 140.17)(4000)^{0.5}/1000$$
$$= 16.14 \text{ kips}$$
$$< P_u$$

Hence the connection is inadequate for combined dead load plus seismic load.

The ratio of edge distance to bolt diameter is

$$d_e/d_b = 8.5/0.75$$
$$= 11.33$$
$$> 10$$

Hence the concrete capacity in shear is given by

$$\phi V_c = 800\phi\lambda A_b(f_c')^{0.5}$$
$$= 800 \times 0.65 \times 1 \times 0.442 \times (4000)^{0.5}/1000$$
$$= 14.54 \text{ kips}$$
$$< V_u$$

Hence the connection is inadequate for the applied dead load.   ======

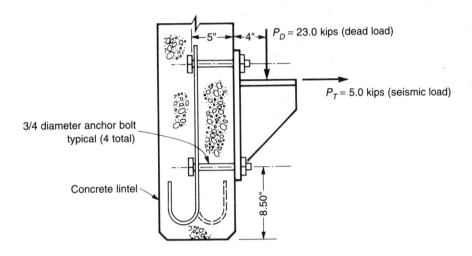

Wall Bracket—Section

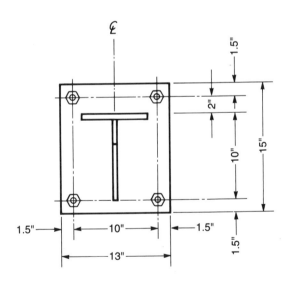

Wall Bracket —Elevation

FIGURE 91C–1

371

STRUCTURAL ENGINEER EXAMINATION – 1990 ══════════════════

PROBLEM C–1 – WT.6.0 POINTS

GIVEN: A concrete corbel projecting from an 18 inches square concrete column as shown.

CRITERIA: Concrete: f'_c = 3000 psi
Reinforcement: f_y = 60,000 psi
Neglect the corbel weight

REQUIRED:
1. Determine whether the corbel's dimensions and location of the vertical load V_u are satisfactory.
2. Determine the maximum vertical load and horizontal tensile load N_{uc} that may be applied simultaneously assuming that N_{uc} = $0.20V_u$.
3. Determine the required size and spacing of horizontal ties.

SOLUTION

1. DIMENSIONAL REQUIREMENTS

Assuming the primary tension reinforcement consists of one inch diameter bars with one and a half inch clear concrete cover, the effective depth to the tension reinforcement measured at the face of the column is

$$d = h - d_c$$
$$= 24 - 1.5 - 0.5$$
$$= 22 \text{ inches}$$

The shear span-to-depth ratio is

$$a/d = 3/22$$
$$< 1$$

and the requirement of UBC Section 1911.9.1 is satisfied.

The corbel depth at the outside edge of the bearing area is

$$h_c = 8 + 3 \times 16/8$$
$$= 14 \text{ inches}$$
$$> d/2$$

and the requirement of UBC Section 1911.9.2 is satisfied.

2. MAXIMUM SHEAR LOAD

The maximum design shear strength is given by UBC Section 1911.9.3.2.1 as

$$\phi V_n = 0.8\phi b_w d$$
$$= 0.8 \times 0.85 \times 18 \times 22$$
$$= 269 \text{ kips}$$

or

$$\phi V_n = 0.2\phi b_w d f'_c$$
$$= 0.2 \times 0.85 \times 18 \times 22 \times 3$$
$$= 202 \text{ kips} \dots \text{ governs}$$

For normal weight concrete, placed monolithically, the coefficient of friction at the shear interface is obtained from UBC Section 1911.7.4.3 as

$$\mu = 1.4\lambda$$

where $\lambda = 1.0 \dots$ for normal weight concrete

and $\mu = 1.4$

The required area of shear–friction reinforcement is given by UBC Formula (11–26) as

$$A_{vf} = V_u/\phi f_y \mu$$
$$= V_u/(0.85 \times 60 \times 1.4)$$
$$= 0.01401 V_u \text{ square inches}$$

The required area of reinforcement to resist the tensile force N_{uc} is obtained from UBC Section 1911.9.3.4 as

$$A_n = N_{uc}/\phi f_y$$
$$= 0.2 V_u/(0.85 \times 60)$$
$$= 0.00392 V_u \text{ square inches}$$

In accordance with UBC Section 1911.9.3.5, the required area of primary tension reinforcement shall not be less than

$$A_s = 2A_{vf}/3 + A_n$$
$$= (2 \times 0.01401/3 + 0.00392)V_u$$
$$= 0.0133 V_u \text{ square inches}$$

The primary tension reinforcement provided consists of three No.8 bars with an area of

$$A_s = 2.37 \text{ square inches}$$

Hence

$$V_u = 2.37/0.0133$$
$$= 179 \text{ kips} \dots \text{ governs}$$
$$< 202 \text{ kips}$$

Using the value $V_u = 179$ kips for the ultimate shear, the applied ultimate moment is given by

$$M_u = V_u a + N_{uc}(h - d)$$
$$= 179[3 + 0.2(24 - 22)]$$
$$= 608 \text{ kip inches}$$

The required area of flexural reinforcement is derived from UBC Section 1910.2 and is

$$A_f = 0.85b_w df'_c[1.0 - (1.0 - 2M_u/0.765bd^2f'_c)^{0.5}]/f_y$$
$$= 0.52 \text{ square inches}$$

In accordance with UBC Section 1911.9.3.5, the required area for the primary tension reinforcement is given by

$$A_s = A_f + A_n$$
$$= 0.52 + 0.00392V_u$$
$$= 0.52 + 0.00392 \times 179$$
$$= 1.22 \text{ square inches}$$
$$< 2.37$$

Hence $\quad V_u = 179 \text{ kips}$

and $\quad N_{uc} = 0.2V_u$
$$= 35.8 \text{ kips}$$

3. CLOSED TIES

The required area of closed ties is given by UBC Section 1911.9.4 as

$$A_h = 0.5(A_s - A_n)$$
$$= 0.5(2.37 - 0.00392 \times 179)$$
$$= 0.83 \text{ square inches}$$

In accordance with UBC Section 1911.9.4 the ties must be distributed over a depth of

$$h_v = 2d/3$$
$$= 2 \times 22/3$$
$$= 14.7 \text{ inches}$$

Using four No.3 ties at a spacing of 3.5 inches, provides a reinforcement area of

$$A_h = 4 \times 2 \times 0.11$$
$$= 0.88 \text{ square inches}$$
$$> 0.83$$

and the reinforcement provided is satisfactory.

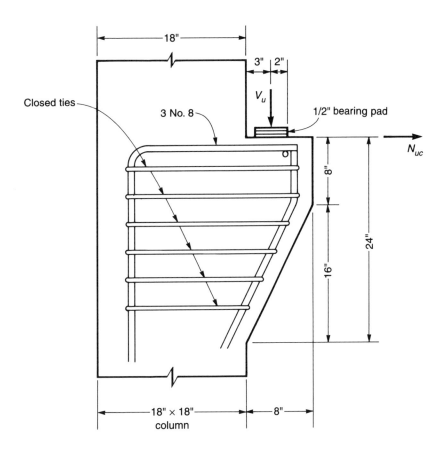

FIGURE 90C–1

STRUCTURAL ENGINEER REGISTRATION

CHAPTER 3

STRUCTURAL CONCRETE DESIGN

SECTION 3.8

PRESTRESSED CONCRETE

PROBLEMS:

1990 C–3
1987 C–2

STRUCTURAL ENGINEER EXAMINATION – 1990

PROBLEM C-3 – WT. 6.0 POINTS

GIVEN:

Design a precast prestressed concrete plank system to span a single bay as shown on plan.

- Live load: 75 psf
- Superimposed dead load: 20 psf

CRITERIA:

- Unit weight of concrete: 150 pounds per cubic foot.
- Precast prestressed plank: 3.5 inches thick solid plank.
- Cast-in-place concrete topping: 2.5 inches thick.
- Shoring: Snug fit at mid-span, adequate to support wet concrete topping.
- Non-composite section properties for a one foot wide section:
  Area, $A = 42.0$ in^2
  Section modulus, $S = 24.5$ in^3
- Composite section properties for one foot wide section:
  Section modulus at top of topping, $S_{2c} = 76.4$ in^3
  Section modulus at bottom of plank, $S_{1c} = 67.8$ in^3
- Effective prestressing force = 25.5 kips per foot

REQUIRED:

Determine the top & bottom fiber stresses at midspan for the following:

(a) Prestressing force + plank dead load
(b) Plank dead load + wet concrete topping (shoring at midspan)
(c) Plank dead load + hardened concrete topping (composite section) at the time immediately after shoring removal.
(d) Superimposed dead load + live load only
(e) Combine all of the above stresses at the top and bottom of composite section.

SOLUTION

a. INITIAL SECTION

The eccentricity of the prestressing force is

$$e \quad = 3.5/2 - 1.25$$
$$= 0.5 \text{ inches}$$

Considering a twelve inch wide section, the top and bottom fibre stresses produced by the prestressing force and self weight are

$$f_2 \quad = M_D/S + P(1/A - e/S)$$
$$= 1.5 \times 3.5 \times 150 \times 20^2/(12 \times 24.5) + 25{,}500(1/42 - 0.5/24.5)$$
$$= 1071 + 607 - 520$$
$$= 1158 \text{ pounds per square inch, compression}$$
$$f_1 \quad = -M_D/S + P(1/A + e/S)$$
$$= -1071 + 607 + 520$$
$$= 56 \text{ pounds per square inch, compression}$$

b. CAST TOPPING

The bending moment produced in the prestressed plank, at the location of the prop, by the concrete topping is

$$M_P \quad = w\ell^2/8$$
$$= 1.5 \times 150 \times 2.5 \times 10^2/12$$
$$= 4688 \text{ pound inches}$$

The reaction produced in the prop is

$$R_p \quad = 1.25w\ell$$
$$= 1.25 \times 150 \times 2.5 \times 10/12$$
$$= 390 \text{ pounds}$$

The top and bottom fibre stresses produced in the prestressed plank by the topping are

$$f_2 \quad = -M_p/S$$
$$= -4688/24.5$$
$$= -191 \text{ pounds per square inch, tension}$$
$$f_1 \quad = M_p/S$$
$$= 191 \text{ pounds per square inch, compression}$$

c. PROP REMOVAL

Removing the shoring is equivalent to applying a downward load of R_p to the composite section at midspan. This produces a moment at midspan of

$$
\begin{aligned}
M_R &= R_p \ell/4 \\
&= 390 \times 20 \times 12/4 \\
&= 23{,}400 \text{ pound inches}
\end{aligned}
$$

The resultant stresses produced at the top of the topping and at the bottom of the prestressed plank are

$$
\begin{aligned}
f_{2c} &= M_R/S_{2c} \\
&= 23{,}400/76.4 \\
&= 306 \text{ pounds per square inch, compression} \\
f_1 &= -M_R/S_{1c} \\
&= -23{,}400/67.8 \\
&= -345 \text{ pounds per square inch, tension}
\end{aligned}
$$

d. SUPERIMPOSED LOAD

The moment produced in the prestressed plank at midspan by the live load and superimposed dead load is

$$
\begin{aligned}
M_s &= w\ell^2/8 \\
&= 1.5(75 + 20) \times 20^2 \\
&= 57{,}000 \text{ pound inches}
\end{aligned}
$$

The resultant stresses produced at the top of the topping and at the bottom of the prestressed plank are

$$
\begin{aligned}
f_{2c} &= M_s/S_{2c} \\
&= 57{,}000/76.4 \\
&= 746 \text{ pounds per square inch, compression} \\
f_1 &= -M_s/S_{1c} \\
&= -57{,}000/67.8 \\
&= -841 \text{ pounds per square inch, tension}
\end{aligned}
$$

e. COMBINED STRESSES

At the top of the topping, the final combined stress is

$$
\begin{aligned}
f_{2c} &= 306 + 746 \\
&= 1052 \text{ pounds per square inch, compression}
\end{aligned}
$$

At the bottom of the prestressed plank, the final combined stress is

$$
\begin{aligned}
f_1 &= 56 + 191 - 345 - 841 \\
&= -939 \text{ pounds per square inch, tension}
\end{aligned}
$$

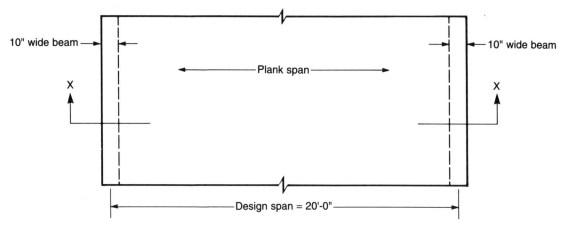

Partial Plan

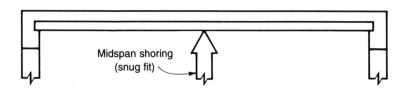

Section X—X

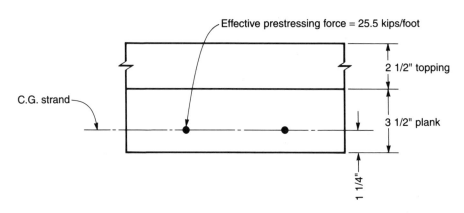

Typical Plank Section

FIGURE 90C-3

STRUCTURAL ENGINEER EXAMINATION – 1987

PROBLEM C-2 – WT. 8.0 POINTS

GIVEN: The post-tensioned concrete beam shown. Assume the beam is hinged at the supports for post-tensioned force calculation purposes.

CRITERIA:

- Concrete: f'_c = 4000 psi
- Prestressing tendon ultimate strength: f_{pu} = 270 ksi _(1862 MPa)_
- Prestressing tendon initial prestress: f_{pi} = 0.70f_{pu}
- Prestressing tendon final prestress: f_{se} = 0.60f_{pu}
- Bonded reinforcement: E_s = 29,000 ksi
            f_y = 60,000 ksi
- Prestressing tendon: ½ inch diameter, area A_t = 0.153 in^2

ASSUME:

BEAM SECTION PROPERTIES:

- Area: A = 704.0 in^2
- Moment of inertia: I = 122,502 in^4
- Section modulus at top of beam: S_t = 6720.8 in^3
- Section modulus at bottom of beam: S_b = 5626.4 in^3

REQUIRED:

1. Calculate the required effective post-tensioning force. Indicate the required number of ½ inch diameter tendons.
2. Calculate the amount of bonded reinforcing required. Indicate the number and size of bonded bars required.
3. What is the required concrete strength at transfer.
4. Calculate the beam deflection due to dead load after allowance for all prestress losses, neglecting bonded reinforcing.
5. Check the beam ultimate moment capacity.

SOLUTION

LOADING

The beam self weight is

$$
\begin{aligned}
w \quad &= \gamma_c A \\
&= 150 \times 704/(144 \times 1000) \\
&= 0.733 \text{ kips per linear foot}
\end{aligned}
$$

The bending moments produced at mid span by the self weight, superimposed dead load, and live load are given by

$$
\begin{aligned}
M_S \quad &= w\ell^2/8 \\
&= 1.5 \times 0.733 \times 45^2 \\
&= 2226 \text{ kip inches} \\
M_L \quad &= 20 \times 15/12 \\
&= 3600 \text{ kip inches} \\
M_D \quad &= 38.4 \times 15/12 \\
&= 6912 \text{ kip inches}
\end{aligned}
$$

SECTION PROPERTIES

The eccentricity of the prestressing force at mid span is

$$
\begin{aligned}
e \quad &= d_p - y_t \\
&= 34 - 18.23 \\
&= 15.77 \text{ inches}
\end{aligned}
$$

Then
$$
\begin{aligned}
R_t \quad &= 1/A - e/S_t \\
&= 1/704 - 15.77/6720.8 \\
&= -0.000926
\end{aligned}
$$

and
$$
\begin{aligned}
R_b \quad &= 1/A + e/S_b \\
&= 1/704 + 15.77/5626.4 \\
&= 0.004223
\end{aligned}
$$

1. REQUIRED PRESTRESSING FORCE

The maximum allowable final compressive stress at the top of the beam, in accordance with UBC Section 1918.4.2, is given by

$$
\begin{aligned}
f_t \quad &= 0.45f_c' \\
&= 0.45 \times 4 \\
&= 1.80 \text{ kips per square inch} \\
&= (M_S + M_L + M_D)/S_t + P_e R_t \\
&= (2226 + 3600 + 6912)/6720.8 - 0.000926 P_e
\end{aligned}
$$

and
$$
\begin{aligned}
P_e \quad &= \text{final prestressing force} \\
&= 103 \text{ kips}
\end{aligned}
$$

The maximum allowable final tensile stress at the bottom of the beam, in accordance with UBC Section 1918.4.2, is given by

$$f_b = -6(f'_c)^{0.5}$$
$$= -6 \times (4000)^{0.5}/1000$$
$$= -0.379 \text{ kips per square inch}$$
$$= -(M_S + M_L + M_D)/S_b + P_e R_b$$
$$= -(2226 + 3600 + 6912)/5626.4 + 0.004223 P_e$$

and
$$P_e = 446 \text{ kips}$$

This value for the effective prestressing force governs and the required number of half inch diameter tendons is

$$n = P_e/A_t f_{se}$$
$$= 446/(0.153 \times 0.6 \times 270)$$
$$= 18 \text{ tendons}$$

The area of prestressed reinforcement is

$$A_{ps} = 18 \times 0.153$$
$$= 2.75 \text{ square inches}$$

2. BONDED REINFORCING

In accordance with UBC Section 1918.4.1, bonded reinforcement is required in the top of the beam if the tensile stress at transfer at mid span exceeds.

$$f_{ti} = 3(f'_{ci})^{0.5}$$

where
$$f'_{ci} = \text{concrete compressive stress at transfer}$$
$$= 3000 \text{ pounds per square inch, assumed.}$$

Then
$$f_{ti} = -3(3000)^{0.5}/1000$$
$$= -0.164 \text{ kips per square inch}$$

The actual stress at transfer is

$$f_{ti} = M_S/S_t + P_i R_t$$
$$= 2226/6720.8 - 18 \times 0.153 \times 0.7 \times 270 \times 0.000926$$
$$= -0.151 \text{ kips per square inch}$$
$$> -0.164$$

Hence bonded reinforcement is not required in the top of the beam at mid span.

At the ends of the beam, no tensile stresses are produced since the location of the tendon coincides with the centroid of the section. Hence no bonded reinforcement is required at the ends of the beam.

In accordance with UBC Section 1918.9.2 the minimum area of bonded reinforcement required at mid span in the bottom of the beam[7] is given by UBC Formula (18–6) as

$$A_{s(min)} = 0.004A$$

where A = concrete area between bottom of beam and section centroid
= $10 \times 20 + 11.77 \times 12$
= 341.2 square inches

Then $A_{s(min)}$ = 0.004×341.2
= 1.365 square inches

In addition[7], sufficient non-bonded reinforcement must be provided to carry all dead load plus twenty five percent of the live load assuming load factors and strength reduction factors of unity. Then the design moment is given by

$$M = M_S + M_D + 0.25M_L$$
$$= 2226 + 6912 + 0.25 \times 3600$$
$$= 10,038 \text{ kip inches}$$

The required reinforcement ratio is given by

$$\rho = 0.85f_c'[1.0 - (1.0 - 2K/0.85f_c')^{0.5}]/f_y$$

where K = M/bd^2
= $10,038/(36 \times 37.5^2)$
= 0.198

and ρ = 0.00341

The required area of bonded reinforcement is then

$$A_s = \rho bd$$
$$= 0.00341 \times 36 \times 37.5$$
$$= 4.60 \text{ square inches}$$

This value governs and may be provided by three No. 11 bars giving an area of

$$A_s = 4.68 \text{ square inches}$$
$$> 4.60$$

3. CONCRETE STRENGTH AT TRANSFER

The maximum allowable compressive stress, at transfer, in the bottom of the beam, in accordance with UBC Section 1918.4.1, is given by

$$f_{bi} = 0.6f_{ci}'$$
$$= -M_S/S_b + P_iR_b$$
$$= -2226/5626 + 18 \times 0.153 \times 0.7 \times 270 \times 0.004223$$
$$= 1.802 \text{ kips per square inch}$$

Hence f_{ci}' = 1.802/0.6
= 3.004 kips per square inch

Similarly the minimum allowable tensile stress at transfer in the top of the beam is

$$f_{ti} = -3(f_c')^{0.5}$$
$$= M_s/S_t + P_iR_t$$
$$= -0.151 \text{ kips per square inch}$$

Hence $\quad f_{ci}' = (151/3)^2$
$$= 2533 \text{ pounds per square inch}$$

Hence the concrete strength at transfer must not be less than

$$f_{ci}' = 3004 \text{ pounds per square inch}$$

4. DEAD LOAD DEFLECTION

The modulus of elasticity of the concrete after all prestress losses have occurred is given by UBC Section 1908.5.1 as

$$E_c = 57(f_c')^{0.5}$$
$$= 57(4000)^{0.5}$$
$$= 3605 \text{ kips per square inch}$$

The prestressing force may be considered equivalent to a balancing load consisting of two concentrated upward forces at the third points of the span, each with a magnitude of

$$W_e = aP_e/x$$
where $\quad a = $ drape of the tendon
$$= 40 - 18.25 - 6$$
$$= 15.75 \text{ inches}$$
and $\quad x = \ell/3$
$$= 45 \times 12/3$$
$$= 180 \text{ inches}$$ _39.1 ^{k}._
Then $\quad W_e = 15.75 \times 446/180$
$$= 41.3 \text{ kips, upward}$$

The net force at the third points is

$$W_n = W_D - W_e \quad 39.1$$
$$= 38.4 - 41.3$$
$$= -2.9 \text{ kips, upward}$$

The initial deflection produced by this net force is

$$\delta_n = 23W_n\ell^3/648EI$$
$$= 23 \times 2.9 \times (45 \times 12)^3/(648 \times 3605 \times 122{,}502)$$
$$= 0.037 \text{ inches, upward}$$

The initial deflection produced by the beam self weight is

$$\delta_s = 22.5 w\ell^4/EI$$
$$= 22.5 \times 0.733 \times 45^4/(3605 \times 122{,}502)$$
$$= 0.153 \text{ inches, downward}$$

Hence the net initial deflection due to the self-weight, dead superimposed load and prestressing force is

$$\delta_i = \delta_s + \delta_n$$
$$= 0.153 - 0.037$$
$$= 0.116 \text{ inches, downward}$$

The multiplier for additional long–term deflection is given by UBC Formula (9–10) as

$$\lambda = \xi/(1 + 50\rho')$$

where ξ = time dependant factor for five years
$$= 2$$

and ρ' = compression reinforcement ratio
$$= 0$$

Then $\lambda = 2$

Hence the final deflection after all prestress losses have occurred is given by

$$\delta_e = \delta_i(\lambda + 1)$$
$$= 0.116(2 + 1)$$
$$= 0.348 \text{ inches}$$

5. ULTIMATE MOMENT CAPACITY

The required factored moment is given by UBC Formula (9–1) as

$$M_u = 1.4(M_S + M_D) + 1.7M_L$$
$$= 1.4(2226 + 6912) + 1.7 \times 3600$$
$$= 18{,}913 \text{ kip inches}$$

Assuming that the compressive stress block, at the nominal strength, is located within the flange, the reinforcement ratio for the prestressed reinforcement is

$$\rho_p = A_{ps}/bd_p$$
$$= 2.75/(36 \times 34)$$
$$= 0.00225$$

Since the span–to–depth ratio of the beam is less than thirty five, the stress in the prestressed tendons, at the nominal strength, is given by UBC Formula (18–4) as

$$
\begin{aligned}
f_{ps} &= f_{se} + 10 + f_c'/100\rho_p \\
&= 0.6 \times 270 + 10 + 4/(100 \times 0.00225) \\
&= 189.78 \text{ kips per square inch} \\
&< f_{py} \quad (= 0.9 \times 270 = 243 \text{ ksi ... satisfactory}) \\
&< f_{se} + 60 \quad (= 222 \text{ ksi ... satisfactory})
\end{aligned}
$$

Then the depth of the equivalent rectangular stress block is given by

$$
\begin{aligned}
a &= (A_{ps}f_{ps} + A_sf_y)/0.85f_c'b \\
&= (2.75 \times 189.78 + 4.68 \times 60)/(0.85 \times 4 \times 36) \\
&= 6.56 \text{ inches} \\
&> h_f
\end{aligned}
$$

Hence the depth of the stress block exceeds the flange thickness.

The reinforcement ratio for the prestressed reinforcement referred to the web width is

$$
\begin{aligned}
\rho_{pw} &= A_{ps}/b_w d_p \\
&= 2.75/(12 \times 34) \\
&= 0.00675
\end{aligned}
$$

The revised stress in the prestressed tendons, at the nominal strength, is

$$
\begin{aligned}
f_{ps} &= f_{se} + 10 + f_c'/100\rho_{pw} \\
&= 0.6 \times 270 + 10 + 4/(100 \times 0.00675) \\
&= 177.93 \text{ kips per square inch}
\end{aligned}
$$

Then the area of reinforcement required to develop the compressive strength of the flange is

$$
\begin{aligned}
A_{pf} &= 0.85f_c'h_f(b - b_w)/f_{ps} \\
&= 0.85 \times 4 \times 6(36 - 12)/177.93 \\
&= 2.75 \text{ square inches}
\end{aligned}
$$

Hence the area of prestressed reinforcement available to develop the web is

$$
\begin{aligned}
A_{pw} &= A_{ps} - A_{pf} \\
&= 2.75 - 2.75 \\
&= 0
\end{aligned}
$$

Then the reinforcement index is derived from UBC Section 1918.8.1 as

$$
\begin{aligned}
R_i &= \omega_{pw} + d(\omega_w - \omega_w')/d_p \\
&= 0 + d(\omega_w - 0)/d_p \\
&= dA_sf_y/f_c'b_w dd_p \\
&= 4.68 \times 60/(4 \times 12 \times 34) \\
&= 0.172
\end{aligned}
$$

The limiting reinforcement index to ensure an under reinforced section is

$$R_i = 0.36\beta_1$$
$$= 0.36 \times 0.85$$
$$= 0.306$$
$$> 0.172$$

Hence the section is under reinforced with all reinforcement contributing to the nominal strength to give a stress block depth of

$$a = A_s f_y / 0.85 f'_c b_w$$
$$= 4.68 \times 60 / (0.85 \times 4 \times 12)$$
$$= 6.88 \text{ inches}$$

Then the design moment strength is given by

$$\phi M_n = \phi[A_{ps}f_{ps}(d_p - h_f/2) + A_s f_y(d - a/2)]$$
$$= 0.9[2.75 \times 177.93(34 - 3) + 4.68 \times 60(37.5 - 3.44)]$$
$$= 22,259 \text{ kip inches}$$
$$> M_u \ldots \text{ satisfactory}$$

The modulus of rupture of the concrete is obtained from UBC Formula (9–9) as

$$f_r = 7.5(f'_c)^{0.5}$$
$$= 7.5(4000)^{0.5}/1000$$
$$= 0.474 \text{ kips per square inch}$$

The cracking moment is that moment which, when applied to the beam after all losses have occurred, will cause cracking at the bottom of the beam. The cracking moment is, then

$$M_{cr} = S_b(P_e R_b + f_r)$$
$$= 5626.4(446 \times 0.004223 + 0.474)$$
$$= 13,264 \text{ kip inches}$$

To provide adequate warning of impending failure, UBC Section 1918.8.3 requires the flexural strength to be at least

$$\phi M_{n(min)} = 1.2 M_{cr}$$
$$= 1.2 \times 13,264$$
$$= 15,917 \text{ kip inches}$$
$$< \phi M_n \ldots \text{ satisfactory}$$

Hence the beam ultimate moment capacity is adequate.

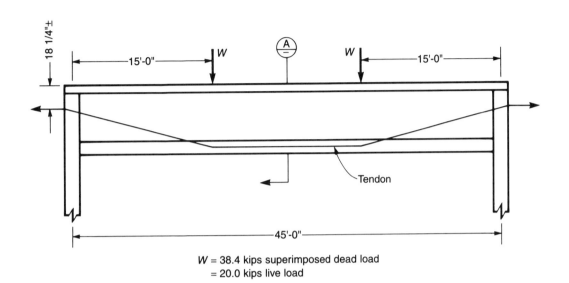

W = 38.4 kips superimposed dead load
= 20.0 kips live load

Post-Tensioned Concrete Beam Elevation

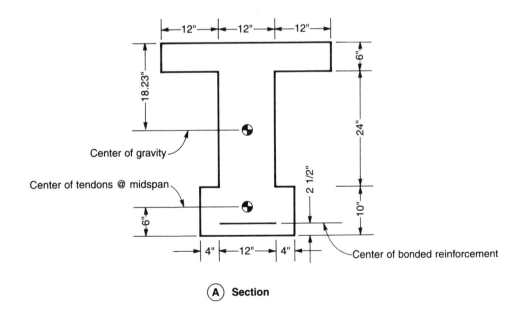

Ⓐ **Section**

FIGURE 87C–2

CHAPTER 3

REFERENCES

1. Cross, H. *The Analysis of continuous frames by distributing fixed end moments.* Transactions American Society of Civil Engineers, Vol 96, 1932. pp 1 – 56.

2. Williams, A. *The analysis of indeterminate structures.* Hart Publishing Company Inc, New York, 1968. pp 114 – 184.

3. American Concrete Institute. *Commentary on Building Code Requirements for Reinforced Concrete (ACI 318 – 83),* Section 10.3. Detroit, MI, 1985.

4. American Concrete Institute. *Commentary on Building Code Requirements for Reinforced Concrete (ACI 318 – 83),* Section 11.11.5. Detroit, MI, 1985.

5. American Concrete Institute and Structural Engineers Association of Southern California. *Report of the task committee on slender walls.* Los Angeles, CA, 1982.

6. Cole, E.E. et al. *Seismic design examples in seismic zones 4 and 2A.* Concrete Reinforcing Steel Institute. Schaumburg, IL, 1993.

7. Bondy, K. *Mild steel required in unbonded prestressed concrete.* SEAOSC Design Seminar. Los Angeles, CA, December 1991.

4

STRUCTURAL TIMBER DESIGN

SECTION 4.1

TIMBER BEAMS

PROBLEMS:

1990 D–3
1988 D–1

STRUCTURAL ENGINEER EXAMINATION – 1990 ══════════════

PROBLEM D-3 – WT. 7.0 POINTS

GIVEN: A two story residential building with framing and gravity loads as shown.

CRITERIA: 90 mph wind zone, exposure C.
Show all calculations.

Materials:
* Joists: Douglas Fir–Larch select structural grade.
* Header H: 5⅛ inch visually graded western species glulam combination 24F-V4 with 1½ inch laminations.

Assumptions:
* Dry condition of use.
* No reduction of live load for tributary area.
* Ignore wind uplift.

REQUIRED:
1. For the notched 2 × 12 joists, determine the following:
   a. The maximum and minimum reactions at each support. (Describe connection requirements at line 2)
   b. The maximum shear and bending stresses in each span and compare with the allowable stresses.
2. For the 5⅛ inch wide glulam header H determine the least standard depth allowed by the UBC. State the required load combinations and show all calculations for flexure, shear and vertical deflection.

SOLUTION

1. NOTCHED JOIST

REACTIONS, SHEARS AND MOMENTS

Dead and live load over both the cantilever and the beam span produce maximum shear and moment in the cantilever, maximum reaction at line 1 and maximum shear in the beam span. These values are given by

$$V_c = (30 + 85 \times 6)16/12$$
$$= 720 \text{ pounds}$$
$$M_c = (30 \times 6 + 85 \times 6^2/2)16/12$$
$$= 2280 \text{ pounds feet}$$
$$R_{1(max)} = (30 \times 20 + 85 \times 6 \times 17 + 50 \times 14 + 50 \times 14^2/2)16/(14 \times 12)$$
$$= 1416 \text{ pounds}$$

$$V_B = R_{1(max)} - V_C - 50 \times 16/12$$
$$= 1416 - 720 - 66$$
$$= 630 \text{ pounds.}$$

Dead load only over both the cantilever and beam span produces the minimum reaction at line 1 which is given by

$$R_{1(min)} = (30 \times 20 + 25 \times 6 \times 17 + 50 \times 14 + 10 \times 14^2/2)16/(14 \times 12)$$
$$= 460 \text{ pounds}$$

Dead load and live load on the cantilever and dead load only on the beam span produce the minimum reaction at line 2 which is given by

$$R_{2(min)} = (30 \times 6 + 85 \times 6^2/2 - 10 \times 14^2/2)16/(14 \times 12)$$
$$= -70 \text{ pounds, uplift}$$

Dead load only on the cantilever and dead load plus live load on the beam span produce maximum reaction at line 2 and maximum positive moment in the beam span. These values are given by

$$R_{2(max)} = (30 \times 6 + 25 \times 6^2/2 - 50 \times 14^2/2)16/(14 \times 12)$$
$$= 407 \text{ pounds, downwards}$$

Hence a joist hanger is required at line 2 with an uplift capacity of 70 pounds and a bearing capacity of 407 pounds.

The maximum sagging moment occurs in the beam span at a distance x from line 2 and is given by

$$M = 407x - 50 \times 16x^2/24$$

Differentiating and equating dM/dx to zero gives

$$x = 6.1 \text{ feet}$$

Hence the maximum positive moment is

$$M = 1242 \text{ pounds feet}$$

Then the maximum negative moment at line 1 governs the flexural design of the beam span and is given by

$$M_B = M_C$$
$$= 2280 \text{ pounds feet}$$

CANTILEVER STRESSES

The relevant properties of the 1½ inches notched section are

$$d' = d - 1.5$$
$$= 11.25 - 1.5$$
$$= 9.75 \text{ inches}$$
$$b = 1.5 \text{ inches}$$

$$A \quad = bd'$$
$$\quad = 14.63 \text{ inches}^2$$
$$S \quad = Ad'/6$$
$$\quad = 23.77 \text{ inches}^3$$

The tabulated stresses in shear and bending are obtained from UBC Table 23-I-A-1 and are

$$F_v \quad = 95 \text{ pounds per square inch}$$
$$F_b \quad = 1450 \text{ pounds per square inch}$$

The size factor is specified in UBC Table 23-I-A-1 footnote No.3 as

$$C_F \quad = 1.1 \text{ ... for a ten inches wide member}$$

The repetitive member factor C_r is not applicable at a support.

The actual bending stress is given by

$$f_b \quad = M_C/S$$
$$\quad = 2280 \times 12/23.77$$
$$\quad = 1151$$
$$\quad < F_b C_F \text{ ... satisfactory}$$

The 1½ inch notch is on the tension side of the beam and, in accordance with UBC Section 2306.4 Formula (6-2) the allowable shear force is

$$V_A \quad = (d'/d)2bd'F_v/3$$
$$\quad = (9.75/11.25)2 \times 1.5 \times 9.75 \times 95/3$$
$$\quad = 803 \text{ pounds}$$
$$\quad > V_C \text{ ... satisfactory}$$

Hence the cantilever stresses are satisfactory.

BEAM SPAN STRESSES

By inspection, the beam span stresses are satisfactory.

2. DEPTH OF HEADER BEAM

LOAD COMBINATIONS

The applicable load combinations are given in UBC Section 1603.6 and allowable increases in stress are given in UBC Section 2304.3.4 as

- Vertical dead plus floor live load without any increase in permissible stresses.
- Dead plus floor live load plus lateral wind load, with an increase of one-third in permissible stress for connections and fasteners, and an increase of sixty percent in permissible stress for members.

APPLIED LOADING

The applied dead load plus live load is given by

$$w_T = 50 + 12R_{1(max)}/16$$
$$= 50 + 12 \times 1416/16$$
$$= 1112 \text{ pounds per linear foot}$$

The applied live load is given by

$$w_L = (60 \times 6 \times 17 + 40 \times 14^2/2)/14$$
$$= 717 \text{ pounds per linear foot}$$

In accordance with UBC Formula (18–1) the design wind pressure is given by

$$p = C_e C_q q_s I_w$$
$$C_e = 1.06 \dots \text{ for exposure C at a height of ten feet, from UBC Table 16–G}$$
$$C_q = 1.2 \dots \text{ for a wall element, from UBC Table 16–H}$$
$$q_s = 20.8 \text{ pounds per square foot} \dots \text{ for a ninety miles per hour wind speed,}$$
$$\text{from UBC Table 16–F}$$
$$I_w = 1.0 \dots \text{ for the importance factor given in UBC Table 16–K}$$

Then
$$p = 1.06 \times 1.2 \times 20.8 \times 1.0$$
$$= 26.46 \text{ pounds per square foot}$$

Thus for a tributary width of

$$t = h/2$$
$$= 10/2$$
$$= 5 \text{ feet}$$

the wind load on the header beam is

$$w_W = pt$$
$$= 26.46 \times 5$$
$$= 132.3 \text{ pounds per linear foot}$$

DEAD PLUS LIVE LOAD

The bending moment and shear force in the header beam due to the dead plus live gravity load are given by

$$M_x = w_T \ell^2/8$$
$$= 1112 \times 18^2/8$$
$$= 45,036 \text{ pounds feet}$$
$$V_x = w_T(\ell - 2d)/2$$
$$= 1112(18 - 3.5)/2 \dots \text{ at a distance } d \text{ from the support (UBC Section 2306.3)}$$
$$= 8062 \text{ pounds}$$

Neglecting lateral instability, a 5.125×18 glulam beam is adequate for these applied loads. However, the compression side of the beam is unsupported, and to account for lateral instability effects and the lateral wind load, a 5.125×21 beam will be selected.

The relevant properties of this beam are

$$
\begin{aligned}
A &= 107.6 \text{ inches}^2 \\
I_x &= 3955 \text{ inches}^4 \\
S_x &= 376.7 \text{ inches}^3 \\
S_y &= 91.9 \text{ inches}^3 \\
\ell_u/d &= 18 \times 12/21 \\
&= 10.3 \\
&> 7
\end{aligned}
$$

The relevant tabulated stresses are obtained from UBC Table 23–I–C–1 and are

$$
\begin{aligned}
F_{bx} &= 2400 \text{ pounds per square inch} \\
F_{vx} &= 165 \\
F_{by} &= 1500 \\
F_{vy} &= 145 \\
E_x &= 1.8 \times 10^6
\end{aligned}
$$

For a single span beam with a uniformly distributed applied load, the effective unbraced length is given by UBC Section 2363.1 as

$$
\begin{aligned}
\ell_e &= 1.63\ell_u + 3d \ ... \text{ for } \ell_u/d > 7 \\
&= 1.63 \times 18 + 3 \times 21/12 \\
&= 34.59 \text{ feet}
\end{aligned}
$$

The allowable stress, adjusted for stability or volume factors, is obtained from calculator program A.3.2 in Appendix A.

The slenderness factor is given by UBC Section 2363.1 as

$$
\begin{aligned}
R_B &= (\ell_e d/b^2)^{0.5} \\
&= 18.22 \\
&< 50 \ ... \text{ satisfactory}
\end{aligned}
$$

The critical buckling design value is given by UBC Section 2363.2 as

$$ F_{bE} = K_{bE}E'/R_B^2 $$

where K_{bE} = Euler buckling coefficient
= 0.609 ... for glued–laminated timber

E' = adjusted tabulated modulus of elasticity
= 1.8×10^6 pounds per square inch

and F_{bE} = 3303 pounds per square inch

The volume factor is given by UBC Section 2312.4.5 as

$$ C_V = k(1291.5/bdL)^{1/x} $$

where k = loading condition coefficient
= 1.00 ... for uniformly distributed load

x = 10 ... for western species timber

and C_V = 0.96

The beam stability factor is given by UBC Section 2363.2 as

$$C_L = (1.0 + F)/1.9 - \{[(1.0 + F)/1.9]^2 - F/0.95\}^{0.5}$$

where
$F_b^* = $ adjusted tabulated bending stress
$\qquad = 2400$ pounds per square inch
$F = F_{bE}/F_b^*$
$\qquad = 1.38$

and
$C_L = 0.91$
$\qquad < C_V$

Then slenderness effects govern and the allowable flexural stress is

$$F_b' = F_b^* C_L$$
$$\qquad = 2186 \text{ pounds per square inch}$$

The actual stress in the header beam is

$$f_{bx} = M_x/S_x$$
$$\qquad = 45{,}036 \times 12/376.7$$
$$\qquad = 1435 \text{ pounds per square inch}$$
$$\qquad < F_b'$$

The shear stress in the header beam is given by

$$f_{vx} = 1.5 V_x/A$$
$$\qquad = 1.5 \times 8062/107.6$$
$$\qquad = 112 \text{ pounds per square inch}$$
$$\qquad < F_{vx}$$

Hence the header beam is satisfactory under dead plus live gravity loads.

WIND LOAD

The bending moment and shear force in the header beam due to the transverse wind loading are given by

$$M_y = w_w \ell^2/8$$
$$\qquad = 132.3 \times 18^2/8$$
$$\qquad = 5362 \text{ pounds feet}$$
$$V_y = w_w(\ell - 2d)/2$$
$$\qquad = 132.3(18 - 0.85)/2$$
$$\qquad = 1134 \text{ pounds feet}$$

The transverse flexural stress in the header beam is

$$f_{by} = M_y/S_y$$
$$\qquad = 12 \times 5362/91.9$$
$$\qquad = 700 \text{ pounds per square inch}$$
$$\qquad < F_{by} \times 1.60 \ldots \text{ UBC Section 2304.3.4}$$

The transverse shear stress in the header beam is

$$
\begin{aligned}
f_{vy} \quad &= 1.5 V_y/A \\
&= 1.5 \times 1134/107.6 \\
&= 16 \text{ pounds per square inch} \\
&< F_{vy} \times 1.60
\end{aligned}
$$

COMBINED DEAD, LIVE AND WIND LOAD

The combined flexural stresses due to vertical and transverse loading, after allowing for the sixty percent increase in permissible stress, must satisfy the interaction equation

$$ f_{bx}/F'_{bx} + f_{by}/F_{by} \le 1.0 \times 1.60 $$

The left hand side of the expression is evaluated as

$$
\begin{aligned}
1435/2113 + 700/1500 \quad &= 1.15 \\
&< 1.60
\end{aligned}
$$

Hence the header beam is satisfactory under the combined loading.

VERTICAL DEFLECTION

From UBC Table 16–D, the allowable deflection due to live load only is given by

$$ \Delta_L \quad = \ell/360 $$

The actual deflection due to the live load is

$$
\begin{aligned}
\delta_L \quad &= 22.5 w_L \ell^4/EI \\
&= 22.5 \times 717 \times 18^4/(1.8 \times 10^6 \times 3955) \\
&= 0.238 \text{ inches} \\
&= \ell/908 \\
&< \Delta_L \ \dots \text{ satisfactory}
\end{aligned}
$$

Adopting a conservative value for K, from UBC Table 16–E, of unity, the allowable deflection due to dead load plus live load is

$$ \Delta_T \quad = \ell/240 $$

The actual deflection due to the dead plus live load is

$$
\begin{aligned}
\delta_T \quad &= 22.5 w_T \ell^4/EI \\
&= 22.5 \times 1112 \times 18^4/(1.8 \times 10^6 \times 3955) \\
&= 0.369 \text{ inches} \\
&= \ell/585 \\
&< \Delta_T \ \dots \text{ satisfactory}
\end{aligned}
$$

Hence the deflection limitations are satisfied.

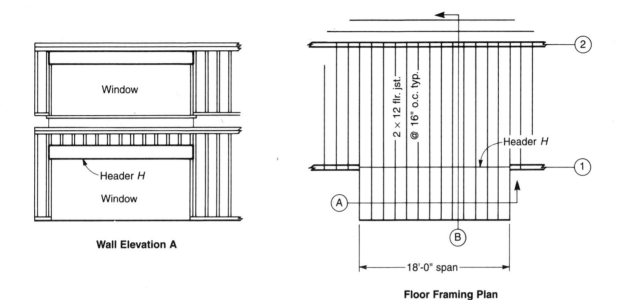

Wall Elevation A

Floor Framing Plan

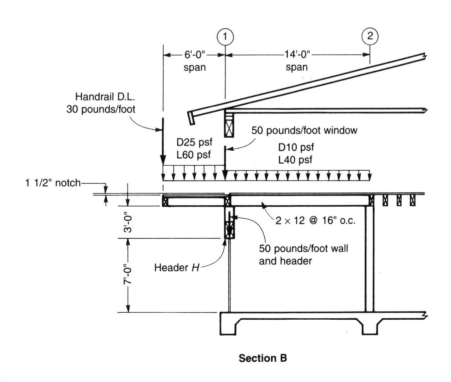

Section B

FIGURE 90D-3

399

STRUCTURAL ENGINEER EXAMINATION – 1988

PROBLEM D-1 – WT. 4.0 POINTS

GIVEN: An existing 2 story wood frame office building is proposed to be converted to a restaurant and dining room, as shown in the sectional elevation.
- Floor LL = assembly occupancy, (no fixed seating)
- Floor DL = 16.8 psf (beam weight not included)

CRITERIA: Joist design span length is 15′
Beam design span length is 14′ – 8″

Materials:
- Lumber is Douglas Fir-Larch.
- Joists are select structural grade.
- Beams are select structural grade.
- Posts are dense No.1 Grade.

Assumptions:
- Check for vertical loading only.
- The top of the column is adequately braced.
- The length of the split on the wide face of the beam is 9½ inches.

REQUIRED: 1. Check the existing 2 × 12 floor joists at 16 inches on center.
              2. Design a sawn wood beam to replace the existing first floor bearing wall.
              3. Design the wood posts to support the new beams. The owner wants to provide carved decorations of 9/16 inch maximum depth into all faces of the column, except for the top 24 inches.
              4. List two items of concern when using a large sawn wood beam as above.

SOLUTION

1. FLOOR JOISTS

ALLOWABLE STRESSES

The tabulated stresses for 2 × 12 Douglas Fir-Larch select structural grade lumber are obtained from UBC Table 23-I-A-1. The joists are at sixteen inches on center and the repetitive member factor C_r with a value of 1.15 is applicable, as specified in UBC Table 23-I-A-1 footnote 4. The size factor is specified in UBC Table 23-I-A-1 footnote No.3 as

$$C_F \quad = 1.00 \text{ ... for a twelve inch wide member}$$

The relevant tabulated stresses are

$$F_b \quad = 1450 \text{ pounds per square inch}$$
$$F_v \quad = 95 \text{ pounds per square inch}$$
$$E_x \quad = 1.9 \times 10^6 \text{ pounds per square inch}$$

400

APPLIED LOADING

The live loading for assembly occupancy with movable seating is given in UBC Table 16–A as

$$w_L \quad = 100.0 \quad \text{pounds per square foot}$$
$$w_D \quad = \underline{16.8} \quad \text{given dead load}$$
$$w_T \quad = 116.8 \quad \text{total load}$$

The total load applied to each joist is

$$w \quad = w_T \times 16/12$$
$$= 155.34 \text{ pounds per linear foot}$$

SECTION PROPERTIES

The relevant properties of a 2×12 sawn lumber joist are

$$A \quad = 16.9 \text{ inches}^2$$
$$S \quad = 31.6 \text{ inches}^3$$
$$I \quad = 178 \text{ inches}^4$$

MEMBER STRESSES

The flexural stress at mid span and the shear stress a distance d from the support are given by

$$f_b \quad = 1.5w\ell^2/S$$
$$= 1.5 \times 155.34 \times 15^2/31.6$$
$$= 1659 \text{ pounds per square inch}$$
$$< F_b C_r \dots \text{ satisfactory}$$
$$f_v \quad = 1.5w(\ell-2d)/2A$$
$$= 1.5 \times 155.34(15 - 2 \times 11.25/12)/(2 \times 16.9)$$
$$= 91 \text{ pounds per square inch}$$
$$< F_v \dots \text{ satisfactory}$$

DEFLECTION

From UBC Table 16–D, the allowable deflection due to live load only is given by

$$\Delta_L \quad = \ell/360$$

The actual deflection due to the live load is

$$\delta_L \quad = 22.5 \times 16w_L\ell^4/12E_xI$$
$$= 22.5 \times 16 \times 100 \times 15^4 /(12 \times 1.9 \times 10^6 \times 178)$$
$$= 0.447 \text{ inches}$$
$$= \ell/402$$
$$< \Delta_L \dots \text{ satisfactory}$$

Adopting a conservative value for K from UBC Table 16–E of unity, the allowable deflection due to dead load plus live load is

$$\Delta_T \quad = \ell/240$$

401

The actual deflection due to the dead plus live load is

$$
\begin{aligned}
\delta_T &= 22.5 w \ell^4 / E_x I \\
&= 22.5 \times 155.34 \times 15^4 / (1.9 \times 10^6 \times 178) \\
&= 0.523 \text{ inches} \\
&= \ell / 344 \\
&< \Delta_T \dots \text{satisfactory}
\end{aligned}
$$

Hence the 2×12 joists are satisfactory.

2. WOOD BEAM

ALLOWABLE STRESSES

The tabulated stresses for Douglas Fir-Larch select structural grade lumber are obtained from UBC Table 23-I-A-4. Footnote No.5 of this Table indicates that the specified shear stress may be increased by one-third when the length of split on the wide face of the beam equals the width of the narrow face. The relevant stresses are

$$
\begin{aligned}
F_b &= 1600 \text{ pounds per square inch} \\
F_v &= 85 \times 1.33 = 113 \text{ pounds per square inch} \\
E_x &= 1.6 \times 10^6 \text{ pounds per square inch}
\end{aligned}
$$

APPLIED LOADING

In accordance with UBC Section 1606, no reduction is allowable in the live loading for assembly occupancy and the live load acting is

$$
\begin{aligned}
w_L &= 100 \times 15 \\
&= 1500 \text{ pounds per linear foot}
\end{aligned}
$$

Assuming the beam self weight is 36 pounds per foot, the dead load acting is

$$
\begin{aligned}
w_D &= 16.8 \times 15 + 36 \\
&= 288 \text{ pounds per linear foot}
\end{aligned}
$$

The applied dead load plus live load is, then

$$
\begin{aligned}
w &= w_L + w_D \\
&= 1788 \text{ pounds per linear foot}
\end{aligned}
$$

REQUIRED SECTION PROPERTIES

The top of the beam is braced by the floor joists and no slenderness factor adjustment is necessary. Using a 16 inch deep beam with a width exceeding five inches, the size factor is given by UBC Table 23-I-A-4 footnote 3 as

$$
\begin{aligned}
C_F &= (12/d)^{1/9} \\
&= (12/15.5)^{1/9} \\
&= 0.97
\end{aligned}
$$

The required section modulus is given by

$$S = 1.5w\ell^2/F_bC_F$$
$$= 1.5 \times 1788 \times 14.67^2/(1600 \times 0.97)$$
$$= 371 \text{ inches}^3$$

The required area is given by

$$A = 1.5 \ w(\ell-2d)/2F_v$$
$$= 1.5 \times 1788(14.67 - 2 \times 15.5/12)/(2 \times 113)$$
$$= 143 \text{ inches}^2$$

The required moment of inertia is given by

$$I = 360 \times 22.5w_L\ell^4/12E_x\ell$$
$$= 30 \times 22.5 \times 1500 \times 14.67^3/1.6 \times 10^6$$
$$= 1998 \text{ inches}^4$$

Hence a 10×16 beam is required and this has the following properties

$$A = 147 \text{ inches}^2$$
$$S = 380 \text{ inches}^3$$
$$I = 2948 \text{ inches}^4$$
$$w = 36 \text{ pounds per linear foot}$$

3. WOOD POST

ALLOWABLE STRESSES

The size factor for a post not exceeding twelve inches in width is given by UBC Table 23-I-A-4 footnote 3 as $C_F = 1.00$. The relevant tabulated stresses for a Douglas Fir-Larch dense No.1 Grade post are obtained from UBC Table 23-I-A-4 and are

$$F_c = F_c^* = 1400 \text{ pounds per square inch}$$
$$E = E' = 1.7 \times 10^6 \text{ pounds per square inch}$$

APPLIED LOADING

The combined dead load plus live load is given by

$$P = 1788 \times 14.67$$
$$= 26{,}230 \text{ pounds}$$

REQUIRED SECTION

Assuming an 8×10 post is required, and deducting $9/16$ inch from all faces for the carved decorations, the relevant properties are

$$d_1 = 7.5 - 2 \times 9/16 = 6.38 \text{ inches}$$
$$d_2 = 9.5 - 2 \times 9/16 = 8.38 \text{ inches}$$
$$A = d_1 \times d_2 = 53.43 \text{ inches}^2$$

The effective unbraced length about both axes, in accordance with UBC Section 2307.3 is

$$\ell_e = K_e\ell$$
$$= 1.0 \times 14.46$$
$$= 14.46 \text{ feet}$$

The slenderness ratio about the weak axis is

$$\ell_e/d = 14.46 \times 12/6.38$$
$$= 27.2$$
$$< 50 \ldots \text{satisfactory}$$

The relevant design factors are obtained from calculator program A.3.3 in Appendix A. The critical buckling design value is given by UBC Section 2307.3 as

$$F_{cE} = K_{cE}E'/(\ell_e/d)^2$$

where
$$K_{cE} = \text{Euler buckling coefficient for columns}$$
$$= 0.30 \ldots \text{for visually graded lumber}$$
$$E' = \text{adjusted tabulated modulus of elasticity}$$
$$= 1.7 \times 10^6 \text{ pounds per square inch}$$

and
$$F_{cE} = 689 \text{ pounds per square inch}$$

The column stability factor is given by UBC Section 2307.3 as

$$C_P = (1.0 + F)/2c - \{[(1.0+ F)/2c]^2 - F/c\}^{0.5}$$

where
$$F_c^* = \text{adjusted tabulated compressive stress}$$
$$= 1400 \text{ pounds per square inch}$$
$$F = F_{cE}/F_c^*$$
$$= 0.49$$
$$c = \text{column parameter}$$
$$= 0.8 \ldots \text{for sawn lumber}$$

and
$$C_P = 0.43$$

The allowable compressive stress is

$$F_c' = F_c^*C_P$$
$$= 600 \text{ pounds per square inch}$$

The allowable load is then

$$P' = F_c' \times A$$
$$= 600 \times 53.43$$
$$= 32,058 \text{ pounds}$$
$$> P$$

Hence the 8 × 10 section is adequate.

4. LARGE BEAM PROBLEMS

Some problems associated with large sawn members are
    (i) Long term shrinkage.
    (ii) Out of straightness and warping.
    (iii) Availability.

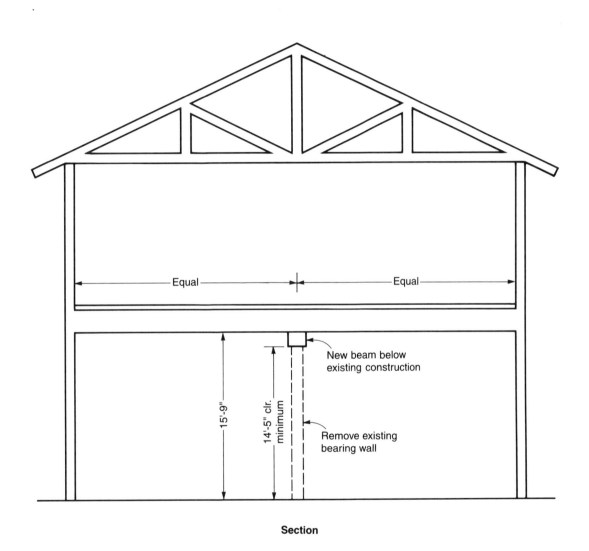

Section

FIGURE 88D–1

STRUCTURAL ENGINEERING REGISTRATION

CHAPTER 4

STRUCTURAL TIMBER DESIGN

SECTION 4.2

TIMBER COLUMNS

PROBLEMS:

1988 D–2
1987 D–5

STRUCTURAL ENGINEER EXAMINATION – 1988 ════════════════

PROBLEM D-2 – WT. 5.0 POINTS

<u>GIVEN</u>:   A wood column located on an interior bay of the exterior face of the first floor of a three story wood framed office building, as shown in Section A in Figure 88D-2(i).

<u>CRITERIA</u>:   Duration of load factor for snow is 1.15
Wind conditions – 80 mph exposure C
$C_q = 1.6$ (partially enclosed structure)

Materials:
- Wood column tabulated stresses: $F_b = 1050$ psi, $F_v = 70$ psi, $F_c = 750$ psi and $E = 1,300,000$ psi.
- Glued-laminated wood beams: combination 24F-V4 visually graded western species.
- Steel: Grade ASTM A-36.
- Bolts: ¾ inch diameter A-307 or Lag bolts, typically.

Assumptions:
- Do not consider seismic loads.
- Top of column is adequately braced in the plane of the wall.

<u>REQUIRED</u> 1. Design the smallest 6 × wood column below the second floor that will carry the appropriate UBC prescribed load combinations.
       2. Calculate the required length of bearing and bearing plate thickness for the connection at the top of the first floor column shown in Section A.
       3. Complete Section B showing missing bolts and dimensions.

SOLUTION

1. WOOD COLUMN

LOAD COMBINATIONS

The applicable load combinations are given in UBC Section 1603.6 and the appropriate stress adjustments for duration of load are given in UBC Sections 2304.3.4 and 2308.2. The combinations are

- Dead load plus floor live load without any increase in permissible stress.
- Dead plus floor live plus snow load with a 15 percent increase in permissible stress.
- Dead plus floor live plus lateral wind load with an increase of one–third in permissible stress for connections and fasteners and an increase of sixty percent in permissible stress for members.
- Dead plus floor live plus snow plus fifty percent wind load with the same increase in permissible stress as for the above case.
- Dead plus floor live plus wind plus fifty percent snow load with the same increase in permissible stress as for the above case.

DEAD PLUS LIVE LOAD

The combined dead plus live load is given by

$$P = 6250 + 27{,}500 + 18{,}000$$
$$= 51{,}750 \text{ pounds}$$

The load duration factor is

$$C_D = 1.00$$

Assuming that an 18 inches wide column is required, the size factor is given by UBC Table 23–I–A–4 footnote 3 as

$$C_F = (12/d)^{1/9}$$
$$= 0.96$$

The effective unbraced length about both axes, in accordance with UBC Section 2307.3, is given by

$$\ell_e = K_e \ell$$
$$= 1.0 \times 9.5$$
$$= 9.5 \text{ feet}$$

The slenderness ratio about the weak axis is

$$\ell_e/b = 9.5 \times 12/5.5$$
$$= 20.7$$
$$< 50 \ldots \text{ satisfactory}$$

The relevant design factors are obtained from calculator program A.3.3 in Appendix A. The critical buckling design value is given by UBC Section 2307.3 as

$$F_{cE} = K_{cE}E'/(\ell_e/d)^2$$

where
K_{cE} = Euler buckling coefficient for columns
= 0.30 ... for visually graded lumber
E' = adjusted tabulated modulus of elasticity
= 1.3×10^6 pounds per square inch

and
F_{cE} = 910 pounds per square inch

The column stability factor is given by UBC Section 2307.3 as

$$C_P = (1.0 + F)/2c - \{[(1.0 + F)/2c]^2 - F/c\}^{0.5}$$

where
F_c^* = adjusted tabulated compressive stress
= $F_C C_D C_F$
= $750 \times 1.0 \times 0.96$
= 720 pounds per square inch

F = F_{cE}/F_c^*
= 1.26

c = column parameter
= 0.8 ... for sawn lumber

and
C_P = 0.77

The allowable compressive stress is

F_c' = $F_c^* C_P$
= 551 pounds per square inch

The required area of the column is given by

A = P/F_c'
= 51,750/551
= 93.9 square inches

DEAD, LIVE AND SNOW LOAD

The combined dead, live and snow load is given by

P = 51,750 + 6250
= 58,000 pounds

The load duration factor is

C_D = 1.15

409

The relevant design values are

$$F_{cE} = K_{cE}E'/(\ell_e/d)^2$$
$$= 910 \text{ pounds per square inch}$$
$$C_P = (1.0 + F)/2c - \{[(1.0+ F)/2c]^2 - F/c\}^{0.5}$$

where F_c^* = adjusted tabulated compressive stress
$$= F_C C_D C_F$$
$$= 750 \times 1.15 \times 0.96$$
$$= 828 \text{ pounds per square inch}$$
$$F = F_{cE}/F_c^*$$
$$= 1.10$$

and $C_P = 0.72$

The allowable compressive stress is

$$F_c' = F_c^* C_P$$
$$= 598 \text{ pounds per square inch}$$

The required area of the column is given by

$$A = P/F_c'$$
$$= 58,000/598$$
$$= 96.9 \text{ square inches, which governs}$$
$$> 93.9 \text{ square inches}$$

Hence a 6×18 column is required and this has the following properties

$$A = 96.3 \text{ inches}^2$$
$$S = 281 \text{ inches}^3$$
$$\ell_u/d = 9.5 \times 12/17.5$$
$$= 6.5$$
$$< 7$$

DEAD, LIVE, SNOW, AND 50 PERCENT WIND

In accordance with UBC Formula (18–1) the design wind pressure is given by

$$p = C_e C_q q_s I_w$$
C_e = 1.31 ... for exposure C, an outward acting force, and a mean roof height of between 30 and 40 feet, from UBC Table 1623–G.
C_q = 1.6 ... outwards, given
q_s = 16.4 pounds per square foot ... for an eighty mile per hour wind speed, from UBC Table 16–F.
I_w = 1.0 ... for the importance factor given in UBC Table 16–K

Then p = $1.31 \times 1.6 \times 16.4 \times 1.0$
$$= 34.37 \text{ pounds per square foot}$$

For a post spacing of five feet, the distributed load on the post due to 50 percent of the wind pressure is

$$w = 5p/2$$
$$= 86.0 \text{ pounds per linear foot}$$

The bending moment and shear force acting on the post are

$$M = w\ell^2/8$$
$$= 1.5 \times 86.0 \times 9.5^2$$
$$= 11,634 \text{ pound inches}$$
$$V = w(\ell - 2d)/2$$
$$= 86.0(9.5 - 2 \times 15.5/12)/2$$
$$= 297 \text{ pounds}$$

The shear stress in the post, for 50 percent of the full wind pressure, is given by

$$f_v = V/A$$
$$= 297/96.3$$
$$= 3 \text{ pounds per square inch}$$
$$< F_v C_D \text{ ... satisfactory}$$

The bending stress in the post, for 50 percent of the full wind pressure, is

$$f_b = M/S$$
$$= 11,634/281$$
$$= 41 \text{ pounds per square inch}$$

The axial stress in the post is given by

$$f_c = P/A$$
$$= 58,000/96.3$$
$$= 602 \text{ pounds per square inch}$$

The load duration factor is

$$C_D = 1.60$$

The relevant design values, for the post acting as a column, are

$$F_{cE} = K_{cE}E'/(\ell_e/d)^2$$
$$= 910 \text{ pounds per square inch}$$
$$C_P = (1.0 + F)/2c - \{[(1.0 + F)/2c]^2 - F/c\}^{0.5}$$

where

$$F_c^* = \text{adjusted tabulated compressive stress}$$
$$= F_C C_D C_F$$
$$= 750 \times 1.60 \times 0.96$$
$$= 1152 \text{ pounds per square inch}$$
$$F = F_{cE}/F_c^*$$
$$= 0.79$$

and

$$C_P = 0.60$$

411

The allowable compressive stress is

$$F_c' = F_c^* C_P$$
$$= 697 \text{ pounds per square inch}$$
$$> f_c \ldots \text{satisfactory}$$

For the post acting as a single span beam with a uniformly distributed applied load, the effective unbraced length is given by UBC Section 2363.1 as

$$\ell_e = 2.06\ell_u \ldots \text{for } \ell_u/d < 7$$
$$= 2.06 \times 9.5$$
$$= 19.57 \text{ feet}$$

The slenderness factor is given by UBC Section 2363.1 as

$$R_B = (\ell_e d/b^2)^{0.5}$$
$$= 11.66$$
$$< 50 \ldots \text{satisfactory}$$

The critical buckling design value is given by UBC Section 2363.2 as

$$F_{bE} = K_{bE}E'/R_B^2$$
where $\quad K_{bE}$ = Euler buckling coefficient
$$= 0.438 \ldots \text{for visually graded timber}$$
$\quad\quad E'$ = adjusted tabulated modulus of elasticity
$$= 1.3 \times 10^6 \text{ pounds per square inch}$$
and $\quad F_{bE}$ = 4191 pounds per square inch

The beam stability factor is given by UBC Section 2363.2 as

$$C_L = (1.0 + F)/1.9 - \{[(1.0+ F)/1.9]^2 - F/0.95\}^{0.5}$$
where $\quad F_b^*$ = adjusted tabulated bending stress
$$= F_c C_D C_F$$
$$= 1050 \times 1.60 \times 0.96$$
$$= 1612 \text{ pounds per square inch}$$
$\quad\quad F$ $= F_{bE}/F_b^*$
$$= 2.60$$
and $\quad C_L$ = 0.97

The allowable flexural stress is

$$F_b' = F_b^* C_L$$
$$= 1566 \text{ pounds per square inch}$$
$$> f_b \ldots \text{satisfactory}$$

The combined compressive and flexural stresses due to vertical and transverse loading must satisfy the interaction equation given in UBC Section 2308.2 as

$$(f_c/F_c')^2 + f_b/F_b'(1.0 - f_c/F_{cE}) \leq 1$$

The left hand side of the expression is evaluated as

$$(602/697)^2 + 41/1560(1.0 - 602/910) = 0.83$$

Hence the post is satisfactory under the combined dead, plus live, plus snow, plus 50 percent wind load.

DEAD, LIVE, WIND, AND 50 PERCENT SNOW

By inspection, this loading combination is less critical than the previous. Hence the post is satisfactory under all loading conditions.

2. BEARING PLATE

The allowable compressive stress perpendicular to the grain for a 24F-V4 glulam beam is obtained from UBC Table 23-I-C-1 as

$$F_{c\perp} = 650 \text{ pounds per square inch.}$$

No increase in this value is permitted for duration of loading and the maximum load acting, due to dead, plus floor live, plus snow load is

$$P = 58,000$$

The required bearing length is, then

$$
\begin{aligned}
\ell &= P/F_{c\perp}b \\
&= 58,000/(650 \times 5.125) \\
&= 17.4 \text{ inches}
\end{aligned}
$$

Allowing for a 0.6 inch shortness in the beam, the required bearing plate length is

$$\ell_p = 18 \text{ inches}$$

Using a plate width of 5.25 inches, the bending moment per inch run in the plate is

$$
\begin{aligned}
m &= 650 \times 5.25^2/8 \\
&= 2240 \text{ pounds inch}
\end{aligned}
$$

The required plate thickness is, then

$$
\begin{aligned}
t &= (6m/F_b)^{0.5} \\
&= (6 \times 2240/27,000) \\
&= 0.71 \text{ inches}
\end{aligned}
$$

Full details are shown in Figure 88D-2(ii).

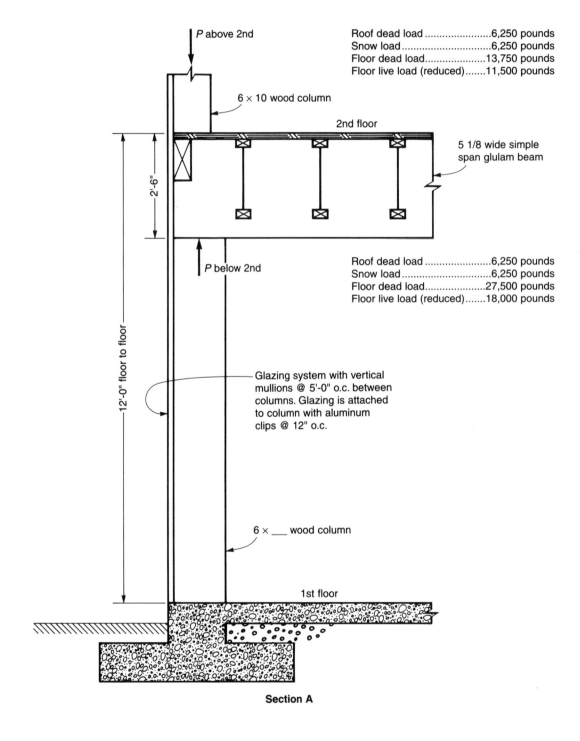

P above 2nd

Roof dead load ....6,250 pounds
Snow load ....6,250 pounds
Floor dead load....................13,750 pounds
Floor live load (reduced).......11,500 pounds

6 × 10 wood column

2nd floor

5 1/8 wide simple span glulam beam

2'-6"

P below 2nd

Roof dead load ....6,250 pounds
Snow load ....6,250 pounds
Floor dead load....................27,500 pounds
Floor live load (reduced).......18,000 pounds

12'-0" floor to floor

Glazing system with vertical mullions @ 5'-0" o.c. between columns. Glazing is attached to column with aluminum clips @ 12" o.c.

6 × ___ wood column

1st floor

Section A

FIGURE 88D–2(i)

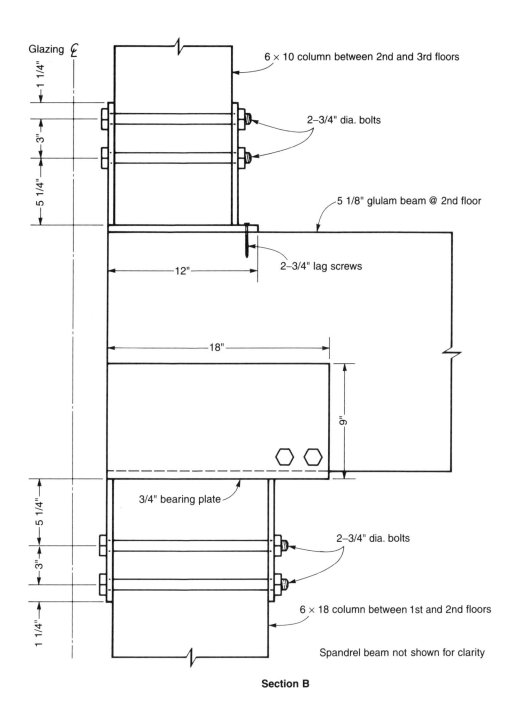

Glazing $\mathcal{C}$

6 × 10 column between 2nd and 3rd floors

2–3/4" dia. bolts

1 1/4"

3"

5 1/4"

5 1/8" glulam beam @ 2nd floor

2–3/4" lag screws

12"

18"

9"

3/4" bearing plate

5 1/4"

3"

2–3/4" dia. bolts

6 × 18 column between 1st and 2nd floors

1 1/4"

Spandrel beam not shown for clarity

Section B

FIGURE 88D–2(ii)

STRUCTURAL ENGINEER EXAMINATION – 1987 ═══════════

PROBLEM D–5 – WT 3.0 POINTS

GIVEN: During a renovation, the contractor must temporarily remove a damaged interior masonry column which supports two precast beams. The structural engineer must design temporary supports for the beam using 4 × 4 wood posts.

CRITERIA: Properties of the 4 × 4 construction grade wood posts on site are:
Actual size $= 3.5" \times 3.5"$
$F_c = 1150$ psi
$E = 1500$ ksi

Assume pinned connection for beams at A, B, and C.

REQUIRED: a. Calculate the load at B given the loads shown.
                         b. Calculate the number of 4 × 4 posts the contractor must place under the beam adjacent to B to temporarily support the beams until a new column can be reinstalled.
                         c. Sketch the location of the posts on a plan or elevation for the contractor's use.

SOLUTION:

a. SUPPORT LOADS

Locate the center of the temporary support twelve inches on either side of grid line B, in order to allow working space for removal and replacement of the damaged masonry column. The load on each temporary support is, then

$$P = 1.125 \times 16^2/(2 \times 15) + 8 \times 16/15$$
$$= 18.13 \text{ kips}$$

b. POSTS REQUIRED

Assuming that the temporary supports will be required for not more than two months, the applicable load duration factor is given by UBC Section 2304.3.42504(c)4 as

$$C_D = 1.15$$

The size factor is obtained from UBC Table 23–I–A–1 as

$$C_F = 1.00$$

The effective unbraced length of each 4×4 post is

$$
\begin{aligned}
\ell_e &= K_e\ell \\
&= 1.0 \times 10 \\
&= 10 \text{ feet}
\end{aligned}
$$

The slenderness ratio about each axis is

$$
\begin{aligned}
\ell_e/b &= 10 \times 12/3.5 \\
&= 34.3 \\
&< 50 \text{ ... satisfactory}
\end{aligned}
$$

The relevant design factors are obtained from calculator program A.3.3 in Appendix A. The critical buckling design value is given by UBC Section 2307.3 as

$$
\begin{aligned}
F_{cE} &= K_{cE}E'/(\ell_e/d)^2 \\
\end{aligned}
$$

where K_{cE} = Euler buckling coefficient for columns
$= 0.30$ ... for visually graded lumber

E' = adjusted tabulated modulus of elasticity
$= 1.5 \times 10^6$ pounds per square inch

and F_{cE} = 382 pounds per square inch

The column stability factor is given by UBC Section 2307.3 as

$$
C_P = (1.0 + F)/2c - \{[(1.0 + F)/2c]^2 - F/c\}^{0.5}
$$

where F_c^* = adjusted tabulated compressive stress
$= F_c C_D C_F$
$= 1750 \times 1.15 \times 1.00$
$= 2013$ pounds per square inch

$F = F_{cE}/F_c^*$
$= 0.19$

c = column parameter
$= 0.8$ ... for sawn lumber

and $C_P = 0.18$

The allowable compressive stress is

$$
\begin{aligned}
F_c' &= F_c^* C_P \\
&= 366 \text{ pounds per square inch}
\end{aligned}
$$

The number of posts required to provide each temporary support is, then

$$
\begin{aligned}
n &= 1000P/AF_c' \\
&= 1000 \times 18.13/(12.25 \times 366) \\
&= 4.04
\end{aligned}
$$

Hence, four posts are required for each temporary support.

c. POST LOCATION

The location of the temporary supports is shown in Figure 87D-5.

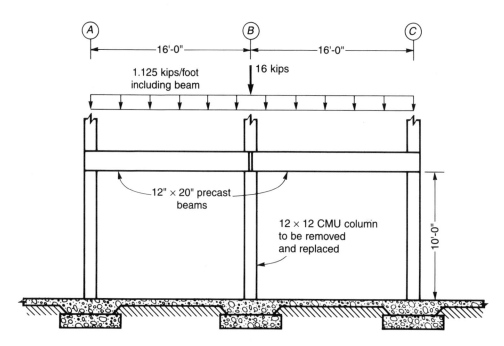

Elevation

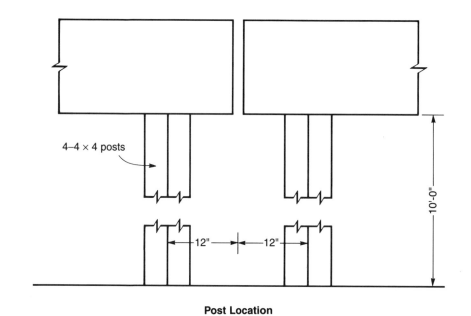

Post Location

FIGURE 87D-5

418

$$\begin{aligned}
C_d &= p/12D \\
\text{where} \quad D &= \text{nail diameter from UBC Table 23–III–MM} \\
&= 1.5/(12 \times 0.162) \\
&= 0.77
\end{aligned}$$

In addition, a seismic load duration factor of 1.33 is applicable, as specified in UBC Section 2304.3.4. Hence the adjusted lateral load is

$$\begin{aligned}
Z' &= 105 \times 1.33 \times 0.77 \quad {}^{C_d} \\
&= 108 \text{ pounds}
\end{aligned}$$

The required number of 16d nails is, then

$$\begin{aligned}
n &= F_b/Z' \\
&= 3200/108 \\
&= 30 \text{ nails}
\end{aligned}$$

In accordance with UBC Section 2326.11.2, joints in the double top plates shall have a minimum offset of 48 inches. Adopting a splice length of 60 inches provides a nail spacing of

$$\begin{aligned}
s &= 60/30 \\
&= 2 \text{ inches}
\end{aligned}$$

The minimum spacing required is specified in UBC Section 2311.3.3 as

$$\begin{aligned}
s_r &= p \\
&= 1.94 \text{ inches} \\
&< s \ldots \text{satisfactory}
\end{aligned}$$

The completed detail is shown in Figure 90D–2.

GRID LINE Y

The unit shear in the diaphragm along grid line Y is given by

$$\begin{aligned}
v_y &= (V_e/2 - 5w_e)/L \\
&= (10{,}000/2 - 5 \times 250)/80 \\
&= 47 \text{ pounds per linear foot}
\end{aligned}$$

The drag force at point A is given by

$$\begin{aligned}
F_A' &= 40v_y \\
&= 40 \times 47 \\
&= 1880 \text{ pounds}
\end{aligned}$$

The chord force at point A is given by

$$\begin{aligned}
F_A &= M_B/35 \\
&= 112{,}000/35 \\
&= 3200 \text{ pounds, which governs.}
\end{aligned}$$

The drag force at point C is given by

$$F'_C = 25v_y$$
$$= 25 \times 47$$
$$= 1175 \text{ pounds}$$

The bending moment at point C is given by

$$M_C = 25V_n/2 - w_n25^2/2$$
$$= 25 \times 11,200/2 - 140 \times 25^2/2$$
$$= 96,250 \text{ pounds feet}$$

The corresponding chord force is

$$F_C = M_C/d$$
$$= 96,250/35$$
$$= 2750 \text{ pounds, which governs}$$

The diaphragm capacity in the east–west direction is 180 pounds per linear foot. Hence to develop the chord force will require a collector length of

$$L_C = F_C/180$$
$$= 2750/180$$
$$= 15.3 \text{ feet}$$

This requires 2×8 blocking between eight joist spaces giving a total collector length of 16 feet. Using a 14 gauge strap, and allowing for the seismic load duration factor of 1.33, the adjusted lateral load capacity of a 16d common nail is obtained from UBC Table 23–III–MM as

$$Z' = 136 \times 1.33$$
$$= 181 \text{ pounds}$$

The required number of 16d nails in both the beam and the collector is, then

$$n = F_C/Z'$$
$$= 2750/181$$
$$= 16 \text{ nails}$$

In the collector, two nails are required in each block and, in the beam, the required nail spacing is 2.0 inches.

Using a 14 gauge by 1.5 inch wide strap provides a tensile capacity of

$$T = 0.6F_ybt \times 1.33$$
$$= 0.6 \times 36,000 \times 1.5 \times 0.075 \times 1.33$$
$$= 3230 \text{ pounds}$$
$$> F_C \ldots \text{ satisfactory}$$

GRID LINE Z

The drag force at point D is given by

$$\begin{aligned} F_D' &= v_e \times 24/2 \\ &= 76.9 \times 12 \\ &= 923 \text{ pounds} \end{aligned}$$

The bending moment at point D is given by

$$\begin{aligned} M_D &= 32V_n/2 - w_n32^2/2 \\ &= 32 \times 11{,}200/2 - 140 \times 32^2/2 \\ &= 107{,}520 \text{ pounds feet} \end{aligned}$$

The corresponding chord force is

$$\begin{aligned} F_D &= M_D/B \\ &= 107{,}502/40 \\ &= 2688 \text{ pounds, which governs} \end{aligned}$$

The drag force at point E is given by

$$\begin{aligned} F_E' &= v_e \times 15/2 \\ &= 76.9 \times 7.5 \\ &= 577 \text{ pounds} \end{aligned}$$

The bending moment at point E is given by

$$\begin{aligned} M_E &= 20V_n/2 - w_n20^2/2 \\ &= 20 \times 11{,}200/2 - 140 \times 20^2/2 \\ &= 84{,}000 \text{ pounds feet} \end{aligned}$$

The corresponding chord force is

$$\begin{aligned} F_E &= M_E/B \\ &= 84{,}000/40 \\ &= 2100 \text{ pounds, which governs} \end{aligned}$$

4. CONNECTION DETAILS

The connection details are shown in Figure 90D–2 ===================

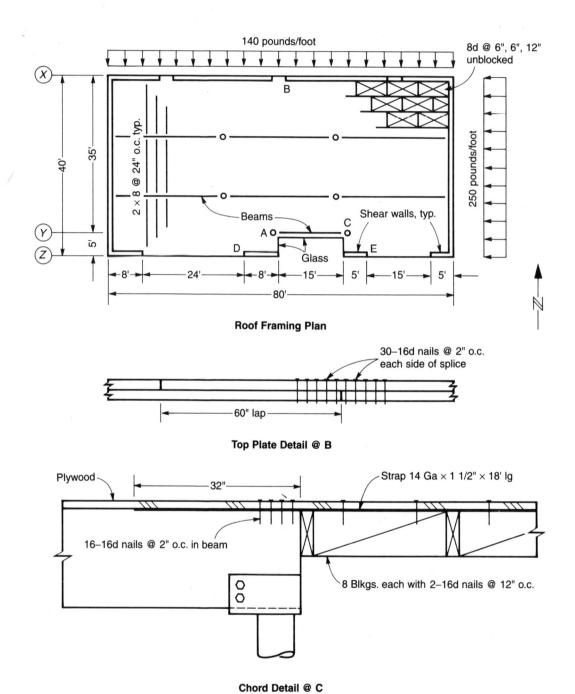

Roof Framing Plan

Top Plate Detail @ B

Chord Detail @ C

FIGURE 90D-2

426

STRUCTURAL ENGINEER EXAMINATION – 1989

PROBLEM D-3 – WT. 7.0 POINTS

GIVEN:
- A single story wood frame building with roof framing plan and section as shown.
- Wall weight is 15 pounds per square foot.
- Roof dead weight is 25 pounds per square foot.
- Snow load is 120 pounds per square foot, including snow drift.
  (The building official has determined that the use of 75 percent of the snow load shall be used for lateral analysis)

CRITERIA:
- Seismic zone 4, C=2.75
- Consider the north–south direction only.

Materials
- Lumber: Douglas Fir–Larch No.1 grade.
- Glued–laminated beams: Douglas Fir–Larch 24F–V4 grade.
- Steel connectors: ASTM A36 grade.
- Use 15/32 inch plywood for walls and ¾ inch plywood for roof, structural I grade.

Assumptions
- Column provides no lateral load resistance

REQUIRED:
1. Design and detail the connection required at location A for the two cases:

   - The 2 × 12 joists line up with the 3 × 12 studs.
   - The 2 × 12 joists do not line up with the 3 × 12 studs.

   Consider the vertical loads and lateral loads perpendicular to the wall along line J.
2. Determine the maximum high roof diaphragm shears and specify the nailing requirement.
3. Design and detail the splice required at location B on line J for the two 3 × top plates.
4. Determine the drag load at location C, and design and detail the required plate splice at the glued–laminated beams.
5. Determine the diaphragm shear at line 1 and design and detail the necessary connections for lateral load transfer to the shear wall.

SOLUTION

1. DETAIL A

LATERAL FORCE

The relevant dead load tributary to the low roof in the north–south direction is due to the south wall, dead load and reduced snow load and is given by

South wall	$= 15 \times 15/2$	$=$	113
Low roof	$= 25 \times 12$	$=$	300
Snow	$= 120 \times 12 \times 0.75$	$=$	1080
Total	$= W$	$=$	1493 pounds per linear foot

The seismic force acting is obtained from UBC Section 1628 Formula (28–1) as

	V	$= (ZIC/R_w)W$
where	Z	$= 0.4$ for zone 4 from UBC Table 16–I
	I	$= 1.0$ from UBC Table 16–L for a standard occupancy structure
	C	$= 2.75$ as given
	R_w	$= 8$ from UBC Table 16–N item 1.1.a
	W	$= 1493$ pounds per linear foot, as calculated
Then	V	$= (0.4 \times 1.0 \times 2.75/8)1493$
		$= 0.138 \times 1493$
		$= 205$ pounds per linear foot

This force is resisted by the 3×12 stud wall, as the low roof can not function as a diaphragm since its aspect ratio of 5 exceeds the limiting value of 4 specified in UBC Table 23–I–I.

VERTICAL FORCE

The vertical force acting at the connection is due to roof dead load and full snow load and is given by

Low roof	$= 25 \times 6$	$= 150$
Snow	$= 120 \times 6$	$= 720$
Total	$= w$	$= 870$ pounds per linear foot

JOISTS AND STUDS IN ALIGNMENT

To support the joists at the ledger, a face mounted hanger is required with a capacity of

$$P = 2w$$
$$= 2 \times 870$$
$$= 1740 \text{ pounds}$$

Allowing for the duration factor for snow load, a Simpson[7] type HUS210 hanger has a capacity of

$$P_a = 3920 \text{ pounds}$$
$$> P \ldots \text{ satisfactory}$$

The vertical load applied to each stud is given by

$$P = 16w/12$$
$$= 16 \times 870/12$$
$$= 1160 \text{ pounds}$$

Allowing for the snow load duration factor, as specified in UBC Section 2304.3.4, the adjusted vertical load is

$$P' = 1160/1.15$$
$$= 1009$$

Using two ⅝ inch diameter bolts, the allowable load is obtained from UBC Table 23–III–J as

$$P_a = 2 \times Z_{s\perp}$$
$$= 2 \times 700$$
$$= 1400$$
$$> P' \ldots \text{ satisfactory}$$

In accordance with UBC Tables 23–III–H and 23–III–I, the minimum bolt spacing in the stud is four times the bolt diameter and the minimum edge distance to the loaded edge of ledger is also four times the bolt diameter. Hence the bolts located as shown in the detail, with the bolt heads countersunk one inch into the ledger, are satisfactory.

A joist and a stud coincide at intervals of four feet. At these locations, the lateral force is given by

$$P = 4V$$
$$= 4 \times 205$$
$$= 820 \text{ pounds}$$

Providing a Simpson anchor type H6 with eight 8d nails in both the joist and the stud gives a capacity of

$$P_a = 915 \text{ pounds}$$
$$> P \ldots \text{ satisfactory}$$

JOISTS AND STUDS NOT IN ALIGNMENT

The lateral seismic force may be transferred from each joist to continuous blocking between the studs. The lateral force at each joist is

$$P = 2V$$
$$= 2 \times 205$$
$$= 410 \text{ pounds}$$

To transfer this force to the blocking requires a Simpson holdown type HD2A, with two ⅝ inch diameter bolts in the joists and a ⅝ inch diameter tie rod to the blocking, providing a capacity of

$$P_a = 1555 \text{ pounds}$$
$$> P \text{ ... satisfactory}$$

The bolts in the joists require a minimum edge distance of seven bolt diameters or 4½ inches. The blocking may be attached to the studs, at each end, with two Simpson angles type L30 with four 10d nails in each, providing a capacity of

$$P_a = 2 \times 292$$
$$= 584 \text{ pounds}$$
$$> P \text{ ... satisfactory}$$

2. DIAPHRAGM NAILING

The dead load tributary to the high roof in the north-south direction is due to the walls on lines J and M, roof dead load, reduced snow load and low roof dead load and is given by

Walls on J and M	$= 15 \times 1 \times 20/2$	$= 300$
High roof self weight	$= 25 \times 60$	$= 1500$
Snow	$= 120 \times 60 \times 0.75$	$= 5400$
Low roof	$= 1493 \times 15/20$	$= \underline{1120}$
Total	$= w$	$= 8320$ pounds per linear foot

The total seismic force in the north-south direction is given by UBC Formula (28-1) as

$$V = 0.138wL$$
$$= 0.138 \times 8320 \times 60$$
$$= 68,890 \text{ pounds}$$

The unit shear along the diaphragm boundaries at line 1 and line 2 is

$$v = V/2B$$
$$= 68,890/(2 \times 60)$$
$$= 574 \text{ pounds per linear foot}$$

The capacity of ¾ inch structural I plywood with ten penny nails is governed by the strength of the nails[3]. Using a case 1 plywood layout, with all edges blocked and with 3 × framing, the required nail spacing is obtained from UBC Table 23–I–J–1 as:

at diaphragm boundaries	= 2½ inches
at panel edges	= 4 inches
at intermediate framing members	= 12 inches

and this provides a shear capacity of 720 pounds per linear foot which exceeds the required value of 574.

3. CHORD FORCE AT B

The bending moment at point B, twenty feet from line 2, due to the north–south seismic force is

$$M_B = VL/9$$
$$= 68,890 \times 60/9$$
$$= 459,267 \text{ pounds feet}$$

The corresponding chord force is

$$F_B = M_B/B$$
$$= 459,267/60$$
$$= 7654 \text{ pounds}$$

2 ½" × 11 ¼" ACTUAL.

The double 3 × 12 top plates provide a tensile capacity, allowing for a bored hole of 13/16 inch diameter, of

$$F_a = AF_tC_D$$
$$= 5(11.25 - 0.813)675 \times 1.33$$
$$= 46,849$$
$$> F_B \text{ ... satisfactory}$$

why need this?

Using a Simpson tie type HST6 at the splice, with twelve ¾ inch diameter bolts, provides a capacity of

$$F_a = 12,465 \text{ pounds}$$
$$> F_B \text{ ... satisfactory}$$

The required member thickness is 4½ inches and the double top plates provide 5 inches. Bolt spacing is three inches, which equals the required minimum of four diameters, and end distance in the top plate is 5¼ inches, which equals the required minimum of seven diameters. Standard washers are required between the top plate and the nuts.

4. DRAG FORCE AT C

The drag force at point C, twenty feet from line M, due to the north–south seismic force is

$$
\begin{aligned}
F_C \quad &= v \times 20 \\
&= 574 \times 20 \\
&= 11{,}480 \text{ pounds}
\end{aligned}
$$

To transfer this force across the glulam beam requires two Simpson holdowns type HD14A, each with four one inch diameter bolts and with a one inch tie bar. Using the necessary bolt spacing and end distance of four inches and seven inches, respectively, in the five inch thick double top plate, provides a capacity of

$$
\begin{aligned}
F_a \quad &= 12{,}230 \text{ pounds} \\
&> F_C \dots \text{satisfactory}
\end{aligned}
$$

5. SHEAR WALL AT LINE 1

The unit shear in the shear wall is given by

$$
\begin{aligned}
q \quad &= 2v \\
&= 2 \times 574 \\
&= 1148 \text{ pounds per linear foot}
\end{aligned}
$$

Using 15/32 inch structural I plywood on both sides of the stud wall, the required spacing of 10d nails is obtained from UBC Table 23-I-K-1 as :

at panel edges	= 3 inches
at interior framing members	= 12 inches

This provides a shear capacity of

$$
\begin{aligned}
q_a \quad &= 2 \times 665 \\
&= 1330 \text{ pounds per linear foot} \\
&> q \dots \text{satisfactory}
\end{aligned}
$$

To transfer the shear from the blocking to the double top plate requires two Simpson angles type L70 at twelve inch centers which provide a shear capacity of

$$
\begin{aligned}
q_a \quad &= 2 \times 592 \\
&= 1184 \text{ pounds per linear foot} \\
&> q \dots \text{satisfactory}
\end{aligned}
$$

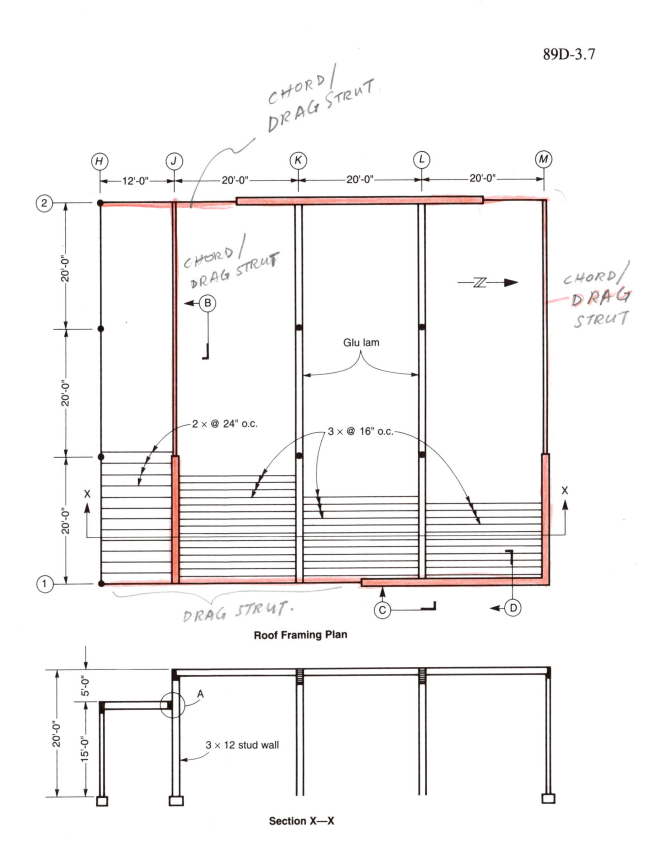

Roof Framing Plan

Section X—X

FIGURE 89D–3(i)

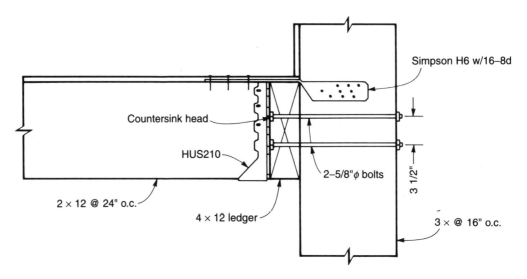

Detail A
Case 1—2 × 12 Joists Line Up with 3 × 12 Studs

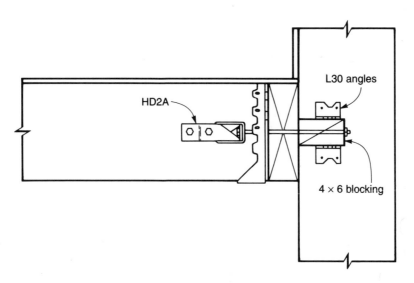

Detail A
Case 2—Joists and Studs Do Not Line Up

FIGURE 89D-3(ii)

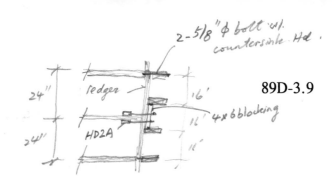

89D-3.9

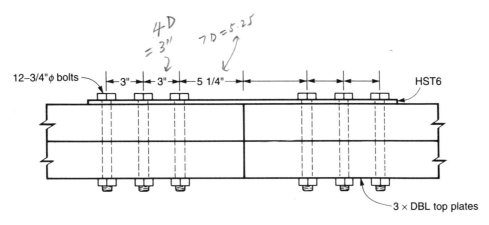

Detail B

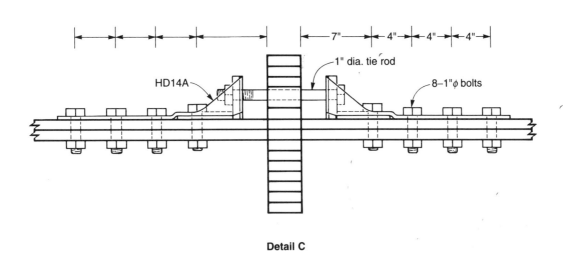

Detail C

FIGURE 89D−3(iii)

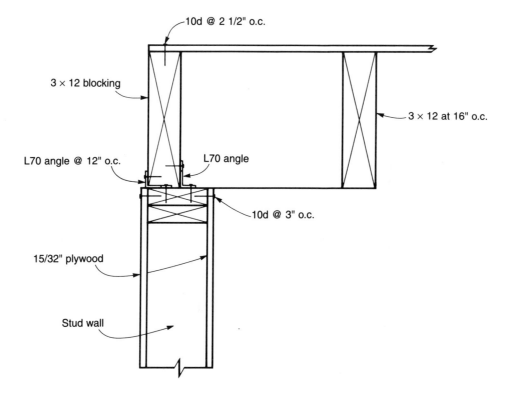

10d @ 2 1/2" o.c.

3 × 12 blocking

3 × 12 at 16" o.c.

L70 angle @ 12" o.c.

L70 angle

10d @ 3" o.c.

15/32" plywood

Stud wall

Detail D

FIGURE 89D–3(iv)

STRUCTURAL ENGINEER EXAMINATION – 1988

PROBLEM D-3 – WT. 5.0 POINTS

GIVEN: A panelized roof system framing plan for a one story building with 6 inches concrete tilt-up shear walls as shown.

CRITERIA: Seismic zone 4
Materials:
- Roof plywood: 15/32 inch structural I grade.
- Glued-laminated beams: Douglas Fir-Larch 24F-V8 grade.
- Purlins and ledger: Douglas Fir-Larch (north) No.1.
- Roof weight: 15 pounds per square foot including beam and girder weight.
- Concrete wall: thickness = 6 inches, weight = 75 pounds per square foot.

Assumptions:
- Wind forces do not govern.
- Nails are common nails.
- Concrete walls span between roof and slab on grade.
- The north and south walls at roof level are designed to span eight feet between glued-laminated beams.
- Shear and chord stresses calculated for the sub-diaphragm need not be added to the main diaphragm stresses.
- Neglect the roof dead load in the subdiaphragm calculations.

REQUIRED: 1. Calculate the nailing required and show where nail spacing may be changed, for forces in the north-south direction only.
2. Determine the maximum diaphragm chord forces on line A and D.
3. Provide a detail of the north and south wall anchorage to the roof diaphragm to resist seismic loads perpendicular to the wall at the 3½ × 13½ glued- laminated beams at eight feet on centers. Do not consider gravity loads in this connection.
4. Calculate the required subdiaphragm nailing and chord forces.
5. Calculate the seismic forces at the 6¾ × 24 glued-laminated beam hinges.

SOLUTION

1. DIAPHRAGM NAILING

LATERAL FORCE

The relevant dead load tributary to the diaphragm in the north–south direction is due to the north and south wall and the roof dead load and is given by

Roof	$= 15 \times 120$	$= 1800$
North wall	$= 75 \times 14^2/(2 \times 12.5)$	$= 588$
South wall		$= \underline{588}$
Total	$= w$	$= 2976$ pounds per linear foot

The seismic force acting is obtained from UBC Section 1628 Formula (28–1) as

$$V = (ZIC/R_w)wL$$

$Z = 0.4$ for zone 4 from UBC Table 16–I

$I = 1.0$ from UBC Table 16–L for a standard occupancy structure

$C = 2.75$ from UBC Section 1628.2.1

$R_w = 6$ from UBC Table 16–N item 1.2.a

$w = 2976$ pounds per linear foot as calculated

$L = 256$ feet, length of building

$$V = (0.4 \times 1.0 \times 2.75/6)2976 \times 256$$
$$= 0.183 \times 761,856$$
$$= 139,420 \text{ pounds}$$

NAILING REQUIREMENTS

The unit shear along the diaphragm boundaries at grid lines 1 and 9 is

$$v_1 = V/2B$$
$$= 139,420/(2 \times 120)$$
$$= 581 \text{ pounds per linear foot}$$

The nail spacing may be changed at the beam locations, as shown in Figure 88D–3, and the unit shear a distance 40 feet from the boundary is given by

$$v_2 = v_1 \times 88/128$$
$$= 399 \text{ pounds per linear foot}$$

At 64 feet from the boundary, the unit shear is

$$v_3 = v_1 \times 64/128$$
$$= 291 \text{ pounds per linear foot}$$

The required nail spacing is obtained from UBC Table 16-I-J-1 with a case 4 plywood layout applicable, all edges blocked and with 3½ inch framing at continuous panel edges parallel to the load. Using 15/32 inch structural I plywood and 10d nails with 1⅝ inch penetration, the nail spacing required in the three diaphragm zones are

Zone	1	2	3
Diaphragm boundaries	2½"	4"	6"
Continuous panel edges	2½"	4"	6"
Other edges	4"	6"	6"
Capacity provided, plf	640	425	320
Capacity required, plf	581	399	291

2. CHORD FORCE

The bending moment at the mid point of the north and south boundaries due to the north-south seismic force is

$$
\begin{aligned}
M &= VL/8 \\
&= 139{,}420 \times 256/8 \\
&= 4{,}461{,}440 \text{ pounds feet}
\end{aligned}
$$

The corresponding chord force is

$$
\begin{aligned}
F &= M/B \\
&= 4{,}461{,}440/120 \\
&= 37{,}179 \text{ pounds}
\end{aligned}
$$

3. WALL ANCHORAGE

The seismic loading on the concrete wall is given by UBC Section 1630 Formula (30-1) as

$$F_p = ZI_pC_pw_p$$

where
Z = 0.4 for zone 4 from UBC Table 16-I
I_p = 1.0 from UBC Table 16-K for a standard occupancy structure
C_p = 0.75 from Table 16-O item 1.1.b
w_p = 75 pounds per square foot, given wall weight

Then
$$
\begin{aligned}
F_p &= 0.4 \times 1.0 \times 0.75 \times 75 \\
&= 0.3 \times 75 \\
&= 22.5 \text{ pounds per square foot}
\end{aligned}
$$

The anchorage force at the roof is given by

$$p = F_p \times 14^2/(2 \times 12.5)$$
$$= 176 \text{ pounds per foot}$$
$$< 200$$

Hence as stipulated in UBC Section 1611, the minimum anchorage force of 200 pounds per linear foot of wall governs and the force on each anchor for a spacing of eight feet is

$$P = 8p$$
$$= 8 \times 200$$
$$= 1600 \text{ pounds}$$

A Simpson anchor type PA18 with a capacity of 2255 pounds is adequate and a suitable detail is shown in Figure 88D-3.

4. SUBDIAPHRAGM DETAILS

The function of the subdiaphragm is to transfer the wall anchorage forces into the main diaphragm[4]. The aspect ratio of the subdiaphragm shown is

$$b/d = 32/20$$
$$= 1.6$$
$$< 4$$

Hence the requirement of UBC Table 23-I-I is satisfied and the unit shear in the subdiaphragm is

$$v = pb/2d$$
$$= 200 \times 32/(2 \times 20)$$
$$= 160 \text{ pounds per linear foot}$$
$$< 320 \ldots \text{ satisfactory}$$

Hence the capacity of the nailing in the main diaphragm is adequate.

The chord force in the subdiaphragm is

$$F = pb^2/8d$$
$$= 200 \times 32^2/(8 \times 20)$$
$$= 1280 \text{ pounds}$$

5. TIE FORCE AT HINGES

The anchor force to the ties at a spacing of 32 feet is

$$P = pb$$
$$= 200 \times 32$$
$$= 6400 \text{ pounds}$$

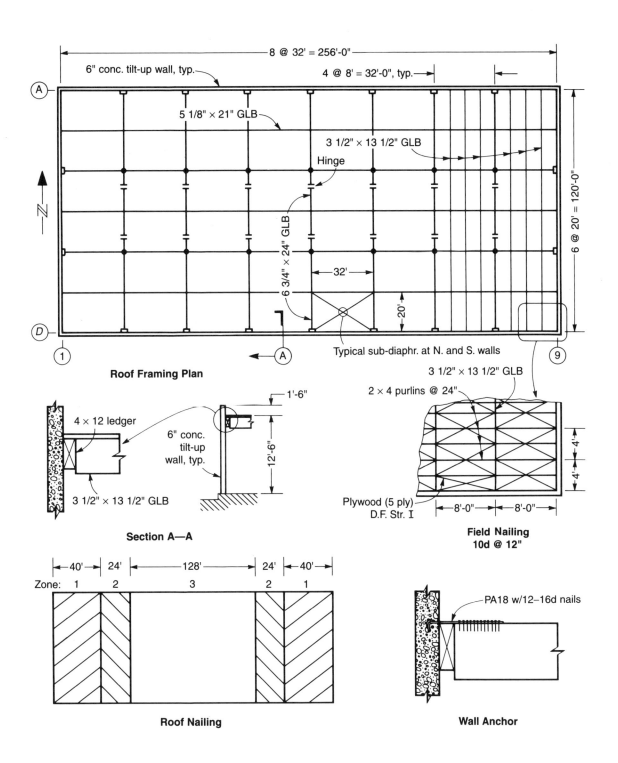

Roof Framing Plan

8 @ 32' = 256'-0"

6" conc. tilt-up wall, typ.

4 @ 8' = 32'-0", typ.

5 1/8" × 21" GLB

3 1/2" × 13 1/2" GLB

Hinge

6 3/4" × 24" GLB

32'

20'

6 @ 20' = 120'-0"

Typical sub-diaphr. at N. and S. walls

Section A—A

4 × 12 ledger

1'-6"

6" conc. tilt-up wall, typ.

12'-6"

3 1/2" × 13 1/2" GLB

Field Nailing 10d @ 12"

3 1/2" × 13 1/2" GLB

2 × 4 purlins @ 24"

Plywood (5 ply) D.F. Str. I

4'

4'

8'-0"

8'-0"

Roof Nailing

40' 24' 128' 24' 40'

Zone: 1 2 3 2 1

Wall Anchor

PA18 w/12–16d nails

FIGURE 88D-3

441

STRUCTURAL ENGINEERING REGISTRATION

CHAPTER 4

STRUCTURAL TIMBER DESIGN

SECTION 4.4

SHEAR WALLS

PROBLEMS:

1987 D-1

STRUCTURAL ENGINEER EXAMINATION – 1987

PROBLEM D–1 – WT. 7.0 POINTS   67 min

GIVEN: A two story wood framed building with plywood shear walls as shown. The total building lateral forces have been determined for each level and are indicated.

CRITERIA: Materials:
- Wood framing: Douglas Fir–Larch No.1 grade.
- Plywood walls: 15/32 inch C–D grade, 32/16 panel index, exposure 1.
- Plywood floor and roof: 19/32 C–D, 32/16 panel index, exposure 1.
- Machine bolts: ASTM A307 grade.
- Nails: 10d common nails at roof and floor plywood and 8d common nails at shear wall plywood.
- Framing connectors: as approved by the ICBO.

Assumptions:
- The shear walls are adequate in the longitudinal direction.
- The end transverse shear walls are adequate.
- The roof and floor framing are adequate for dead load plus live load.
- The foundations are adequate for dead load, live load, and lateral loads.

REQUIRED: a. Determine the lateral forces resisted by shear walls A and B.
b. For the transverse direction of the second floor diaphragm, determine the shear diagram and the nailing requirements.
c. Design shear wall A for the critical lateral load determined above. Include nailing requirements and boundary members.
d. Show all design details at locations 2 and 3 and section 1 of shear wall A which are required to adequately transfer lateral loads into and out of the shear wall. If framing connectors are utilized, show their configuration and required minimum load capacity.

SOLUTION

a. SHEAR WALL FORCES

SHEAR WALL A

Due to wind load, the applied force in shear wall A is

$$V_A = 121 \times 40$$
$$= 4840 \text{ pounds}$$

Due to seismic load, the applied force in shear wall A is

$$V_A = 200 \times 40$$
$$= 8000 \text{ pounds, which governs.}$$

SHEAR WALL B

Due to seismic load, the applied force in shear wall B is

$$V_B = 124 \times 40 + 8000 \times 40/50$$
$$= 11{,}360 \text{ pounds}$$

Due to wind load, the applied force in shear wall B is

$$V_B = 244 \times 40 + 4840 \times 40/50$$
$$= 13{,}632 \text{ pounds, which governs for connections and fasteners.}$$

b. SECOND FLOOR DIAPHRAGM

SHEAR DIAGRAM

Shear wall B effectively subdivides the second floor diaphragm into two simply supported segments. These are shown in Figure 87D-1(ii) as span 12 and span 23. Wind loading governs and the wind loads acting on the diaphragm are indicated and produce the support reactions

$$R_{12} = 244 \times 15 \qquad\qquad\qquad = 3660 \text{ pounds}$$
$$R_{21} \qquad\qquad\qquad\qquad\qquad = 3660 \text{ pounds}$$
$$R_{23} = 244 \times 25 + 4840 \times 40/50 = 9972 \text{ pounds}$$
$$R_{32} = 244 \times 25 + 4840 \times 10/50 = 7068 \text{ pounds}$$

The resultant shear force diagram is shown in Figure 87D-1(ii).

NAILING REQUIREMENTS

The diaphragm unit shears are given by

$$q_{12} = R_{12}/40 \qquad = 3660/40 \qquad = 92 \text{ pounds per linear foot}$$
$$q_{21} = R_{21}/40 \qquad = 3660/40 \qquad = 92 \text{ pounds per linear foot}$$
$$q_{23} = R_{23}/40 \qquad = 9972/40 \qquad = 249 \text{ pounds per linear foot}$$
$$q_{32} = R_{32}/40 \qquad = 7068/40 \qquad = 177 \text{ pounds per linear foot}$$
$$q_{42} = 7532/40 \qquad\qquad\qquad = 188 \text{ pounds per linear foot}$$
$$q_{43} = 2692/40 \qquad\qquad\qquad = 67 \text{ pounds per linear foot}$$

The required nail spacing is obtained from UBC Table 23-I-J-1 with a case 1 plywood layout applicable, all edges blocked, and with 3½ inch framing. Using 19/32 inch grade C-D plywood and 10d nails with 1⅝ inch penetration, the nail spacing required is:

at diaphragm boundaries	= 6 inches
at all other edges	= 6 inches
at intermediate framing	= 12 inches

and this provides a shear capacity of 360 pounds per linear foot which exceeds the required capacity of 249 pounds per linear foot.

444

c and d. SHEAR WALL A

NAILING REQUIREMENTS

The aspect ratio of shear wall A is

$$\ell/h = 20/12$$
$$= 1.7$$
$$< 3.5 \ldots \text{satisfactory}$$

This conforms to the requirements of UBC Table 23-I-I, for plywood panels nailed at all edges, and the unit shear in wall A for seismic loading is given by

$$q = V_A/\ell$$
$$= 8000/20$$
$$= 400 \text{ pounds per linear feet}$$

The nail spacing required to provide a shear capacity of 400 pounds per linear foot using 15/32 inch grade C-D plywood on one side only and 8d nails with 1½ inch minimum penetration, may be obtained from UBC Table 23-I-K-1. With two inch nominal Douglas Fir-Larch vertical studs at sixteen inches on center and all panel edges backed with two inch nominal blocking, the required nail spacing is:

at all panel edges = 3 inches
at intermediate framing members = 12 inches

and this provides a shear capacity of 490 pounds per linear foot.

FLOOR ANCHORAGE

Provide a 2 × 4 sill plate with ⅝ inch diameter lag screws, five inches long, to anchor the shear wall to the four inch nominal Douglas Fir-Larch supporting joist. The combined thickness of the floor diaphragm and the sill plate is

$$t = 1.5 + 0.625$$
$$= 2.125 \text{ inches}$$

The total penetration of the lag screw into the main member, after allowing for an ⅛-inch thick washer is given by

	p	= T – E + S – t – 0.125
where	T	= thread length given in UBC Table 23-III-UU
	E	= length of tapered tip given in UBC Table 23-III-UU
	S	= unthreaded shank length given in UBC Table 23-III-UU
Then	p	= 2.594 + 2 – 2.125 – 0.125
		= 2.344 inches

The penetration depth factor is given in UBC Section 2337.3.3 and Formula (37-5) as

	C_d	= p/8D
where	D	= diameter of lag screw
		= 0.625
Hence	C_d	= 2.344/(8 × 0.625)
		= 0.469

The load duration factor for seismic load is obtained from UBC Section 2304.3.4 as

$$C_D \quad = 1.33$$

The nominal design lateral load for a ⅝-inch diameter lag screw in a 1½ inch thick side member of Douglas Fir–Larch and a main member of Douglas Fir–Larch is given by UBC Table 23-III-T as

$$Z_\| \quad = 920 \text{ pounds}$$

The adjusted design lateral load is

$$
\begin{aligned}
Z_\|' \quad &= Z_\| \times C_D \times C_d \\
&= 920 \times 1.33 \times 0.469 \\
&= 574 \text{ pounds}
\end{aligned}
$$

The required lag screw spacing is then

$$
\begin{aligned}
s \quad &= Z_\|' \times 12/q \\
&= 574 \times 12/400 \\
&= 17.2 \text{ inches}
\end{aligned}
$$

Hence a spacing of sixteen inches is satisfactory, as the edge distance of 1.75 inches provided exceeds the requirements of UBC Section 2337.4 which is 1.5D. The floor anchorage details are shown in Section 1.

END STUDS

Neglecting vertical loads acting on the wall, the maximum compression or tension on the end studs is

$$
\begin{aligned}
P \quad &= qh \\
&= 400 \times 12 \\
&= 4800 \text{ pounds}
\end{aligned}
$$

Providing 4 × 6 Douglas Fir–Larch visually graded sawn lumber end posts, braced in the strong direction, the compressive or tensile stress in the end posts is

$$
\begin{aligned}
f_c \quad &= P/A \\
&= 4800/19.25 \\
&= 249 \text{ pounds per square inch}
\end{aligned}
$$

The effective unbraced length about the weak axis is

$$
\begin{aligned}
\ell_e \quad &= K_e \ell \\
&= 1.0 \times 12 \\
&= 12 \text{ feet}
\end{aligned}
$$

The corresponding slenderness ratio is

$$
\begin{aligned}
\ell_e/b \quad &= 12 \times 12/3.5 \\
&= 41.4 \\
&< 50 \ldots \text{ satisfactory}
\end{aligned}
$$

The relevant design factors are obtained from calculator program A.3.3 in Appendix A. The critical buckling design value is given by UBC Section 2307.3 as

$$F_{cE} = K_{cE}E'/(\ell_e/d)^2$$

where
K_{cE} = Euler buckling coefficient for columns
= 0.30 ... for visually graded lumber
E' = adjusted tabulated modulus of elasticity
= 1.7×10^6 pounds per square inch

and
F_{cE} = 298 pounds per square inch

The column stability factor is given by UBC Section 2307.3 as

$$C_P = (1.0 + F)/2c - \{[(1.0+ F)/2c]^2 - F/c\}^{0.5}$$

where
F_c^* = adjusted tabulated compressive stress
= $F_C C_D C_F$
= $1450 \times 1.33 \times 1.10$
= 2121 pounds per square inch
F = F_{cE}/F_c^*
= 0.14
c = column parameter
= 0.8 ... for sawn lumber

and
C_P = 0.136

The allowable compressive stress is

F_c' = $F_c C_D C_P C_F$
= $1450 \times 1.33 \times 0.136 \times 1.10$
= 289 pounds per square inch
> f_c ... satisfactory

By inspection, the tensile stress in the end post is also less than the allowable value. Hence the end studs are adequate.

HOLD-DOWN ANCHORS

A Simpson[7] type HD2A hold-down with two ⅝ inch diameter bolts must be installed on each side of the supporting column and on each side of the shear wall end stud. The capacity in the four inch nominal member is

T = 2×2775
> P ... satisfactory

The minimum end distance, in accordance with UBC Table 23-III-H, is

ℓ_n = 7D
= 7×0.625
= 4.38 inches

and a ⅝ inch diameter tie rod is required between hold-downs. The hold-down details are shown in Detail 3.

TOP DRAG STRUT

Provide a double 2×4 top plate to shear wall A. The drag force developed at the end of the top plate is

$$
\begin{aligned}
F \quad &= 10V_A/40 \\
&= 10 \times 8000/40 \\
&= 2000 \text{ pounds}
\end{aligned}
$$

A Simpson type MST136 strap tie with thirty-six 10d $\times$ 1½ inch nails has a capacity of

$$
\begin{aligned}
F_a \quad &= 2090 \text{ pounds} \\
&> F \ldots \text{satisfactory}
\end{aligned}
$$

The stress in the top plate is

$$
\begin{aligned}
f_t \quad &= F/A \\
&= 2000/5.25 \\
&= 381 \text{ pounds per square inch}
\end{aligned}
$$

The allowable tensile stress for Douglas Fir–Larch No.1 grade lumber is

$$
\begin{aligned}
F_t \quad &= 675 \text{ pounds per square inch} \\
&> f_t \ldots \text{satisfactory}
\end{aligned}
$$

The drag strut connection details are shown in Detail 2.

BOTTOM JOIST CONNECTION

The drag force transmitted across the column cap to the supporting joists is

$$
V_A \quad = 8000 \text{ pounds}
$$

A Simpson type CC46 column cap with two ⅝ inch diameter bolts in each joist, after allowing for a ½ inch gap between the joists, provides an end distance of four inches which is less than the specified seven bolt diameters for full design load. The capacity of the bolts in the nominal four inch joists with 3 gauge metal side plates, as given by UBC Table 23-III-O, after allowing for a load duration factor of 1.33 is

$$
\begin{aligned}
P_L \quad &= 2 \times 2250 \times 1.33 \times 4.0/4.375 \\
&= 5472 \text{ pounds}
\end{aligned}
$$

A Simpson type MST48 strap tie with eight ½ inch diameter bolts provides a capacity of

$$
F_A \quad = 2970 \text{ pounds}
$$

Hence, the total capacity of the column cap and the strap tie is

$$
\begin{aligned}
P_L + F_A \quad &= 5472 + 2970 \\
&= 8442 \text{ pounds} \\
&> V_A \ldots \text{satisfactory}
\end{aligned}
$$

Details of the bottom joist connection are shown in Detail 3.

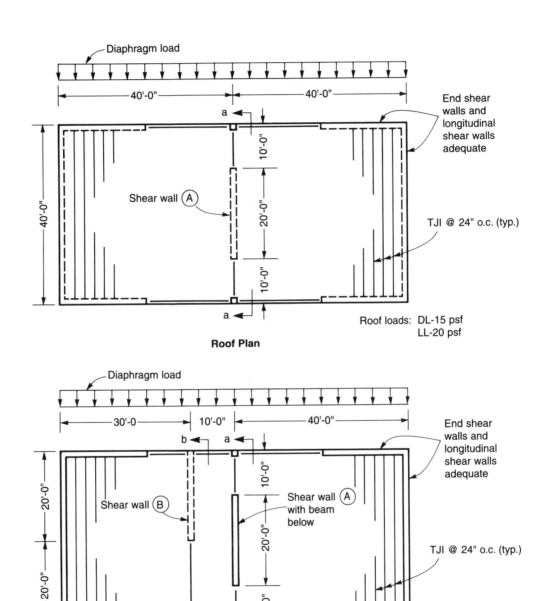

Roof Plan

Roof loads: DL-15 psf
LL-20 psf

Second Floor Plan

2nd floor loads: DL-15 psf
LL-50 psf
Partitions-20 psf

FIGURE 87D–1(i)

449

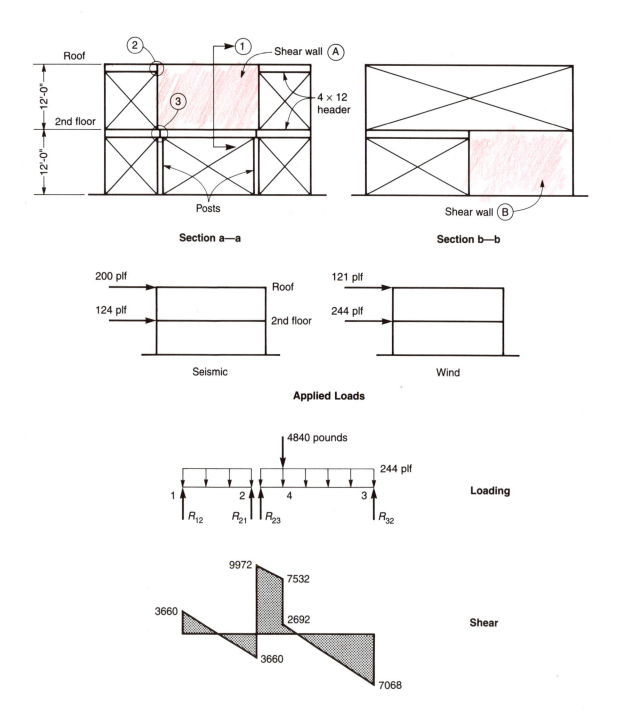

FIGURE 87D-1(ii)

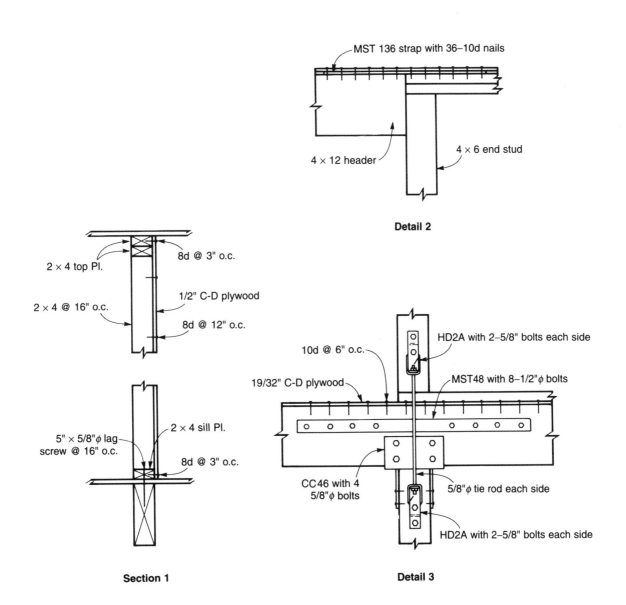

MST 136 strap with 36–10d nails

4 × 12 header

4 × 6 end stud

Detail 2

2 × 4 top Pl.

8d @ 3" o.c.

2 × 4 @ 16" o.c.

1/2" C-D plywood

8d @ 12" o.c.

10d @ 6" o.c.

HD2A with 2–5/8" bolts each side

19/32" C-D plywood

MST48 with 8–1/2"φ bolts

5" × 5/8"φ lag screw @ 16" o.c.

2 × 4 sill Pl.

8d @ 3" o.c.

CC46 with 4 5/8"φ bolts

5/8"φ tie rod each side

HD2A with 2–5/8" bolts each side

Section 1

Detail 3

FIGURE 87D–1(iii)

STRUCTURAL ENGINEERING REGISTRATION

CHAPTER 4

STRUCTURAL TIMBER DESIGN

SECTION 4.5

CONNECTION DESIGN

PROBLEMS:

1988 D–4
1987 D–4

STRUCTURAL ENGINEER EXAMINATION – 1988 ===============

PROBLEM D–4 – WT. 6.0 POINTS

GIVEN: The 2nd floor framing plan of a two story wood building is shown. Drag strut loads are as follows:

Line Y = 260 pounds per linear foot.
Line 2 = 400 pounds per linear foot.

CRITERIA: Materials:
- Bolts: A307 grade.

Assumptions:
- There are no nailers on the top of any of the steel beams because of wood shrinkage.
- The tops of the joists are a ½ inch above the tops of the steel beams. Joist hangers welded to the beams are adequate for vertical loads but are not adequate to stay the top flange.
- Neglect the provisions of UBC Section 1631.2.9.6.

REQUIRED: Provide calculations and complete the following details:
1.   Section A, at the W16 × 40 without the drag strut load, calculations are not required for this section.
2.   Section B, at the W16 × 40 with a drag strut load of 260 pounds per linear foot.
3.   Section C, the drag strut detail connection to the wall at line Y.
4.   Section D, the drag strut detail connection to the wall at line 2.
5.   Comment on 3 areas of special concern for this type of floor construction.

SOLUTION

1. CONNECTION AT STEEL BEAM

To support the 2 × 14 floor joists a hanger, welded to the top flange of the W16 × 40 beam, is required. Assuming a maximum load of 1600 pounds, a Simpson[7] type LB214 hanger is suitable, with the hanger shortened by ½ inch. To provide lateral support to the top flange of the steel beam, bracing is necessary at a maximum spacing of

$$L_c = 7.4 \text{ feet}$$

The required bracing capacity is

$$
\begin{aligned}
P &= 2 \times 0.66 F_y A_f / 100 \\
&= 2 \times 0.66 \times 36{,}000 \times 7 \times 0.505/100 \\
&= 1680 \text{ pounds}
\end{aligned}
$$

A Simpson type FTA5 twisted strap connector with four ¾ inch diameter bolts is suitable. The allowable lateral load in a two inch nominal Douglas Fir-Larch joist is

$$
\begin{aligned}
P_L &= 1865 \text{ pounds} \\
&> P \dots \text{satisfactory}
\end{aligned}
$$

The clear span of the strap must be reduced to seven inches to fit over the flange of the W16 × 40. The strap must be provided at every fifth joist, at a spacing of 80 inches, and welded to the top of the flange. The details are shown in Figure 88D-4(ii), Section A.

2. CONNECTION AT DRAG STRUT

It may be assumed that the unit shears on either side of grid line Y are proportional to the respective diaphragm areas and inversely proportional to the respective diaphragm lengths. Hence the unit shears shown in Figure 88D-4(i) are proportional to the width of the diaphragm on either side of grid line Y and are given by

$$
\begin{aligned}
v_1 &= 260 \times 40/(40 + 30) \\
&= 149 \text{ pounds per linear foot} \\
v_2 &= 260 \times 30/(40 + 30) \\
&= 111 \text{ pounds per linear foot}
\end{aligned}
$$

Provide 2 × 4 blocking between all joists on either side of the drag strut and 15/32 inch C-D grade plywood floor sheathing with 8d common nails into the blocking. The nail spacing required to transfer the unit shear to the blocking is obtained from UBC Table 23-I-G after adjusting for seismic loading and diaphragm construction in accordance with UBC Sections 2304.3.4 and 2340.3.6. The required spacing is

$$
\begin{aligned}
s &= 12 Z C_D C_{di}/v_1 \\
&= 12 \times 76 \times 1.33 \times 1.1/149 \\
&= 9 \text{ inches}
\end{aligned}
$$

On the other side of the drag strut a spacing of twelve inches is adequate.

To transfer the unit shear from the blocking to the drag strut, provide a 5 inch × 12 gauge angle connector with two ¾ inch diameter bolts in the blocking and a three inch long ⅛ inch fillet weld to the beam flange. With a connector at every ninth blocking, at a spacing of twelve feet, the required capacity of a connector is

$$\begin{aligned} P \quad &= 149 \times 12 \\ &= 1788 \text{ pounds} \end{aligned}$$

Neglecting the effect of the steel side plate, the capacity of the two bolts is obtained from UBC Table 23–I–F as

$$\begin{aligned} P_a \quad &= 2pC_D \\ &= 2 \times 720 \times 1.33 \\ &= 1915 \text{ pounds} \\ &> P \ldots \text{ satisfactory} \end{aligned}$$

The capacity of the weld in the 0.105 inch thick connector is

$$\begin{aligned} P_w \quad &= 928 \times 2 \times 1.33 \times 3 \times 0.105/0.125 \\ &= 6220 \text{ pounds} \\ &> P \ldots \text{ satisfactory} \end{aligned}$$

The shear capacity of the connector is

$$\begin{aligned} P_n \quad &= 0.4F_y \times 1.33t(b - 2\phi) \\ &= 0.4 \times 36,000 \times 1.33 \times 0.105(5 - 2 \times 0.8125) \\ &= 6787 \\ &> P \ldots \text{ satisfactory} \end{aligned}$$

The details are shown in Figure 88D–4(ii), Section B.

3. DRAG STRUT DETAIL ON LINE Y

The drag force at grid line 2 is

$$\begin{aligned} P \quad &= 260 \times 52 \\ &= 13,520 \text{ pounds} \end{aligned}$$

The double top plate of the shear wall forms a member with an area of

$$\begin{aligned} A_m \quad &= 5.5(2.5 + 1.5) \\ &= 22 \text{ square inches} \end{aligned}$$

A 4 inch × 5 gauge twisted strap tie, connected to the double top plate with six ¾ inch diameter bolts in two rows of three each and connected to the beam web with a 3/16 inch fillet weld, has an area of

$$\begin{aligned} A_s \quad &= 4 \times 0.21 \\ &= 0.84 \text{ square inches} \end{aligned}$$

The ratio of top plate to strap area is

$$A_m/A_s = 26.2$$

The reduction factor for three fasteners in a row is given by UBC Table 23–III–E as

$$C_g \quad = 0.98$$

Allowing for the effect of the steel side plate, the allowable lateral load on one $7/8$ inch diameter bolt in four inch thick Douglas Fir–Larch material is obtained from UBC Table 23–III–K as

$$Z \quad = 2035 \text{ pounds}$$

After allowing for the load duration factor and group action factor, the capacity of the six bolts is

$$\begin{aligned}
P_A \quad &= 6ZC_DC_g \\
&= 6 \times 2035 \times 1.33 \times 0.98 \\
&= 15{,}915 \text{ pounds} \\
&> P \ldots \text{satisfactory}
\end{aligned}$$

The capacity of two three inch long $3/16$ inch fillet welds is

$$\begin{aligned}
P_w \quad &= 928 \times 3 \times 1.33 \times 6 \\
&= 22{,}216 \text{ pounds} \\
&> P \ldots \text{satisfactory}
\end{aligned}$$

The tensile capacity of the 5 gauge plate, based on its net area, is given by

$$\begin{aligned}
P_n \quad &= 0.5F_u \times 1.33t(b - 2\phi) \\
&= 0.5 \times 58{,}000 \times 1.33 \times 0.21(4 - 2 \times 0.9375) \\
&= 17{,}212 \text{ pounds} \\
&> P \ldots \text{satisfactory}
\end{aligned}$$

The tensile stress in the double top plate is

$$\begin{aligned}
f_t \quad &= P/A \\
&= 13{,}520/4(5.5 - 2 \times 0.9375) \\
&= 932 \text{ pounds per square inch}
\end{aligned}$$

Provide a Douglas Fir–Larch No.2 grade top plate with an allowable stress, obtained from UBC Tables 23–I–A–1, 23–I–A–6, and 23–I–A–1 footnote 3 of

$$\begin{aligned}
F_t' \quad &= F_tC_DC_F \\
&= 575 \times 1.33 \times 1.3 \\
&= 994 \text{ pounds per square inch} \\
&> f_t \ldots \text{satisfactory}
\end{aligned}$$

The details are shown in Figure 88D–4(iii), Section C

4. DRAG STRUT DETAIL ON LINE 2

The drag force at grid line Y is

$$P = 400 \times 40 = 16,000 \text{ pounds}$$

The double 2×6 top plate of the shear wall forms a member with an area of

$$A_m = 5.5(1.5 + 1.5) = 16.5 \text{ square inches}$$

Provide a holdown type connection with a $1\frac{1}{8}$ inch diameter tie rod and eight $\frac{7}{8}$ inch diameter bolts in each holdown. The 5 inch by $\frac{3}{8}$ inch holdown base plate has an area of

$$A_s = 5 \times 0.375 = 1.875 \text{ square inches}$$

The ratio of top plate to base plate area is

$$A_m/A_s = 8.8$$

The corresponding reduction factor for four fasteners in a row is given by UBC Table 23–III–E as

$$C_g = 0.92$$

The allowable lateral load on one $\frac{7}{8}$ inch diameter bolt in three inch thick Douglas Fir–Larch material, with steel side plates, is obtained from UBC Table 23–III–K as

$$Z = 1670 \text{ pounds}$$

After allowing for the load duration factor and the group action factor, the capacity of the eight bolts is

$$P_A = 8ZC_DC_g = 8 \times 1670 \times 1.33 \times 0.96 = 17,058 \text{ pounds} > P \text{ ... satisfactory}$$

The tensile capacity of the holdown base plate is governed by its net area and is given by

$$P_n = 0.5F_u \times 1.33t(b - 2\phi) = 0.5 \times 58,000 \times 1.33 \times 0.375(5 - 2 \times 0.9375) = 45,200 \text{ pounds} > P \text{ ... satisfactory}$$

The tensile capacity of the $1\frac{1}{8}$ inch diameter tie bar is given by AISC Table 1–B as

$$P_t = 19,000 \text{ pounds} > P \text{ ... satisfactory}$$

The tensile stress in the double top plate is

$$f_t = P/A$$
$$= 16{,}000/3(5.5 - 2 \times 0.9375)$$
$$= 1471 \text{ pounds per square inch}$$

Provide a Douglas Fir-Larch select structural grade top plate with an allowable stress, obtained from UBC Table 23-I-A-1, 23-I-A-6, and 23-I-A-1 footnote 3 of

$$F'_t = F_t C_D C_F$$
$$= 1000 \times 1.33 \times 1.3$$
$$= 1729 \text{ pounds per square inch}$$
$$> f_t \ldots \text{ satisfactory}$$

The details are shown in Section D, Figure 88D-4(iii), with bolt spacing conforming to the requirements of UBC Tables 23-III-H and 23-III-I.

5. ITEMS OF CONCERN

Items causing concern in this type of construction are the additional expense incurred compared with normal construction, the necessity to produce special joist hangers and holdowns, and the danger of causing a fire during the welding operations.

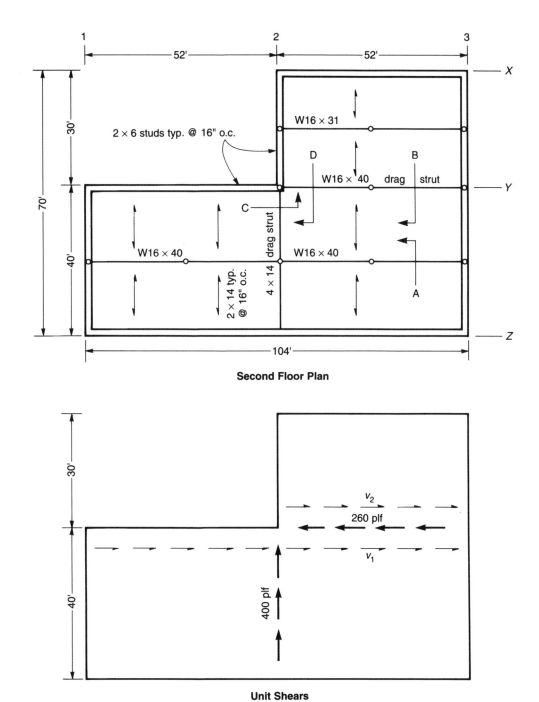

Second Floor Plan

Unit Shears

FIGURE 88D–4(i)

459

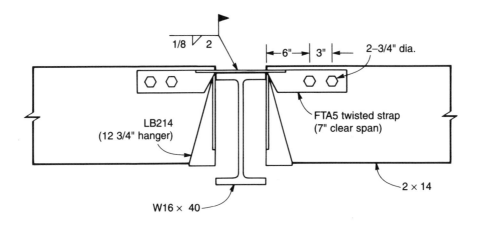

Section A

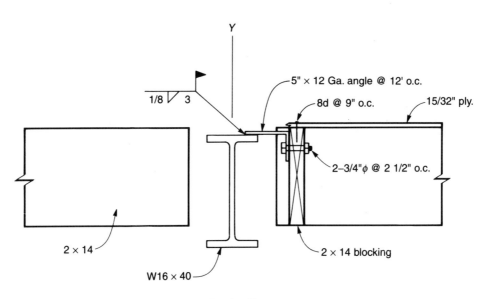

Section B

FIGURE 88D–4(ii)

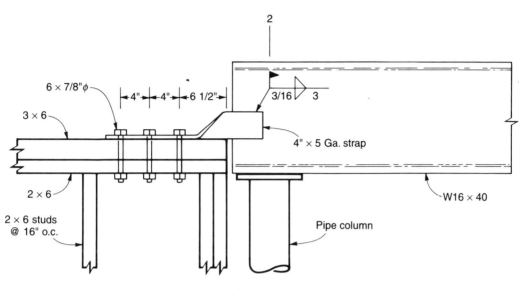

Section C

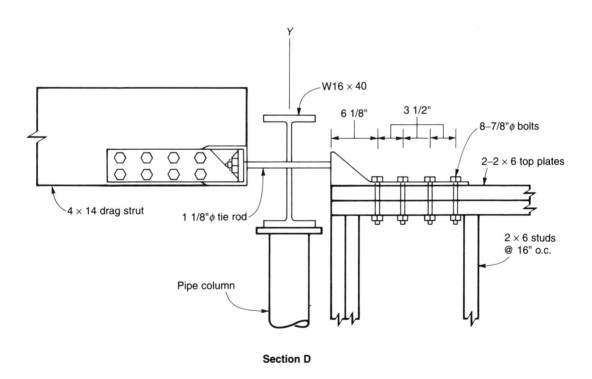

Section D

FIGURE 88D-4(iii)

461

STRUCTURAL ENGINEER EXAMINATION – 1987 ═══════════════

PROBLEM D–4 – WT. 3.0 POINTS

GIVEN: Roof Load R: dead load = 5.0 kips, live load = 3.0 kips.
                             Seismic collector force H = 20.0 kips.
                             The beam is a $5\frac{1}{8}$" × 15" Glulam.

CRITERIA: Species: Douglas Fir–Larch.
                             Combination: 24F–V4.
                             Bolts: A307 Machine Bolts.
                             Steel Plates: A36 grade.
                             Assume bearing stress due to load R is uniform over the bearing surface.

REQUIRED: a. Bolt diameter.
                           b. Dimension A (minimum in inches).
                           c. Dimension B (minimum in inches).
                           d. Dimension C (minimum in inches).
                           e. Dimension D (minimum in inches).
                           f. Maximum horizontal shear stress in the glulam beam due to the gravity load.

SOLUTION

a. BOLT DIAMETER

For a ⅞ inch diameter bolt, the allowable lateral load in double shear on a 5⅛ inch glued–laminated beam, with ¼ inch side plates, is obtained from UBC Table 23–III–Q as

$$Z \quad = 4260 \text{ pounds}$$

The area of the glued–laminated beam is

$$A_m \quad = 76.9 \text{ square inches}$$

The area of the two side plates is

$$\begin{aligned} A_s \quad &= 2 \times 0.25 \times 8.0 \\ &= 4.0 \text{ square inches} \end{aligned}$$

The ratio of the areas is

$$A_m/A_s = 19.2$$

The corresponding group action factor for two fasteners in a row is given by UBC Table 23–III–E as

$$C_g \quad = 1.00$$

The load duration factor, for seismic loads, is given by UBC Section 2304.3.4 as

$$C_D \quad = 1.33$$

Then the adjusted lateral load factor for the four bolts is

$$\begin{aligned} P'_L \quad &= 4 \times ZC_D \\ &= 4 \times 4260 \times 1.33 \\ &= 22{,}663 \text{ pounds} \\ &> H \ \dots \text{ satisfactory} \end{aligned}$$

b. DIMENSION A

Allow a half inch clearance to the end of the beam giving

$$\ell_A \quad = 0.50 \text{ inches}$$

c. DIMENSION B

The minimum end distance in tension for a ⅞ inch diameter bolt, in parallel–to–grain loading, is given by UBC Table 23–III–H as

$$\begin{aligned} \ell_n \quad &= 7D \\ &= 7 \times 0.875 \\ &= 6.125 \text{ inches} \end{aligned}$$

d. DIMENSION C

The minimum spacing for a ⅞ inch diameter bolt, in parallel–to–grain loading, is given by UBC Table 23–III–I as

$$\ell_c = 4D$$
$$= 4 \times 0.875$$
$$= 3.5 \text{ inches}$$

e. DIMENSION D

For compression perpendicular to the grain, no duration of load modification factor is applicable and the allowable stress is given by UBC Table 23–I–C–1 as

$$F_{c\perp} = 650 \text{ pounds per square inch}$$

The required minimum bearing length is, then

$$\ell_b = R/5.125F_{c\perp}$$
$$= 8000/(5.125 \times 650)$$
$$= 2.4 \text{ inches}$$

Hence the minimum required dimension D is

$$\ell_D = \ell_b + \ell_A$$
$$= 2.4 + 0.5$$
$$= 2.9 \text{ inches}$$

f. SHEAR STRESS

Dead load plus live load, with a load duration factor of 1.25, governs. Allowing for the self weight of the beam, the shear force a distance from the support equal to the depth of the beam is

$$V = R - 18.2 \times 15/12$$
$$= 8000 - 18.2 \times 15/12$$
$$= 7977 \text{ pounds}$$

The shear stress in the beam is, then

$$f_v = 1.5V/A$$
$$= 1.5 \times 7977/76.9$$
$$= 156 \text{ pounds per square inch}$$

The allowable shear stress is obtained from UBC Table 23–I–C–1 as

$$F_v = F_v C_D$$
$$= 165 \times 1.25$$
$$> f_v \dots \text{ satisfactory}$$

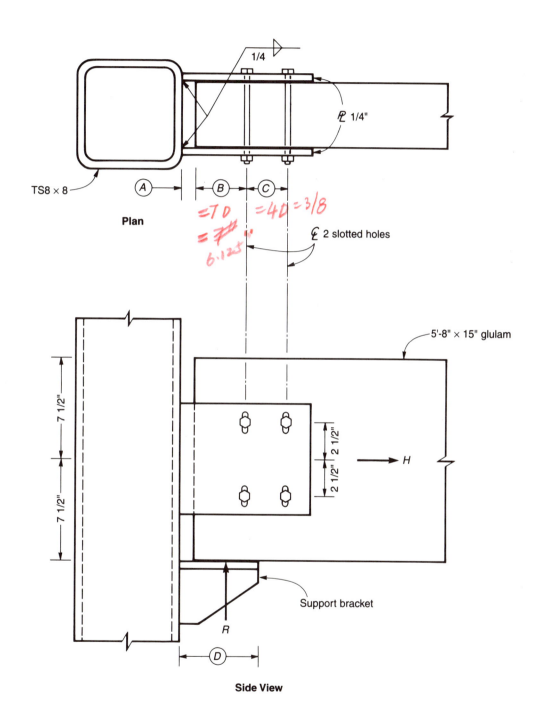

STRUCTURAL ENGINEERING REGISTRATION

CHAPTER 4

STRUCTURAL TIMBER DESIGN

SECTION 4.6

RETROFIT APPLICATIONS

PROBLEMS:

1991 D-1

STRUCTURAL ENGINEER EXAMINATION – 1991 ═══════════

PROBLEM D-1 – WT. 8.0 POINTS

GIVEN: An existing two story building with unreinforced brick walls and wood floor and roof is to be strengthened to improve its seismic resistance. Plans and details are as shown.

CRITERIA: Seismic zone 4.
Standard occupancy building.

Materials:
- Use the following allowable stresses for all new and existing wood: F_b = 1250 pounds per square inch, F_v = 95 pounds per square inch, $F_{c\perp}$ = 625 pounds per square inch.
- Bolts are ASTM A307.
- Brick wall density is 120 pounds per cubic foot.

Assumptions:
- The new steel braced frame on line '1' and the new shotcrete shear wall on line 'Z' are adequate.
- The existing brick walls on lines 'W' and '5' are adequate for shear and are adequate for out-of-plane bending if properly tied at the floor and roof.

REQUIRED: Refer to detail A for questions a through e.

    a.   Find the required spacing of nails 'a'. Use 16d common nails.
    b.   Check the adequacy of bolts 'b' to transfer diaphragm shear to the brick wall. (Check wood only, bolts are adequate in the brick).
    c.   Find the tension in bolts 'b' due to seismic out-of-plane wall loads.

Assume the tension in bolts 'b' is 1700 pounds per bolt to answer parts d and e.

    d.   Find the number of nails 'd' required through the plywood to each block. Use 16d common nails. The strap shown is adequate.
    e.   The blocking acts as a continuous cross-tie for a sub-diaphragm four feet wide. find the design shear in the sub-diaphragm, assuming adequate continuous cross-ties are provided across the building on lines 'X' and 'Y'.

Refer to detail B for questions f through h. Assume the seismic out-of-plane wall load is 500 pounds per linear foot.

f. Check block 'f' for shear and bending due to out-of-plane wall loads. To allow for construction tolerance, assume the bolt may be as much as 3" from the center of the joist space. (See the plan view in detail B.) Neglect the reduction in the cross section due to the bolt hole and neglect the hanger length.

g. Find the required size of square plate washer 'g'. Assume the ¼" thickness is adequate. Account for the reduced bearing area due to a ⅞" diameter hole drilled in the block.

h. Find the sub-diaphragm design shear assuming a single continuous cross-tie is provided along line '3' from line 'W' to line 'Z'. The typical joist laps at lines 'X' and 'Y' have no tension capacity.

SOLUTION

a. DIAPHRAGM NAILING

The seismic load tributary to the second floor diaphragm is given by

$$V = 530 \times 80$$
$$= 42,400 \text{ pounds}$$

The unit shear in the diaphragm along grid line 5, assuming a flexible diaphragm, is

$$q = V/(2L)$$
$$= 42,400/(2 \times 45)$$
$$= 471 \text{ pounds per linear foot}$$

The allowable lateral load on a 16d common nail, with a penetration of 1.94 inches is obtained from UBC Table 23-I-G as

$$Z = 105 \text{ pounds}$$

This value must be adjusted by the following factors[3,5,6].

C_D = load duration factor
= 1.33 for seismic loading from UBC Section 2304.3.4
C_{di} = diaphragm factor
= 1.30 in accordance with NDS Commentary[8] Section 12.3.6 when a load duration factor of 1.33 is used for seismic loading
1.0 for 3 inch nominal framing[5,6]
1.0 for a single line of nails with spacing exceeding two inches[6]

The actual penetration of the nail into the framing is obtained from the nail length, which is given in UBC Table 23-I-G as 3.5 inches, and the thickness of the existing sheathing and the new plywood diaphragm.

Thus the actual penetration is

$$p \quad = 3.5 - 1.5 - 0.5 \quad = 1.5 \text{ inches}$$

and, in accordance with UBC Section 2340.3.4 the penetration depth factor is given by Formula (40–5) as

$$C_d \quad = p/12D$$

where $\quad D \quad = $ nail diameter

$$= 0.162 \text{ inches from UBC Table 23–III–II}$$

Then $\quad C_d \quad = 1.5/(12 \times 0.162) = 0.77$

Hence, the adjusted lateral load is

$$
\begin{aligned}
Z' \quad &= Z \times C_D \times C_{di} \times C_d \\
&= 105 \times 1.33 \times 1.3 \times 0.77 \\
&= 140 \text{ pounds}
\end{aligned}
$$

The required spacing is then

$$
\begin{aligned}
s \quad &= Z' \times 12/q \\
&= 140 \times 12/471 \\
&= 3.6 \text{ inches}
\end{aligned}
$$

The strength of the ½ inch structural I plywood diaphragm is fully developed by a 10d nail[3]. The allowable shear on ½ inch structural I plywood using three inch nominal framing members and 10d nails at 4 inch centers is obtained from UBC Table 23–I–J–1, for case 1 loading, as

$$
\begin{aligned}
q_a \quad &= 480 \text{ pounds per linear foot} \\
&> q
\end{aligned}
$$

Hence the 3.6 inch spacing of the 16d nails is satisfactory.

b. LEDGER BOLT

The lateral seismic force acting on the bolt spaced at 3.5 feet centers is

$$
\begin{aligned}
F \quad &= 3.5q \\
&= 3.5 \times 471 \\
&= 1649 \text{ pounds}
\end{aligned}
$$

The allowable, parallel-to-grain load on a ¾ inch diameter anchor bolt in the two and a half inch thick Douglas Fir-Larch ledger connected to a masonry wall is, in accordance with UBC Section 2311.2, obtained as half the double shear value tabulated in UBC Table 23–I–F for a five inch thick member. Hence the allowable seismic load is

$$
\begin{aligned}
Z'_{\parallel} \quad &= 1.33 \times 3220/2 \\
&= 2141 \text{ pounds} \\
&> F
\end{aligned}
$$

Hence the bolts are satisfactory.

c. WALL ANCHORAGE

The anchorage force is due to that portion of the wall dead load which is tributary to the second floor diaphragm. The tributary dead load is

$$W_p = 170 \times 7 + 130 \times 6$$
$$= 1970 \text{ pounds per linear foot}$$

The pull-out force between the wall and the diaphragm is given by UBC Formula (30-1) as

$$F_p = ZI_pC_pW_p$$

where
$Z = 0.4$ for zone 4 from UBC Table 16-I
$I_p = 1.0$ from UBC Table 16-K for a standard occupancy structure
$C_p = 0.75$ from UBC Table 16-O item I.1.b.
$W_p = 1970$ as calculated

Then
$$F_p = 0.4 \times 1 \times 0.75 \times 1970$$
$$= 591 \text{ pounds per linear foot}$$

and this value governs as it exceeds the minimum value of 200 pounds per linear foot specified in UBC Section 1611. The force on each anchor bolt, for a spacing of three feet six inches, is given by

$$P = 3.5F_p$$
$$= 3.5 \times 591$$
$$= 2069 \text{ pounds}$$

d. SUBDIAPHRAGM NAILING

For a pull-out force of 1700 pounds, the force to be transferred to each block is given by

$$P = 1700/3$$
$$= 567 \text{ pounds.}$$

The allowable lateral load on a 16d common nail with a penetration of 1.5 inches, and allowing for the increase of one-third for seismic loading, is given by

$$Z' = 105 \times 1.33 \times 0.77$$
$$= 108 \text{ pounds}$$

Hence, the number of nails required in each block is given by

$$n = P/Z'$$
$$= 567/108$$
$$= 6 \text{ nails per block}$$

e. SUBDIAPHRAGM SHEAR

For an anchorage force of 1700 pounds and an anchor spacing of three feet six inches, the pull-out force between the wall and the diaphragm is

$$F_p = 1700/3.5$$
$$= 486 \text{ pounds per linear foot}$$

For a subdiaphragm of dimensions four feet by fifteen feet, the unit design shear is given by

$$q = 15F_p/(2 \times 4)$$
$$= 15 \times 486/(2 \times 4)$$
$$= 911 \text{ pounds per linear foot}$$

f. ANCHOR BLOCK

The pull-out force between the wall and the diaphragm is given as

$$F_p = 500 \text{ pounds per linear foot}$$

The force on each anchor bolt, for a spacing of thirty-two inches, is given by

$$P = F_p \times 32/12$$
$$= 500 \times 32/12$$
$$= 1333 \text{ pounds}$$

The maximum bending moment on the anchor block is

$$M = P\ell/4$$
$$= 1333 \times 14.5/4$$
$$= 4833 \text{ pound inches}$$

The corresponding maximum bending stress in the block is given by

$$f_b = M/S$$
$$= 4833/17.6$$
$$= 275 \text{ pounds per square inch}$$
$$< 1.33F_b \ldots \text{ satisfactory}$$

The maximum shear force on the block occurs with the bolt located three inches from the center of the block to give a shear force of

$$V = P \times 10.25/14.5$$
$$= 1333 \times 10.25/14.5$$
$$= 942 \text{ pounds}$$

The shear stress in the block is given by

$$f_v = 1.5V/A$$
$$= 1.5 \times 942/19.25$$
$$= 73 \text{ pounds per square inch}$$
$$< 1.33F_v \ldots \text{ satisfactory}$$

g. BEARING PLATE

In accordance with UBC Section 2306.6 no increase is allowable in bearing stress for seismic forces and, assuming a plate size of 1.5 inches, a size factor of 1.25 is applicable. Hence, the net bearing area required is given by

$$
\begin{aligned}
A_n &= P/1.25F_{c\perp} \\
&= 1333/(1.25 \times 625) \\
&= 1.706 \text{ square inches}
\end{aligned}
$$

The area of the ⅞ inch diameter hole in the bearing plate is

$$
A_h = 0.601 \text{ square inches}
$$

The gross area of bearing plate required is given by

$$
\begin{aligned}
A &= A_n + A_h \\
&= 1.706 + 0.601 \\
&= 2.307 \text{ square inches}
\end{aligned}
$$

Hence the required bearing plate size is

$$
\begin{aligned}
a &= (A)^{0.5} \\
&= (2.307)^{0.5} \\
&= 1.5 \text{ inches}
\end{aligned}
$$

h. SUBDIAPHRAGM SHEAR

For a subdiaphragm of dimensions fifteen feet by forty feet, the unit design shear is given by

$$
\begin{aligned}
q &= 40F_p/(2 \times 15) \\
&= 40 \times 500/(2 \times 15) \\
&= 667 \text{ pounds per linear foot}
\end{aligned}
$$

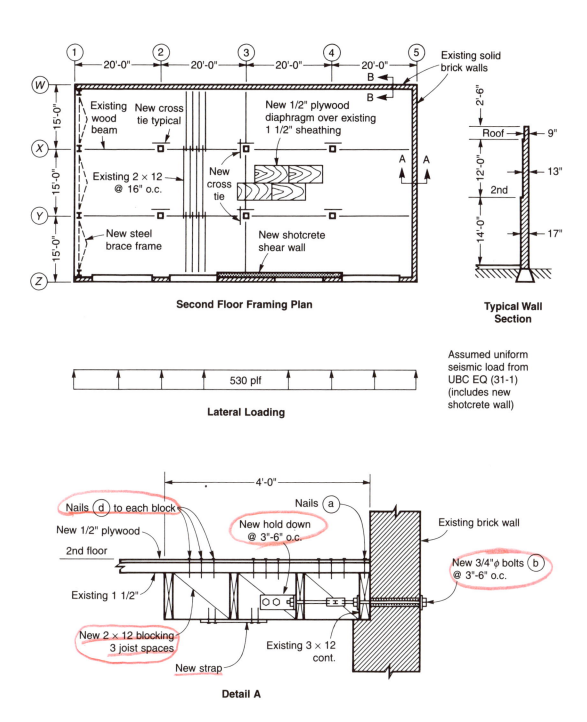

Second Floor Framing Plan

Typical Wall Section

Lateral Loading

Assumed uniform seismic load from UBC EQ (31-1) (includes new shotcrete wall)

Detail A

FIGURE 91D/1(i)

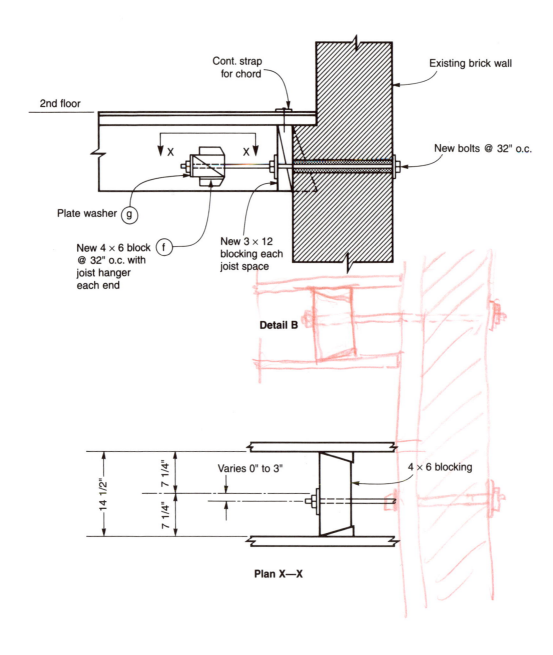

Cont. strap
for chord

Existing brick wall

2nd floor

New bolts @ 32" o.c.

Plate washer (g)

New 4 × 6 block (f)
@ 32" o.c. with
joist hanger
each end

New 3 × 12
blocking each
joist space

Detail B

14 1/2"

7 1/4"

7 1/4"

Varies 0" to 3"

4 × 6 blocking

Plan X—X

FIGURE 91D/1(ii)

STRUCTURAL ENGINEERING REGISTRATION

CHAPTER 4

STRUCTURAL TIMBER DESIGN

SECTION 4.7

ROOF FRAMING

PROBLEMS:

1991 D–3
1989 D–1

STRUCTURAL ENGINEER EXAMINATION – 1991

PROBLEM D-3 – WT. 6.0 POINTS

GIVEN: Roof framing plan, detail A, and loading diagram.

CRITERIA: Materials:
- Glued–laminated timber: combination symbol 24F–V4.
- Sawn lumber: Douglas Fir-Larch, select structural grade.

Assumptions:
- Neglect wind and seismic forces.
- Do not use the UBC Appendix Chapter 16, Division I provisions, for snow loading determination.
- Moisture content of the timber in service does not exceed 19 percent.
- Self weight of beams and/or girders is included in the given roof dead load.

REQUIRED: Show all calculations for the following:
1. Check adequacy of the rafters to resist bending and shear forces.
2. Check adequacy of the glued–laminated girders to resist bending forces if the total load from the post that supports the ridge and hip beams = 9.0 kips.

SOLUTION:

1. RAFTERS

BASIC DESIGN STRESSES

The tabulated basic design stresses for 2 × 10 Douglas Fir-Larch select structural grade lumber are obtained from UBC Table 23–I–A–1 and are

$$F_b = 1450 \text{ pounds per square inch}$$
$$F_v = 95 \text{ pounds per square inch}$$

MODIFIED STRESSES

The moisture content in service does not exceed 19 percent and, in accordance with UBC Section 2304.3.10, no modification to the basic design values is necessary for moisture service conditions. In accordance with UBC Table 23-I-A-1 footnote 3 a size factor adjustment C_F of 1.1 is applicable for 2 × 10 lumber. In addition, the tops of the rafters are braced by the plywood sheathing and no stability factor adjustment is necessary. A snow load duration factor C_D of 1.15 is applicable in accordance with UBC Section 2304.3.4. A repetitive member factor C_r of 1.15 is applicable in accordance with UBC Table 23-I-A-1 footnote 4. In accordance with UBC Table 23-I-A-6 the modified stresses are

$$
\begin{aligned}
F_b' &= F_b C_D C_F C_r \\
&= 1450 \times 1.15 \times 1.1 \times 1.15 \\
&= 2109 \text{ pounds per square inch} \\
F_v' &= F_v C_D \\
&= 95 \times 1.15 \\
&= 109 \text{ pounds per square inch}
\end{aligned}
$$

APPLIED LOADING

The roof slope is

$$
\begin{aligned}
\theta &= \tan^{-1}(9/12) \\
&= 36.87° \\
&> 20°
\end{aligned}
$$

The allowable reduction in snow load is given by UBC Section 1605.4 as

$$
\begin{aligned}
R_S &= S/40 - \tfrac{1}{2} \\
&= 70/40 - \tfrac{1}{2} \\
&= 1.25 \text{ pounds per square foot per degree of pitch over } 20°
\end{aligned}
$$

The reduced snow load is

$$
\begin{aligned}
w_S &= S - R_S(\theta° - 20°) \\
&= 70 - 1.25(36.87-20) \\
&= 48.9 \text{ pounds per square foot}
\end{aligned}
$$

The total load acting on the horizontal plane on one rafter is

$$
\begin{aligned}
w_T &= (w_D + w_S)16/12 \\
&= (20 + 48.9)\, 16/12 \\
&= 91.9 \text{ pounds per linear foot}
\end{aligned}
$$

The bending moment at mid span and the shear force at the support in a rafter due to the total load are

$$
\begin{aligned}
M &= w_T \ell_h^2/8 \\
&= 91.9 \times 15^2/8 \\
&= 2585 \text{ pound feet}
\end{aligned}
$$

$$V = (w_T \ell_h \cos\theta)/2$$
$$= (91.9 \times 15 \times \cos 36.87°)/2$$
$$= 551 \text{ pounds}$$

SECTION PROPERTIES

The relevant properties of a 2×10 sawn lumber joist are

$$A = 13.9 \text{ inches}^2$$
$$S = 21.39 \text{ inches}^3$$

MEMBER STRESSES

The flexural stress at mid span is

$$f_b = M/S$$
$$= 2585 \times 12/21.39$$
$$= 1450 \text{ pounds per square inch}$$
$$< F_b' C_D \dots \text{satisfactory}$$

The shear stress a distance d from the support is

$$f_v = 1.5(V - dw_T \cos^2\theta)/A$$
$$= 1.5(551 - 9.25 \times 91.9\cos^2 36.87°/12)/13.9$$
$$= 55 \text{ pounds per square inch}$$
$$< F_v' C_D \dots \text{satisfactory}$$

2. GLULAM GIRDER

MEMBER PROPERTIES

The relevant section properties of the $5\frac{1}{8} \times 24$ glulam beam are

$$A = 123 \text{ inches}^2$$
$$S = 492 \text{ inches}^3$$
$$w = \text{self weight}$$
$$= 29 \text{ pounds per linear foot}$$

The relevant tabulated basic design stresses are obtained from UBC Table 23-I-C-1 and are

$$F_b = 2400 \text{ pounds per square inch}$$
$$F_v = 165 \text{ pounds per square inch}$$
$$E = 1.8 \times 10^6 \text{ pounds per square inch}$$

The effective span, in accordance with UBC Section 2306.1, is

$$\ell = 30 - 2 + 1$$
$$= 29 \text{ feet}$$

Assume that rotation and lateral movement at the ends of the glulam beam are prevented by nailing the plywood sheathing to the sloping end of the beam. Then the unsupported length of the compression edge of the beam is

$$\ell_u = 26.3 \text{ feet}$$
$$\ell_u/d = 26.3/2$$
$$= 13.2$$
$$> 7$$

For a single span beam with a concentrated load at mid span, the effective unbraced length is given by UBC Section 2363.1 as

$$\ell_e = 1.37\ell_u + 3d \dots \text{ for } \ell_u/d > 7$$
$$= 1.37 \times 26.3 + 3 \times 2$$
$$= 42.03 \text{ feet}$$

The allowable stress, adjusted for stability or volume factors, is obtained from calculator program A.3.2 in Appendix A.

The slenderness factor is given by UBC Section 2363.1 as

$$R_B = (\ell_e d/b^2)^{0.5}$$
$$= 21.47$$
$$< 50 \dots \text{ satisfactory}$$

$$\sqrt{\frac{(42.03 \times 12) \times 24}{5.125^2}}$$

The critical buckling design value is given by UBC Section 2363.2 as

	F_{bE}	$= K_{bE}E'/R_B^2$
where	K_{bE}	= Euler buckling coefficient
		= 0.609 ... for glued–laminated timber
	E'	= adjusted tabulated modulus of elasticity
		$= 1.8 \times 10^6$ pounds per square inch $\times 1.15$
and	F_{bE}	= 2379 pounds per square inch

The volume factor is given by UBC Section 2312.4.5 as

	C_V	$= k(1291.5/bdL)^{1/x}$
where	k	= loading condition coefficient
		= 1.09 ... for a central concentrated load
	b	= width of member
		= 5.125 inches
	d	= depth of member
		= 24 inches
	L	= distance between points of zero moment
		= 29 feet
	x	= 10 ... for western species timber
and	C_V	= 0.98

changed

$$C_V = k\left(\frac{12}{d}\right)^{1/x}\left(\frac{1.125}{b}\right)^{1/x}\left(\frac{21}{L}\right)^{1/x} \leq 1.0$$

The beam stability factor is given by UBC Section 2363.2 as

$$C_L = (1.0 + F)/1.9 - \{[(1.0 + F)/1.9]^2 - F/0.95\}^{0.5}$$

where
$$F_b^* = \text{adjusted tabulated bending stress}$$
$$= F_b C_D$$
$$= 2400 \times 1.15$$
$$= 2760 \text{ pounds per square inch}$$

$$F = F_{bE}/F_b^*$$
$$= 0.86$$

and
$$C_L = 0.75$$
$$< C_V$$

Then slenderness effects govern and the allowable flexural stress is

$$F_b' = F_b^* C_L$$
$$= 2069 \text{ pounds per square inch}$$

APPLIED LOADS

The bending moment at mid span and the shear force at the support due to dead load plus snow load are

$$M = W\ell/4 + w\ell^2/8$$
$$= 9000 \times 29/4 + 29 \times 29^2/8$$
$$= 65{,}250 + 3049$$
$$= 68{,}299 \text{ pound feet}$$

$$V = W/2 + w\ell/2$$
$$= 9000/2 + 29 \times 29/2$$
$$= 4920 \text{ pounds}$$

MEMBER STRESSES

The flexural stress at mid span is

$$f_b = M/S$$
$$= 68{,}299 \times 12/492$$
$$= 1666 \text{ pounds per square inch}$$
$$< F_b' \dots \text{satisfactory}$$

The shear stress a distance d from the support is

$$f_v = 1.5(V - dw)/A$$
$$= 1.5(4920 - 2 \times 29)/123$$
$$= 59 \text{ pounds per square inch}$$
$$< F_v C_D \dots \text{satisfactory}$$

The girder is adequate for the applied loads. ═══════════════

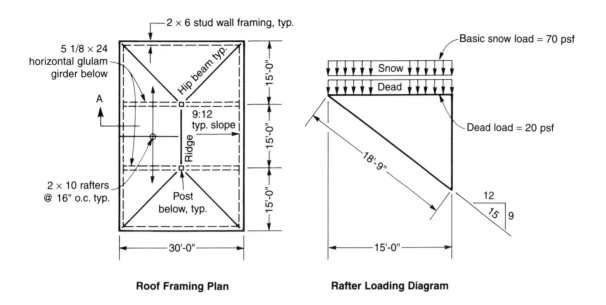

Roof Framing Plan

Rafter Loading Diagram

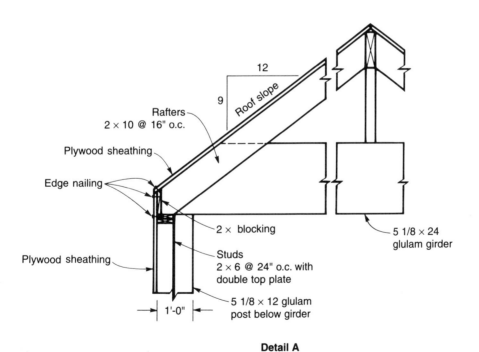

Detail A

FIGURE 91D-3

481

STRUCTURAL ENGINEER EXAMINATION – 1989 ═══════════════

PROBLEM D–1 – WT. 6.0 POINTS

GIVEN:	A building cross section as shown.

CRITERIA:

- Wind zone 70 mph, exposure C (use method 1).
- Use $C_e = 1.40$ (average) and $C_q = 0.7$ (in and out).

Materials:
- Rafters: Douglas Fir–Larch No.2 grade.
- Glued–laminated beam: Douglas Fir–Larch grade 22F–V3.
- $E = 1,600,000$ pounds per square inch.

Assumptions:
- Roof dead load = 12.0 pounds per square foot on the inclined surface, including weight of the rafter.
- Live load as specified in the UBC.
- Assume pin connections at ridge and bases.
- No increase is allowable for repetitive use.
- Assume the buttress walls are rigid and adequate.
- Assume that the rafters are adequately blocked.

REQUIRED:

1. Determine the required minimum size of 2 × rafters at 24 inch on center (neglect the component parallel to the roof surface).
2. Determine the vertical (V_A and V_B) and the horizontal (H_A and H_B) glulam beam reactions due to:
    a. Dead load.
    b. Wind load.
3. Check the adequacy of the glulam beam due to dead load plus wind:
    a. Maximum shear stress.
    b. Maximum combined stress.

SOLUTION

1. RAFTERS

LOAD COMBINATIONS

The applicable load combinations are given in UBC Section 1603.6 as

(i)   Dead load plus roof live load with an increase of twenty-five percent in permissible stress, in accordance with UBC Section 2304.3.4.

(ii) Dead load plus wind load with an increase of sixty percent in permissible stress for member design, in accordance with UBC Section 2304.3.4.

BASIC DESIGN STRESSES

The tabulated basic design stresses for a 2 × Douglas Fir-Larch No.2 rafter are obtained from UBC Table 23-I-A-1 and are

$$F_b \quad = 875 \text{ pounds per square inch}$$
$$F_v \quad = 95 \text{ pounds per square inch}$$

In accordance with UBC Table 23-I-A-1 footnote 3, and assuming 2 × 10 lumber, a size factor adjustment C_F of 1.1 is applicable. The tops of the rafters are braced by the plywood sheathing and no slenderness factor adjustment is necessary for downward loads. In addition, the bottoms of the rafters are adequately braced by means of blocking and no adjustment is necessary for stress reversal. Hence only downward loading need be considered. A load duration factor C_D of 1.25 is applicable for roof live load, and a load duration factor of 1.60 is applicable for wind load, in accordance with UBC Section 2304.3.4.

APPLIED LOADING

The dead load acting along the longitudinal axis of a rafter is

$$w_D \quad = 12 \times 2\cos 40°$$
$$= 18.38 \text{ pounds per linear foot}$$

For a tributary area of less than 200 square feet and a rise of 10 inches per foot, the applicable live load is obtained from UBC Table 16-C as 16 pounds per square foot of horizontal projection. The live load acting along the longitudinal axis of a rafter is

$$w_L \quad = 16 \times 2\cos^2 40$$
$$= 18.78 \text{ pounds per linear foot}$$

The wind stagnation pressure for a seventy miles per hour wind speed is obtained from UBC Table 16-F as

$$q_s \quad = 12.6 \text{ pounds per square foot}$$

Then in accordance with UBC Formula (18–1) the design wind pressure, normal to the roof surface, is given by

$$p = C_e C_q q_s I_w$$
$$= 1.4 \times 0.7 \times 12.6 \times 1.0$$
$$= 12.35 \text{ pounds per square foot}$$

The wind load acting along the longitudinal axis of a rafter is

$$w_W = 2p$$
$$= 2 \times 12.35$$
$$= 24.70 \text{ pounds per linear foot}$$

Allowing for the load duration factor for roof live load and the size factor, the equivalent dead plus wind load on the rafter is

$$w = (w_D + w_W)/C_D C_F$$
$$= (18.38 + 24.70)/(1.60 \times 1.1)$$
$$= 24.48 \text{ pounds per linear foot}$$

The equivalent dead plus roof live load on the rafter is

$$w_T = (w_D + w_L)/C_D C_F$$
$$= (18.38 + 18.78)/(1.25 \times 1.1)$$
$$= 27.03 \text{ pounds per linear foot.}$$

and this value governs.

The bending moment and shear force in a rafter due to the dead plus roof live load are given by

$$M = w\ell^2/8$$
$$= 12 \times 27.03(20 - 8.75/12)^2/8$$
$$= 15,055 \text{ pound inches}$$
$$V = w(\ell - 2d)/2$$
$$= 27.03(19.27 - 1.54)/2$$
$$= 240 \text{ pounds}$$

The required section modulus for a rafter is given by

$$S = M/F_b$$
$$= 15,055/875$$
$$= 17.21 \text{ inches}^3$$

Selecting a 2×10 member, with a section modulus of 21.39, the shear stress in a rafter is given by

$$f_v = 1.5 \, V/A$$
$$= 1.5 \times 240/13.88$$
$$= 26 \text{ pounds per square inch}$$
$$< F_v \dots \text{satisfactory}$$

Hence a 2×10 member is satisfactory.

2. BEAM REACTIONS

The length of each beam is

$$\ell = 59/\cos 40° = 77 \text{ feet}$$

The total dead load, including self weight, on each beam is given by

$$W_D = 77(12 \times 20 + 81) = 24{,}717 \text{ pounds}$$

The vertical and horizontal reactions due to the dead load are

$$V_A = V_B = W_D = 24{,}717 \text{ pounds, acting upwards.}$$
$$H_A = H_B = 59W_D/(2 \times 49.5) = 14{,}730 \text{ pounds, acting inward.}$$

The total wind load on each beam is given by

$$W_W = 77 \times 20 \times 12.35 = 19{,}019 \text{ pounds}$$

acting inward on the windward slope and outward on the leeward slope. The vertical and horizontal components of this load are

$$V_W = W_W \cos 40° = 14{,}569 \text{ pounds}$$
$$H_W = W_W \sin 40° = 12{,}225 \text{ pounds.}$$

The horizontal and vertical reactions due to the wind load are

$$H_A = H_B = H_W = 12{,}225 \text{ pounds, acting to the left.}$$
$$V_A = (59V_W - 49.5H_W)/(2 \times 59) = 2156 \text{ pounds, acting upwards.}$$
$$V_B = 2156 \text{ pounds, acting downwards.}$$

3. GLULAM BEAM

The left hand glulam beam has the higher shear and moment due to the combined dead and wind loading. The top of the beam is continuously braced by the rafters and plywood sheathing and no slenderness factor adjustment is necessary for downward loads. For axial loads, the rafters are not braced in the plane of bending (the strong axis) and stability effects must be considered. The right hand glulam beam has the higher axial stress.

485

BASIC DESIGN STRESSES

The tabulated basic design stresses for a Douglas Fir–Larch 22F–V3 glulam beam are obtained from UBC Table 25–C–1–A and are

$$F_b \quad = 2200 \text{ pounds per square inch}$$
$$F_v \quad = 165 \text{ pounds per square inch}$$
$$F_c \quad = 1500 \text{ pounds per square inch}$$
$$E \quad = 1.6 \times 10^6 \text{ pounds per square inch}$$

SECTION PROPERTIES

The relevant properties of an $8\frac{3}{4} \times 39$ glulam beam are

$$A \quad = 341.3 \text{ inches}^2$$
$$S \quad = 2218 \text{ inches}^3$$
$$C_D \quad = \text{load duration factor for wind}$$
$$\quad = 1.60$$

APPLIED LOADS

The horizontal and vertical reactions at support A due to the combined dead and wind loads are

$$R_H \quad = 14{,}730 - 12{,}225$$
$$\quad = 2505 \text{ pounds, inward}$$
$$R_V \quad = 24{,}717 + 2156$$
$$\quad = 26{,}873 \text{ pounds}$$

At the mid span of the left hand beam, the moment, shear, and axial forces are given by

$$M_c \quad = 59W_D/8 + 77W_W/8$$
$$\quad = 59 \times 24{,}717/8 + 77 \times 19{,}019/8$$
$$\quad = 365{,}346 \text{ pounds feet}$$
$$Q_c \quad = 0$$
$$P_c \quad = R_V \sin 40° + R_H \cos 40° - W_D(\sin 40°)/2$$
$$\quad = 26{,}873 \sin 40° + 2505 \cos 40° - (24{,}717 \sin 40°)/2$$
$$\quad = 11{,}249 \text{ pounds.}$$

At support A, the moment, shear, and axial forces are given by

$$M_A \quad = 0$$
$$Q_A \quad = R_V \cos 40° - R_H \sin 40°$$
$$\quad = 26{,}873 \cos 40° - 2505 \sin 40°$$
$$\quad = 18{,}976 \text{ pounds}$$
$$P_A \quad = R_V \sin 40° + R_H \cos 40°$$
$$\quad = 19{,}193 \text{ pounds}$$

At the support B, the axial force is given by

$$P_B \quad = 22{,}561 \sin 40° + 26{,}955 \cos 40°$$
$$\quad = 35{,}151 \text{ pounds}$$

MEMBER STRESSES

The flexural stress at mid span of the left hand beam is

$$f_b = M_c/S$$
$$= 365,346 \times 12/2218$$
$$= 1976 \text{ pounds per square inch}$$

The volume factor for the beam is given by UBC Section 2312.4.5 as

	C_V	$= k(1291.5/bdL)^{1/x}$
where	k	= loading condition coefficient
		= 1.00 ... for uniformly distributed load
	b	= width of member
		= 8.75 inches
	d	= depth of member
		= 39 inches
	L	= distance between points of zero moment
		= 77 feet
	x	= 10 ... for western species timber
and	C_V	= 0.74

The allowable flexural stress is

$$F_b' = F_b C_D C_V \text{ ... UBC Table 23-I-A-6 footnotes 1 and 2}$$
$$= 2200 \times 1.6 \times 0.74$$
$$= 2605 \text{ pounds per square inch}$$
$$> f_b \text{ ... satisfactory}$$

The shear stress at support A is

$$f_v = 1.5Q_R/A$$
$$= 1.5 \times 18,976/341.3$$
$$= 83 \text{ pounds per square inch}$$
$$< F_v C_D \text{ ... satisfactory}$$

The axial stress at support B is

$$f_c = P_B/A$$
$$= 35,151/341.3$$
$$= 103 \text{ pounds per square inch}$$
$$< F_c C_D \text{ ... satisfactory}$$

The axial stress at the mid span of the left hand beam is

$$f_c = P_c/A$$
$$= 11,429/341.3$$
$$= 33 \text{ pounds per square inch}$$

For axial effects, the beam is braced about its weak axis and the effective unbraced length about the strong axis, in accordance with UBC Section 2307.3, is given by

$$\ell_e = K_e\ell$$
$$= 1.0 \times 77$$
$$= 77 \text{ feet}$$

The slenderness ratio is

$$\ell_e/d = 77 \times 12/39$$
$$= 23.7$$
$$< 50 \dots \text{satisfactory}$$

The relevant design factors are obtained from calculator program A.3.3 in Appendix A. The critical buckling design value is given by UBC Section 2307.3 as

$$F_{cE} = K_{cE}E'/(\ell_e/d)^2$$

where K_{cE} = Euler buckling coefficient for columns
$$= 0.418 \dots \text{for glued-laminated timber}$$
E' = adjusted tabulated modulus of elasticity
$$= 1.6 \times 10^6 \text{ pounds per square inch}$$
and F_{cE} = 1191 pounds per square inch

The column stability factor is given by UBC Section 2307.3 as

$$C_P = (1.0 + F)/2c - \{[(1.0+ F)/2c]^2 - F/c\}^{0.5}$$

where F_c^* = adjusted tabulated compressive stress
$$= F_C C_D$$
$$= 1500 \times 1.6$$
$$= 2400 \text{ pounds per square inch}$$
F = F_{cE}/F_c^*
$$= 0.50$$
c = column parameter
$$= 0.9 \dots \text{for glued-laminated timber}$$
and C_P = 0.46

The allowable compressive stress is

$$F_c' = F_c^* C_P \dots \text{UBC Table 23-I-A-6}$$
$$= 1098 \text{ pounds per square inch}$$
$$> f_c \dots \text{satisfactory}$$

COMBINED STRESSES

The combined compressive and flexural stresses due to vertical and transverse loading must satisfy the interaction equation given in UBC Section 2308.2 as

$$(f_c/F_c')^2 + f_b/F_b'(1.0 - f_c/F_{cE}) \leq 1$$

The left hand side of the expression is evaluated as

$$(33/1098)^2 + 1976/3379(1.0 - 33/1191) = 0.60$$

Hence the glulam beam is satisfactory under the combined dead plus wind loading. $=\!=\!=\!=$

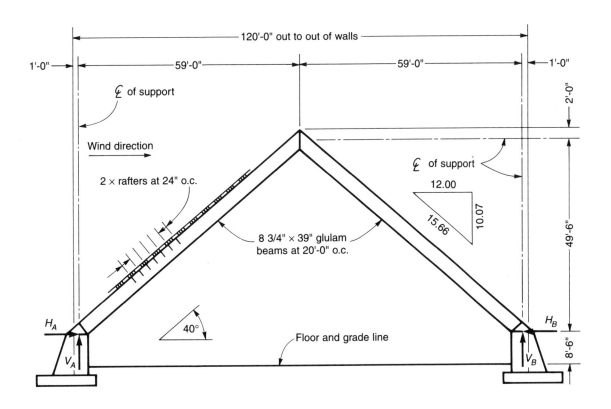

FIGURE 89D-1

CHAPTER 4

REFERENCES

1.   American Institute of Timber Construction. *Timber construction manual*. Englewood, CO, 1985.

2.   American Plywood Association. *Diaphragm*s. APA Design/Construction Guide. Tacoma, WA, 1989. pp 9–23.

3.   American Plywood Association. *Anchorage of concrete and masonry walls*. Commentary on reference 5. Tacoma, WA, 1985.

4.   Coil, J. *Subdiaphragms*. SEAOSC Design Seminar. Los Angeles, CA, March 1991.

5.   Sheedy, P. Anchorage of Concrete and Masonry Walls. *Building Standards*. October, 1983 and April, 1985. International Conference of Building Officials, Whittier, CA.

6.   Tissell, J.R. and Elliot, J.R. *Plywood diaphragms*. Research Report 138. American Plywood Association, Tacoma, WA, 1986.

7.   Simpson Strong-Tie Company. *Wood construction connectors*. Catalog C-96. Pleasanton, CA, January, 1996.

8.   American Forest and Paper Association. *Commentary on the National Design Specification for Wood Construction, Eleventh Edition (ANSI/NFoPA NDS-1991)*. Washington, DC, 1991.

5

STRUCTURAL MASONRY DESIGN

SECTION 5.1

MASONRY BEAMS

PROBLEMS:

1988 D–5

STRUCTURAL ENGINEER EXAMINATION – 1988

PROBLEM D–5 – WT. 5.0 POINTS

GIVEN::   A reinforced masonry lintel beam of a building which is to be renovated to support a structural steel tube column is shown in plan A and section B.

- Total horizontal wind load   = 260 pounds per linear foot (see section B).
- Steel tube column axial load = 14,000 pounds (dead load + live load).

CRITERIA: Materials:
- Masonry: hollow concrete units, grouted solid.
  - f'_m = 1500 pounds per square inch. *(10.3 MPa)*
  - weight = 135 pounds per cubic foot.
- Grout: f'_c = 2000 pounds per square inch. *(13.8 MPa)*
- Reinforcement: ASTM A 615, grade 60.

Assumptions:
- Special inspection was provided during construction and will be provided during renovation.
- Neglect seismic loads.

REQUIRED: Provide calculations to check the adequacy of the lintel beam.

SOLUTION

ALLOWABLE STRESS

The allowable stresses, with special inspection provided, in accordance with UBC Section 2107 are:

$$
\begin{aligned}
F_b &= 0.33f'_m &&= 500 \text{ pounds per square inch} \\
F_v &= (f'_m)^{0.5} &&= 38.7 \text{ pounds per square inch} \\
F_s & &&= 24{,}000 \text{ pounds per square inch} \\
E_m &= 750f'_m &&= 1125 \text{ kips per square inch} \\
E_s & &&= 29{,}000 \text{ kips per square inch}
\end{aligned}
$$

VERTICAL LOADING

The relevant section properties of the lintel are

$$
\begin{aligned}
b &= 11.63 \text{ inches} \\
d &= 52 \text{ inches} \\
L &= 19 \text{ feet} \\
A_s &= 0.88 \text{ square inches} \\
L/b &= 19 \times 12/11.63 \\
&< 32
\end{aligned}
$$

Hence, in accordance with UBC Section 2106.1.7, the lateral support of the beam is adequate.

The lintel frames into masonry walls at either end and, for vertical loads, may be considered fixed ended. The lintel self weight is

$$w = 0.135 \times 4.67$$
$$= 0.63 \text{ kips per linear foot}$$

The bending moment, produced by this self–weight, at the support is

$$M_s = wL^2/12$$
$$= 0.63 \times 19^2/12$$
$$= 18.97 \text{ kip feet}$$

The bending moment, produced by the column load, at a support is

$$M_c = WL/8$$
$$= 14 \times 19/8$$
$$= 33.25 \text{ kip feet}$$

The total moment, due to vertical loads, at a support is

$$M_y = M_s + M_c$$
$$= 18.97 + 33.25$$
$$= 52.22 \text{ kip feet}$$

Using calculator program A.2.1 in Appendix A and neglecting compression steel, the beam stresses, in accordance with UBC Section 2107.2.15 are

$$n = E_s/E_m \qquad\qquad = 25.8$$
$$p = A_s/bd \qquad\qquad = 0.146 \text{ percent}$$
$$k = (n^2p^2 + 2np)^{0.5} - np \qquad = 0.239$$
$$j = 1 - k/3 \qquad\qquad = 0.920$$
$$f_b = 24M_y/jkbd^2 \qquad\qquad = 181 \text{ pounds per square inch}$$
$$< F_b \ldots \text{ satisfactory}$$
$$f_s = 12M_y/jdA_s \qquad\qquad = 14{,}879 \text{ pounds per square inch}$$
$$< F_s \ldots \text{ satisfactory}$$

The total moment, due to the vertical loads, at mid span is

$$M_y' = M_s/2 + M_c$$
$$= 42.74 \text{ kip feet}$$
$$< M_y \ldots \text{ satisfactory}$$

The shear force at each support is given by

$$V = wL/2 + W/2$$
$$= 0.63 \times 19/2 + 14/2$$
$$= 12.99 \text{ kips}$$

The shear stress at each support is given by UBC Formula (7 – 37) as

$$f_v \quad = V/bjd$$
$$= 12{,}990/(11.63 \times 0.92 \times 52)$$
$$= 23.3 \text{ pounds per square inch}$$
$$< F_v \ldots \text{satisfactory}$$

Hence no shear reinforcement is necessary provided that, in accordance with UBC Section 2107.2.8, the shear stress at a distance of one–sixteenth the clear span beyond the point of inflection does not exceed

$$F_v' \quad = 20 \text{ pounds per square inch}$$

The shear force at this point is given by

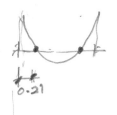

$$V' \quad = V - wL(0.21 - 1/16)$$
$$= 12.99 - 0.63 \times 19(0.21 - 1/16)$$
$$= 11.22 \text{ kips}$$

The corresponding shear stress is given by

$$f_v' \quad = V'/bjd$$
$$= 20.17 \text{ pounds per square inch}$$
$$\approx F_v' \ldots \text{satisfactory}$$

LATERAL LOADING

The relevant section properties of the lintel are

$$b \quad = 56 \text{ inches}$$
$$d \quad = 9.25 \text{ inches}$$
$$L \quad = 19 \text{ feet}$$
$$A_s \quad = 0.88 \text{ square inches}$$
$$L/b \quad = 19 \times 12/56$$
$$< 32$$

Hence the lateral support of the beam is adequate.

For lateral loads, the lintel may be considered as simply supported and the bending moment, produced by the wind load, at mid span is

$$M_x \quad = qL^2/8$$
$$= 0.26 \times 19^2/8$$
$$= 11.73 \text{ kip feet}$$

Using calculator program A.2.1 in Appendix A and neglecting compression reinforcement, the beam stresses, in accordance with UBC Section 2107.2.15 are

$$
\begin{aligned}
n &= E_s/E_m & &= 25.8 \\
p &= A_s/bd & &= 0.170 \text{ percent} \\
k &= (n^2p^2 + 2np)^{0.5} - np & &= 0.255 \\
j &= 1 - k/3 & &= 0.915 \\
f_b &= 24M_x/jkbd^2 & &= 251 \text{ pounds per square inch} \\
& & &< 1.33F_b \ldots \text{satisfactory} \\
f_s &= 12M_x/jdA_s & &= 18{,}901 \text{ pounds per square inch} \\
& & &< 1.33F_s \ldots \text{satisfactory}
\end{aligned}
$$

The shear force at each support is given by

$$
\begin{aligned}
V &= qL/2 \\
&= 0.26 \times 19/2 \\
&= 2.47 \text{ kips}
\end{aligned}
$$

The shear stress at each support is given by

$$
\begin{aligned}
f_v &= V/bjd \\
&= 2470/(56 \times 0.915 \times 9.25) \\
&= 5.2 \text{ pounds per square inch} \\
&< F_v \ldots \text{satisfactory}
\end{aligned}
$$

Hence no shear reinforcement is necessary.

BIAXIAL STRESSES

Assuming that the column live load is not a roof load, the vertical and lateral loading conditions may be combined in accordance with UBC Section 1603.6 and allowable stresses may be increased by one-third in accordance with UBC Section 1603.5. For combined masonry flexural stresses, the maximum resultant stress at mid span is

$$
\begin{aligned}
\Sigma f_b &= 181 \times 42.74/52.22 + 251 \\
&= 399 \text{ pounds per square inch} \\
&< 1.33F_b \ldots \text{satisfactory}
\end{aligned}
$$

For combined reinforcement stresses, the resultant maximum stress at mid span is

$$
\begin{aligned}
\Sigma f_s &= 14{,}879 \times 42.74/52.22 + 18{,}901 \\
&= 31{,}079 \text{ pounds per square inch} \\
&< 1.33F_s \ldots \text{satisfactory}
\end{aligned}
$$

Hence the reinforcement provided in the lintel is adequate.

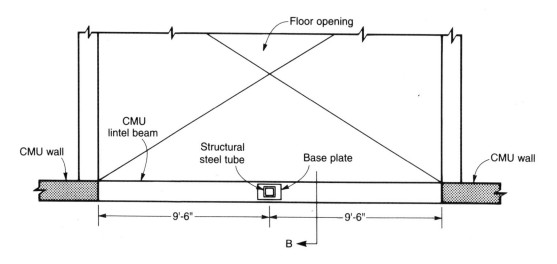

Plan A

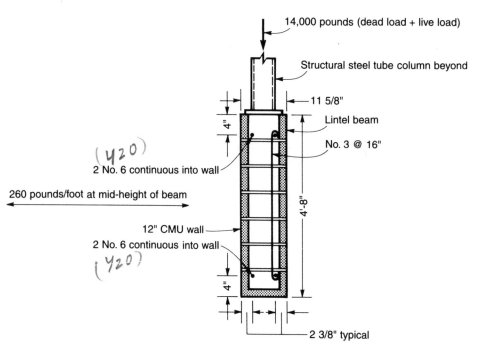

Section B

FIGURE 88D-5

496

STRUCTURAL ENGINEER REGISTRATION

CHAPTER 5

STRUCTURAL MASONRY DESIGN

SECTION 5.2

MASONRY COLUMNS

PROBLEMS:

1987 A–2

STRUCTURAL ENGINEER EXAMINATION – 1987 =========

PROBLEM A–2 – WT. 4.0 POINTS

GIVEN: Shown is a partial elevation of a masonry wall. The building houses an essential facility.

CRITERIA: Roof dead load = 500 pounds per linear foot.
Roof live load = 350 pounds per linear foot.
Basic wind speed 70 mph, exposure category B, using method two.
Seismic zone 4.
Masonry: eight inch hollow concrete units, grouted solid.
f'_m = 1500 pounds per square inch with special inspection..
weight = 135 pounds per square foot.
Reinforcement: f_y = 40,000 psi.
Lintel cannot be braced for lateral loads.

REQUIRED: 1. Check the two No.6 bottom lintel bars.
Assume $M = w\ell^2/8$ for gravity loads.
2. Check the pilaster for four No.6 vertical bars. Assume wind controls using a uniform lateral load.
3. Provide a detail of the pilaster reinforcement required at section X.
4. Comment on the need for top bar T.

SOLUTION

ALLOWABLE STRESSES

The allowable stresses, with special inspection provided, in accordance with Section 2406 are

$$F_b = 0.33f'_m = 500 \text{ pounds per square inch}$$
$$F_s = 0.5f_y = 20,000 \text{ pounds per square inch}$$
$$F_{sc} = 0.4f_y = 16,000 \text{ pounds per square inch}$$
$$E_m = 750f'_m = 1125 \text{ kips per square inch}$$
$$E_s = 29,000 \text{ kips per square inch}$$

1. LINTEL REINFORCEMENT

VERTICAL LOADING

The relevant section properties of the lintel are

$$b = 7.63 \text{ inches}$$
$$d = 58 \text{ inches}$$
$$L = 16 \text{ feet}$$
$$A_s = 0.88 \text{ square inches}$$
$$L/b = 16 \times 12/7.63$$
$$< 32$$

Hence, in accordance with UBC Section 2106.1.7, the lateral support of the beam is adequate.

The total dead load acting on the lintel is

$$w_D = 0.5 + 0.075 \times 5.33$$
$$= 0.9 \text{ kips per linear foot}$$

The total dead plus live load acting on the lintel is

$$w = w_D + w_L$$
$$= 0.9 + 0.35$$
$$= 1.25 \text{ kips per linear foot.}$$

In accordance with the problem statement, the bending moment at mid span due to the total load is

$$M_y = wL^2/8$$
$$= 1.25 \times 16^2/8$$
$$= 40 \text{ kip feet.}$$

Using calculator program A.2.1. in Appendix A and neglecting compression steel, the beam stresses, in accordance with UBC Section 2107.2.15 are

499

$$
\begin{aligned}
n &= E_s/E_m & &= 25.8 \\
p &= A_s/bd & &= 0.199 \text{ percent} \\
k &= (n^2p^2 + 2np)^{0.5} - np & &= 0.273 \\
j &= 1 - k/3 & &= 0.909 \\
f_b &= 24M_y/jkbd^2 & &= 151 \text{ pounds per square inch} \\
& & &< F_b \ldots \text{ satisfactory} \\
f_s &= 12M_y/jdA_s & &= 10{,}346 \text{ pounds per square inch} \\
& & &< F_s \ldots \text{ satisfactory}
\end{aligned}
$$

The bending moment at mid span due to the beam self weight plus roof dead load is

$$
\begin{aligned}
M_y' &= M_y w_D/w \\
&= 40 \times 0.9/1.25 \\
&= 28.8 \text{ kip feet}
\end{aligned}
$$

The corresponding stresses are

$$
\begin{aligned}
f_b &= 151 M_y'/M_y & &= 109 \text{ pounds per square inch} \\
f_s &= 10{,}346 M_y'/M_y & &= 7449 \text{ pounds per square inch}
\end{aligned}
$$

LATERAL LOADING

The relevant section properties of the lintel for lateral loads are obtained by assuming that the upper two feet of the lintel is supported by the roof diaphragm and the lower two feet of the lintel acts as a horizontal beam. Then

$$
\begin{aligned}
b &= 24 \text{ inches} \\
d &= 5.25 \text{ inches} \\
L &= 16 \text{ feet} \\
A_s &= 0.44 \text{ square inches}
\end{aligned}
$$

SEISMIC LOAD

The tributary weight on the lintel is

$$
\begin{aligned}
w_p &= 0.075 \times 2 \\
&= 0.15 \text{ kips per linear foot}
\end{aligned}
$$

The seismic load acting on the lintel is given by UBC Section 1630 Formula (30–1) as

$$
\begin{aligned}
& & F_p &= ZI_p C_p w_p \\
\text{where} & & Z &= 0.4 \text{ for zone 4 from UBC Table 16–I} \\
& & I_p &= 1.50 \text{ from UBC Table 16–K for an essential facility} \\
& & C_p &= 0.75 \text{ from UBC Table 16–O item 1.1.b} \\
& & w_p &= 0.15 \text{ kips per linear foot} \\
\text{Then} & & F_p &= 0.4 \times 1.50 \times 0.75 \times 0.15 \\
& & &= 0.45 \times 0.15 \\
& & &= 0.068 \text{ kips per linear foot}
\end{aligned}
$$

WIND LOAD

The wind pressure acting on the lintel is given by UBC Section 1618 Formula (18–1) as

	p	$= C_e C_q q_s I_w$
where	C_e	= 0.62, for exposure B and a height of 0 to 15 feet, from Table 16–G
	C_q	= 1.3 from UBC Table 16–H for method 2
	q_s	= 12.6 pounds per square foot, for a seventy miles per hour wind speed, from UBC Table 16–F
	I_w	= 1.15 from UBC Table 16–K for an essential facility.
Then	p	$= 0.62 \times 1.3 \times 12.6 \times 1.15$
		= 11.68 pounds per square foot.

The tributary width appropriate to the lintel is

$$\ell = 2 + 12/2$$
$$= 8 \text{ feet}$$

The wind load acting on the lintel is

$$F_w = p\ell$$
$$= 11.68 \times 8/1000$$
$$= 0.093 \text{ kips per linear foot ... governs}$$
$$> F_p$$

Hence the governing lateral moment at mid span is

$$M_x = F_w L^2/8$$
$$= 0.093 \times 16^2/8$$
$$= 2.98 \text{ kip feet}$$

Using calculator program A.2.1 in Appendix A and neglecting compression steel, the beam stresses, in accordance with UBC Section 2107.2.15 are

n	= 25.8
p	= 0.349 percent
k	= 0.344
j	= 0.885
f_b	= 355 pounds per square inch
	$< 1.33F_b$ ... satisfactory
f_s	= 17,484 pounds per square inch
	$< 1.33F_s$ ... satisfactory

BIAXIAL STRESSES

In accordance with UBC Section 1603.6, the lateral wind load may be combined with the vertical dead load and the allowable stresses may be increased by one–third in accordance with UBC Section 1603.5. The combined masonry flexural stress is

$$\Sigma f_b \quad = 109 + 355$$
$$= 464 \text{ pounds per square inch}$$
$$< 1.33F_b \text{ ... satisfactory}$$

The combined reinforcement stress is

$$\Sigma f_s \quad = 7449 + 17,484$$
$$= 24,933 \text{ pounds per square inch}$$
$$< 1.33F_s \text{ ... satisfactory}$$

Hence the lintel reinforcement is adequate.

2. PILASTER REINFORCEMENT

VERTICAL LOADING

The section properties of the pilaster are

$$t \quad = \text{effective pilaster thickness}$$
$$= 15.63 \text{ inches}$$
$$h' \quad = \text{effective pilaster height}$$
$$= 16 \text{ feet}$$
$$A_{sc} \quad = \text{reinforcement area}$$
$$= 1.76 \text{ square inches}$$
$$A_e \quad = \text{effective pilaster area}$$
$$= t^2$$
$$= 244 \text{ square inches}$$

The effective unbraced height of the pilaster for vertical loads is sixteen feet. The simply supported span for lateral loads may also be taken as sixteen feet, from roof to floor level. The load combination of dead load plus wind load governs and the total dead load acting at a height above the base of eight feet is

$$P \quad = w_D(8 + 1.33 + 4) + w_{pilaster} \times 4 \times t/12$$
$$= 0.9(8 + 1.33 + 4) + 0.15 \times 4 \times 1.30$$
$$= 12.8 \text{ kips}$$

The radius of gyration of the pilaster is

$$r \quad = 0.29t$$
$$= 0.29 \times 15.63$$
$$= 4.53 \text{ inches}$$

The slenderness ratio of the pilaster is

$$h'/r = 16 \times 12/4.53$$
$$= 42.4$$
$$< 99 \ldots \text{UBC Formula (7–13) is applicable}$$

The allowable pilaster load is given by UBC Formula (7–13) as

$$P_a = (0.25f'_m A_e + 0.65A_{sc}F_{sc})[1.0 - (h'/140r)^2]$$
$$= (0.25 \times 1.5 \times 244 + 0.65 \times 1.76 \times 16)[1.0 - (42.4/140)^2]$$
$$= (91.5 + 18.3)0.91$$
$$= 83.1 + 16.6$$
$$= 99.7 \text{ kips}$$
$$> P \ldots \text{satisfactory}$$

Then the applied vertical load carried by the masonry is

$$P_m = P \times 83.1/P_a$$
$$= 12.8 \times 83.1/99.7$$
$$= 10.7 \text{ kips}$$

LATERAL LOADING

The section properties of the pilaster for lateral loading are

$$b = 15.63 \text{ inches}$$
$$d = 13 \text{ inches}$$
$$L = 16 \text{ feet}$$
$$A_s = 0.88 \text{ square inches}$$
$$L/b = 16 \times 12/15.63$$
$$< 32$$

Hence in accordance with UBC Section 2106.1.7, the lateral support of the pilaster is adequate for lateral loads. The tributary width, appropriate to the pilaster, for wind loading is

$$\ell = 8 + 1.33 + 4$$
$$= 13.33 \text{ feet}$$

Neglecting the increase in wind pressure on the tributary area above a height of fifteen feet, the wind load acting on the pilaster is

$$F = p\ell$$
$$= 11.68 \times 13.33/1000$$
$$= 0.156 \text{ kips per linear foot}$$

The moment at mid height due to the wind loading is

$$M = FL^2/8$$
$$= 0.156 \times 16^2/8$$
$$= 4.99 \text{ kip feet}$$

Using calculator program A.2.1 in Appendix A and neglecting compression steel, the flexural stresses in the pilaster, in accordance with UBC Section 2107.2.15 are

$$
\begin{aligned}
n &= 25.8 \\
p &= 0.433 \text{ percent} \\
k &= 0.374 \\
j &= 0.875 \\
f_b &= 139 \text{ pounds per square inch} \\
&< 1.33F_b \text{ ... satisfactory} \\
f_s &= 5980 \text{ pounds per square inch} \\
&< 1.33F_s \text{ ... satisfactory}
\end{aligned}
$$

COMBINED STRESSES

In accordance with UBC Section 2107.3.4 Formula (7-42), the combined axial and flexural stresses must satisfy the interaction equation

$$P_m/P_{ma} + f_b/F_b \leq 1.33$$

The left hand side of the equation is evaluated as

$$
\begin{aligned}
10.7/83.1 + 139/500 &= 0.13 + 0.28 \\
&= 0.41 \\
&< 1.33 \text{ ... satisfactory}
\end{aligned}
$$

REINFORCEMENT LIMITS

The vertical reinforcement provided in the pilaster is

$$
\begin{aligned}
A_{sc} &= A_e A_{sc}/A_e \\
&= A_e \times 1.76/244 \\
&= 0.007A_e \\
&> 0.005A_e \\
&< 0.04A_e
\end{aligned}
$$

Hence, the requirements of UBC Section 2107.2.13.1 are satisfied and the pilaster reinforcement is adequate.

3. PILASTER DETAIL

In seismic zone 4, UBC Section 2106.1.12.4 requires that the spacing of column ties shall not exceed eight inches and this requirement governs over that specified in UBC Section 2106.3.6. For No.6 vertical bars, the minimum diameter of tie bar is specified as ¼ inch. A detail of the pilaster is shown in Figure 87A-2.

4. TOP REINFORCEMENT

In seismic zone 4, UBC Section 2106.1.12.3 requires the provision of horizontal reinforcement at the top of all walls of not less than 0.2 square inches in cross-sectional area. In addition, at the level of the diaphragm, reinforcement is required to resist the chord forces.

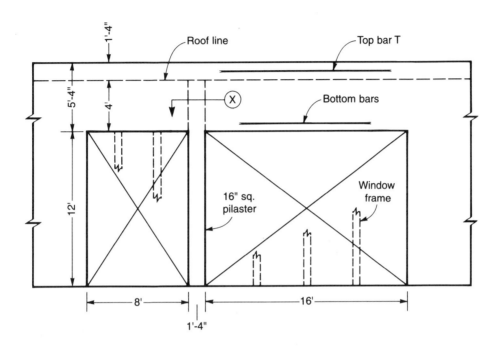

Wall Elevation

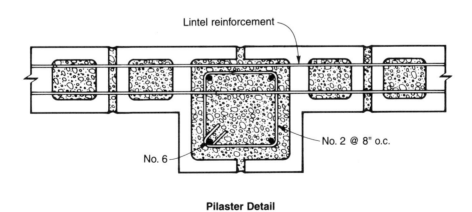

Pilaster Detail

FIGURE 87A-2

505

STRUCTURAL ENGINEER REGISTRATION

CHAPTER 5

STRUCTURAL MASONRY DESIGN

SECTION 5.3

MASONRY WALLS

PROBLEMS:

1990 D–4
1989 D–4

STRUCTURAL ENGINEER EXAM – 1990

PROBLEM D–4 – WT. 6.0 POINTS *(~58 min)*

GIVEN: An 8 inch reinforced concrete masonry bearing wall as shown in elevation and section.

The diaphragm chord force, due to wind = 5000 pounds.

The uniform diaphragm shear = 1500 pounds per linear foot.

The uniform vertical roof load = 2000 pounds per linear foot. (Dead load + reduced live load, non–snow)

Uniform wind pressure = 30 pounds per square foot. (Inward or outward)

CRITERIA: Materials:
- Masonry: solid grouted, hollow concrete units.
    f'_m = 1500 pounds per square inch.
    Weight = 84 pounds per square foot.
- Grout: f'_c = 2000 pounds per square inch.
- Reinforcement: ASTM A615, grade 60.

Assumptions:
- Special inspection shall be provided
- Wind load governs

REQUIRED:
1.   Design the vertical reinforcement at Section B. Use No.4 bars at the wall centerline.
2.   Design the horizontal reinforcement. Use No.4 bars. (For this part, assume j = 0.9)
3.   Show reinforcing steel on a wall elevation and section.

SOLUTION

1. VERTICAL REINFORCEMENT

The relevant section properties of a one foot length of wall are

$$b \quad = \quad 12 \text{ inches}$$
$$t \quad = \quad 7.63 \text{ inches}$$
$$d \quad = \quad 3.81 \text{ inches}$$

ALLOWABLE STRESSES

The allowable stresses, with special inspection provided, in accordance with UBC Section 2107 are

$$F_b \quad = 0.33f'_m \quad = 500 \text{ pounds per square inch}$$
$$F_s \quad\quad\quad\quad\quad = 24{,}000 \text{ pounds per square inch}$$
$$E_m \quad = 750f'_m \quad = 1125 \text{ kips per square inch}$$
$$E_s \quad\quad\quad\quad\quad = 29{,}000 \text{ kips per square inch}$$

WIND PLUS ROOF LOAD

FLEXURAL EFFECTS

In accordance with UBC Section 1603.6, the relevant load combination is wind load plus roof dead load. Assume that the roof live load is negligible compared with the roof dead load and that the given roof load of 2000 pounds per linear foot may be considered as dead load. The horizontal reaction produced at the base of the wall by the wind load is

$$R_w \quad = 30 \times 16 \times 6/14$$
$$= 205.7 \text{ pounds}$$

The horizontal reaction produced at the base of the wall by the eccentric roof load is

$$R_r \quad = 2000 \times 0.5/14$$
$$= 71.4 \text{ pounds}$$

The total reaction at the base is

$$R \quad = R_w + R_r$$
$$= 205.7 + 71.4$$
$$= 277.1 \text{ pounds}$$

The maximum moment occurs in the wall at a height y from the base and is given by

$$M \quad = Ry - 30y^2/2$$
$$= 277.1y - 15y^2$$

Differentiating and equating dM/dy to zero gives

$$y \quad = 277.1/30$$
$$= 9.2 \text{ feet}$$

Hence the maximum moment in the wall is

$$M \quad = 277.1 \times 9.2 - 15 \times 9.2^2$$
$$= 1280 \text{ pounds feet}$$

Assuming the lever arm ratio is

$$j \quad = 0.88$$

the required reinforcement area, allowing for the one–third increase in allowable stress in accordance with UBC Section 1603.5, is

$$A_s \quad = 12M/jdF_s$$
$$= 12 \times 1280/(0.88 \times 3.81 \times 24{,}000 \times 1.33)$$
$$= 0.143 \text{ square inches per foot}$$

Providing No.4 bars at 16 inches on center gives an area of

$$A_s \quad = 0.147 \text{ square inches per foot}$$

Using calculator program A.2.1 in Appendix A, the section stresses, in accordance with UBC Section 2107.2.15 are

n	$= E_s/E_m$	$= 25.8$
p	$= A_s/bd$	$= 0.32$ percent
k	$= (n^2p^2 + 2np)^{0.5} - np$	$= 0.333$
j	$= 1.0 - k/3$	$= 0.889$
f_b	$= 24M/jkbd^2$	$= 596$ pounds per square inch
		$< 1.33F_b$ ... satisfactory
f_s	$= 12M/jdA_s$	$= 30{,}845$ pounds per square inch
		$< 1.33F_s$ ... satisfactory

AXIAL EFFECTS

The weight of the masonry wall above the point of maximum moment is

$$P_w \quad = 84(16 - 9.2)$$
$$= 571 \text{ pounds per linear foot}$$

The weight of the roof is

$$P_r \quad = 2000 \text{ pounds per linear foot}$$

The total applied axial load at the point of maximum moment is

$$P \quad = P_w + P_r$$
$$= 571 + 2000$$
$$= 2571 \text{ pounds per linear foot.}$$

509

The corresponding axial stress in the wall is given by UBC Formula (7–8) as

$$f_a \quad = P/A_e$$
$$= 2571/(12 \times 7.63)$$
$$= 28 \text{ pounds per square inch}$$

The radius of gyration of the wall is obtained from UBC Table 21–H–1 as

$$r \quad = 2.19 \text{ inches}$$

The slenderness ratio of the wall is

$$h'/r \quad = 14 \times 12/2.19$$
$$= 76.7$$
$$< 99 \ldots \text{UBC Formula (7–11) is applicable}$$

Then in accordance with UBC Section 2107.2.5, the allowable stress may be obtained from UBC Formula (7–11) as

$$F_a \quad = 0.25f_m'[1.0 - (h'/140r)^2]$$
$$= 262 \text{ pounds per square inch}$$
$$> f_a \ldots \text{satisfactory}$$

COMBINED STRESSES

In accordance with UBC Section 2107.3.4 Formula (7–42), the interaction expression for combined axial and compressive stresses, allowing for the one–third increase in permissible stresses, is given by

$$f_a/F_a + f_b/F_b \quad \leq 1.33$$

The left hand side of the expression is evaluated as

$$28/262 + 596/500 \quad = 0.11 + 1.19$$
$$= 1.30$$
$$< 1.33 \ldots \text{satisfactory}$$

Hence the masonry wall is adequate for combined wind and roof load.

ROOF LOAD ONLY

FLEXURAL EFFECTS

The maximum moment occurs at the top of the wall and is given by

$$M \quad = 2000 \times 0.5$$
$$= 1000 \text{ pounds feet}$$

The stresses produced by this moment are

$$f_b = 596 \times 1000/1280$$
$$= 465 \text{ pounds per square inch}$$
$$< F_b$$
$$f_s = 30{,}845 \times 1000/1280$$
$$= 24{,}097$$
$$\approx 24{,}000 \ldots \text{ satisfactory}$$

AXIAL EFFECTS

The axial stress at the top of the wall is given by UBC Formula (7-8) as

$$f_a = P/A_e$$
$$= 2000/(12 \times 7.63)$$
$$= 22$$
$$< F_a \ldots \text{ satisfactory}$$

COMBINED STRESSES

The interaction expression for combined axial and compressive stresses is given by UBC Formula (7-42) as

$$f_a/F_a + f_b/F_b \leq 1.0$$

The left hand side of the expression is evaluated as

$$22/262 + 465/500 = 0.08 + 0.93$$
$$= 1.01$$
$$\approx 1.0 \ldots \text{ satisfactory}$$

Hence the wall is adequate for the applied loads.

DOWEL BARS

The unit applied shear on the wall is given as

$$v = 1500 \text{ pounds per foot}$$

Providing a No.4 dowel bar at eight inches on center gives a shear resistance, in accordance with UBC Table 21-F, of

$$v_a = 1.33 \times 850 \times 12/8$$
$$= 1696 \text{ pounds per foot}$$
$$> v \ldots \text{ satisfactory}$$

OVERTURNING

The overturning moment acting on the wall, due to the applied in–plane shear, is given by

$$M_o \quad = vLh'$$
$$= 1500 \times 30 \times 14$$
$$= 630{,}000 \text{ pounds feet}$$

The restoring or stabilizing moment is given by

$$M_r \quad = (2000 + 84 \times 16)30^2/2$$
$$= 1{,}504{,}800$$
$$> 1.5M_o \ldots \text{satisfactory}$$

Hence no additional anchorage reinforcement is required.

2. HORIZONTAL REINFORCEMENT

CHORD REINFORCEMENT

The area of chord reinforcement required is given by

$$A_s \quad = H/1.33F_s$$
$$= 5000/(1.33 \times 24{,}000)$$
$$= 0.16 \text{ square inches}$$

One No.4 bar is required with an area of

$$A_s' \quad = 0.196 \text{ square inches}$$
$$> A_s \ldots \text{satisfactory}$$

SHEAR REINFORCEMENT[1,2]

The unit applied shear is given as

$$v \quad = 1500 \text{ pounds per foot}$$

Since the shear is caused by wind loads, the factor of 1.5 specified by UBC Section 2107.1.7 is not required and the shear stress in the masonry wall is given by UBC Formula (7–37) as

$$f_v \quad = vd/tjd$$
$$= 1500 \times 30/(7.63 \times 0.9 \times 30 \times 12)$$
$$= 18.2 \text{ pounds per square inch}$$

The allowable stress is obtained by applying UBC Section 2107.2.9.

$$\text{The ratio} \quad M/Vd \quad = vdh'/vd^2$$
$$= h'/d$$
$$= 14/30$$
$$= 0.47$$
$$< 1.0 \ldots \text{UBC Formula (7–19) is applicable}$$

Applying UBC Formula (7–19), the allowable stress, with special inspection provided, is the smaller value given by

$$F_v = 1.33(80 - 45M/Vd)$$
$$= 1.33(80 - 45 \times 0.47)$$
$$= 1.33 \times 59 \text{ pounds per square inch}$$

or
$$F_v = 1.33 \times (f'_m)^{0.5}(4 - M/Vd)/3$$
$$= 1.33 \times (1500)^{0.5}(4 - 0.47)/3$$
$$= 1.33 \times 45.6 \text{ pounds per square inch } \ldots \text{ governs}$$
$$> f_v \ldots \text{ satisfactory}$$

Hence the masonry takes all the shear force and nominal horizontal reinforcement[3], only, is required as detailed in UBC Section 2106.1.12.4 for a structure located in seismic zone 4. The maximum allowable reinforcement spacing is forty eight inches and the minimum specified horizontal and vertical reinforcement areas are

$$A_{sh} = A_{sv}$$
$$= 0.0007t'b$$
$$= 0.0007 \times 8 \times 12$$
$$= 0.067 \text{ square inches per foot}$$

Providing No.4 horizontal bars at thirty two inches on center gives a reinforcement area of

$$A'_{sh} = 0.073 \text{ square inches per foot}$$
$$> A_{sh} \ldots \text{ satisfactory}$$

The sum of the horizontal and vertical reinforcement areas provided is

$$A'_{sh} + A'_{sv} = 0.073 + 0.147$$
$$= 0.22 \text{ square inches}$$

The required sum is

$$A_{sh} + A_{sv} = 0.002t'b$$
$$= 0.002 \times 8 \times 12$$
$$= 0.19 \text{ square inches}$$
$$< 0.22 \ldots \text{ satisfactory}$$

In addition, in seismic zone 4, UBC Section 2106.1.12.3 requires the provision of horizontal reinforcement at the top and bottom of all walls of not less than 0.2 square inches in cross–sectional area. Providing one No.4 bar gives an area of

$$A'_s = 0.196 \text{ square inches}$$
$$\approx 0.2 \ldots \text{ satisfactory}$$

3. REINFORCEMENT DETAILS

The required reinforcement details are shown in Figure 90D–4(ii). ══════════════

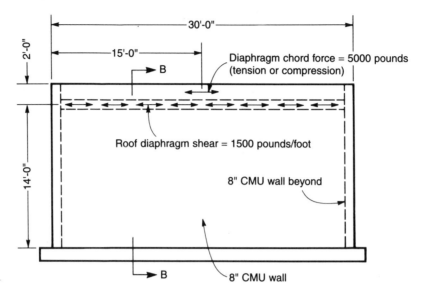

Wall Elevation

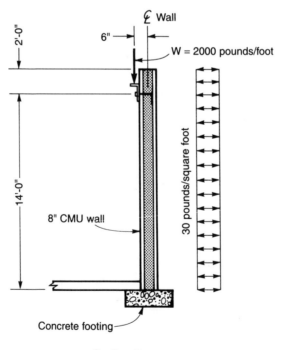

Section B

FIGURE 90D–4(i)

514

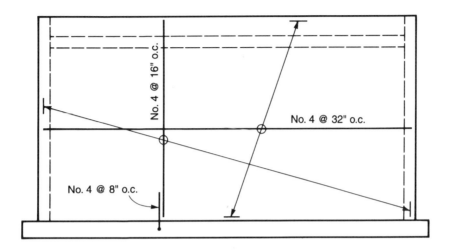

Wall Elevation

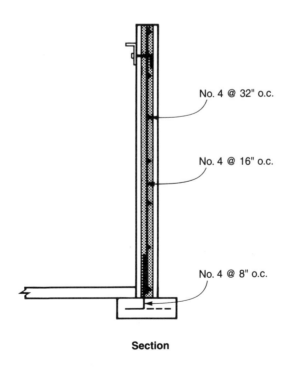

Section

FIGURE 90D-4(ii)

STRUCTURAL ENGINEER EXAMINATION – 1989

PROBLEM D–4 – WT. 6.0 POINTS*

GIVEN: A two–story building, as shown in Figure 89D-4(i), consists of plywood floor and roof diaphragms supported by eight inch concrete masonry walls.

CRITERIA:
- The structure is located in seismic zone 3.
- The seismic loads, derived by using $R_w = 6$, are as shown in Figure 89D-4(i).
- The total uniform dead load at the wall footing = 2200 pounds per linear foot.
- The uniform live load from the second floor, at the wall footing, = 200 pounds per linear foot.

Materials:
- Masonry: Solid grouted, eight inch hollow concrete units, with special inspection and laid in running bond.
  $f'_m = 3000$ pounds per square inch.
- Reinforcement: $f_y = 60,000$ pounds per square inch.

Assumptions:
- The roof and floor diaphragms may be considered flexible.
- The wall is adequate for flexure plus axial loading.
- The nominal shear strength of the wall is less than the shear corresponding to the nominal flexural strength.
- $d = 0.8 \times$ wall length for shear stress computations.

REQUIRED:
1. Determine the maximum design forces in the shear wall indicated. Consideration of out of plane loads is not required.
2. Determine if the reinforcement shown in Figure 89D-4(i) is adequate for the applied shear force.
3. Showing all calculations, determine if confinement reinforcement is required.

---

* To conform to current code requirements, the data has been modified.

SOLUTION *strength design was applied here.*

1. DESIGN FORCES *strength design*

In the bottom story, the factored shear force is given by

$$V_u = 1.4(5 + 6)$$
$$= 15.4 \text{ kips}$$

The factored bending moment at the bottom of the wall due to a response modification factor R_w of 6 is

$$M_u = 1.4(5 \times 22 + 6 \times 12)$$
$$= 255 \text{ kip feet}$$

Using a response modification factor of 1.5, as required by UBC Section 2108.2.5.6 for determining if boundary members are required, gives a factored bending moment of

$$M_u' = 255 \times 6/1.5$$
$$= 1020 \text{ kip feet}$$

In accordance with the problem statement, the wall is adequate for axial loading. To determine the confinement requirements, the factored axial load is required and this is given by

$$P_u = 1.4(2.2 \times 12 + 0.2 \times 12)$$
$$= 40.32 \text{ kips}$$

$U = 0.90 \pm 1.4 E$ should be considered.

2. SHEAR REQUIREMENTS

$$M/Vd = 255/(15.4 \times 0.8 \times 12)$$
$$= 1.72$$
$$> 1.0$$

Then from UBC Table 21–K the value of the masonry shear strength coefficient C_d is

$$C_d = 1.2$$

Since the nominal shear strength of the wall is less than the shear corresponding to the nominal flexural strength, the shear capacity of the masonry may be utilized. The nominal shear strength provided by the masonry is given by UBC Formula (8–46) as

$$V_m = C_d A_{mv}(f'_m)^{0.5}$$
$$= 1.2 \times 144 \times 7.63(3000)^{0.5}/1000$$
$$= 72.2 \text{ kips}$$

The design shear strength of the masonry is

$2108.1.4.3.2$

$$\phi V_m = 0.6 \times 72.2$$
$$= 43.3 \text{ kips}$$
$$> V_u \dots \text{ satisfactory}$$

Hence the masonry is adequate and nominal horizontal reinforcement, only, is required.

REINFORCEMENT DETAILS

The minimum allowable horizontal reinforcement area is specified in UBC Section 2106.1.12.4 as

$$A_{sh} = 0.0007A_g$$
$$= 0.0007 \times 7.63 \times 12$$
$$= 0.064 \text{ square inches per foot}$$

No.4 horizontal bars at 32 inches on center provide a reinforcement area of

$$A'_{sh} = 12 \times 0.2/32$$
$$= 0.075 \text{ square inches per foot}$$
$$> A_{sh} \dots \text{ satisfactory}$$

The maximum allowable spacing of the horizontal reinforcement is given by UBC Section 2106.1.12.4 as

$$s = 48 \text{ inches}$$

The spacing provided is

$$s' = 32 \text{ inches}$$
$$< s \dots \text{ satisfactory}$$

3. BOUNDARY ZONE REQUIREMENTS

CONFINEMENT REINFORCEMENT

Confinement reinforcement is required in a shear wall[4], in accordance with UBC Section 2108.2.5.6, when the compressive strains exceed 0.0015. At this level of strain, stress in the masonry may be assumed essentially elastic and the strain distribution across the section and the forces developed are shown in Figure 89D–4(ii). The strain is determined using factored forces with a value of 1.5 for the response modification factor R_w.

STRAIN DISTRIBUTION

The masonry stress corresponding to a strain of 0.001 is given[2] as

$$f_m = 0.40 f'_m$$

For a masonry strain of 0.0015 the corresponding stress may be taken as

$$f_m = 0.60 f'_m$$

The force in a reinforcing bar is given by

$$
\begin{aligned}
F &= e A_s E_s \times 0.0015/c \\
  &= e \times 0.79 \times 29{,}000 \times 0.0015/c \\
  &= 34.365 e/c
\end{aligned}
$$

where
e = distance of a reinforcing bar from the neutral axis
A_s = area of a reinforcing bar
c = depth of neutral axis

and
$$
\begin{aligned}
F_{max} &= 60 A_s \\
        &= 47.4 \text{ kips}
\end{aligned}
$$

Assuming the depth to the neutral axis is

$$c = 30.3 \text{ inches}$$

then
$$F = 1.134 e$$

TENSILE FORCES

The tensile forces in the reinforcement are

T_1		= 47.4 kips
T_2		= 47.4 kips
T_3		= 47.4 kips
T_4	= 1.13 × 41.7	= 47.3 kips
T_5	= 1.13 × 13.7	= 15.5 kips

The sum of the tensile forces is

$$\Sigma T = 205.0 \text{ kips}$$

COMPRESSIVE FORCES

The compressive force in the reinforcement is

$$C_6 \quad = 1.13 \times 6.3 \quad = 7.2 \text{ kips}$$
$$C_7 \quad = 1.13 \times 26.3 \quad = 29.8 \text{ kips}$$
$$C_m \quad = 0.5f_m tc$$

where $\quad f_m \quad$ = maximum stress in the masonry
$$= 0.60f_m'$$
$$= 1800 \text{ kips per square inch}$$

and $\quad C_m \quad = 0.5 \times 1.8 \times 7.63 \times 30.3$
$$= 208 \text{ kips}$$

The sum of the compressive forces is

$$\Sigma C \quad = 245.0 \text{ kips}$$

AXIAL LOAD CAPACITY

The nominal axial load capacity for this strain distribution is

$$P \quad = \Sigma C - \Sigma T$$
$$= 40.0$$
$$\approx P_u \dots \text{ satisfactory}$$

Hence the assumed neutral axis depth is satisfactory.

MOMENT CAPACITY

The nominal moment capacity for this strain distribution is obtained by summing moments about the mid–depth of the section and is given by

$$M \quad = 68(T_1 + C_7) + 48(T_2 + C_6) + 28(T_3 - T_5) + 61.9C_m$$
$$= 21,639 \text{ kip inches}$$
$$= 1803 \text{ kip feet}$$
$$> M_u' \dots \text{ satisfactory}$$

Hence confinement reinforcement is not necessary, as the factored moment M_u' determined with a value of 1.5 for R_w produces a compressive strain of less than 0.0015.

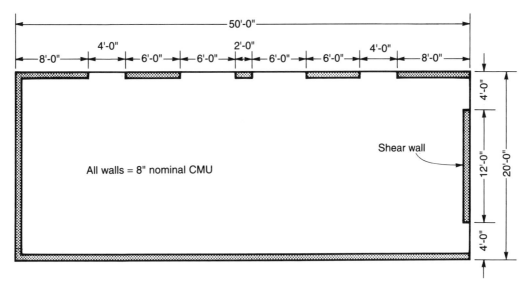

Floor Plan

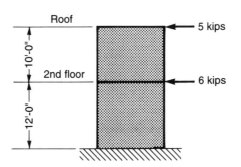

Seismic Loads on Shear Wall

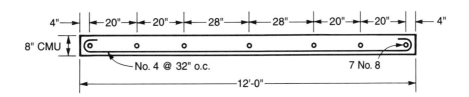

Reinforcement Details

FIGURE 89D–4(i)

521

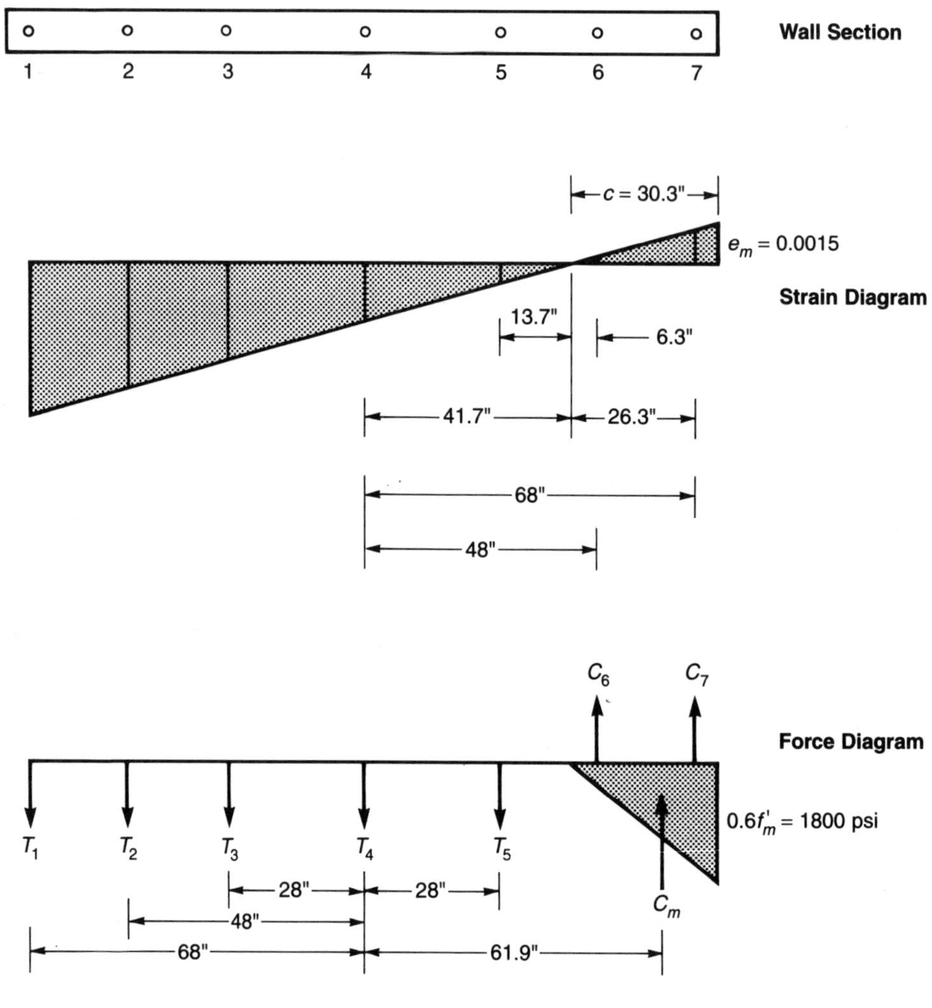

Wall Section

Strain Diagram

$c = 30.3"$

$e_m = 0.0015$

13.7"

6.3"

41.7"

26.3"

68"

48"

Force Diagram

C_6

C_7

$0.6f'_m = 1800$ psi

T_1

T_2

T_3

T_4

T_5

C_m

28"

28"

48"

68"

61.9"

FIGURE 89D–4(ii)

CHAPTER 5

REFERENCES

1. Schneider, R.R. & Dickey, W.L. *Reinforced Masonry Design.* Prentice–Hall, Inc., Englewood Cliffs, NJ, 1987

2. Brandow, G.E. Hart, G.C. and Virdee, A. *Design of reinforced masonry structures.* Concrete Masonry Association of California and Nevada, Citrus Heights, CA, 1995.

3. Masonry Institute of America. *Reinforcing Steel in Masonry.* Los Angeles, CA, 1991.

4. Concrete Masonry Association of California and Nevada. *Design of shear walls.* Masonry Chronicles, fall 1995. Concrete Masonry Association of California and Nevada, Citrus Heights, CA, 1995.

NOTE ON THE CALCULATOR PROGRAMS

The calculator programs in the following section are written for the HP–48G calculator. The same programs are available for the HP–28S calculator and may be obtained on application to the publisher.

STRUCTURAL DESIGN PROGRAMS FOR THE HP–48G CALCULATOR

SECTION A.1:   **CONCRETE DESIGN PROGRAMS**

A.1.1: TRD: Design reinforced concrete beam with tension reinforcement only.

A.1.2: TRA: Analysis of reinforced concrete beam with tension reinforcement only.

A.1.3: CRD: Design reinforced concrete beam with compression reinforcement.

A.1.4: CRA: Analysis of reinforced concrete beam with compression reinforcement.

A.1.5: TBD: Design of reinforced concrete T–beam.

A.1.6: TBA: Analysis of reinforced concrete T–beam.

SECTION A.2:   **MASONRY DESIGN PROGRAMS**

A.2.1: EAS: Analysis of masonry beam to determine stresses.

A.2.2: EAM: Analysis of masonry beam to determine allowable moment.

A.2.3: EDC: Design of masonry beam with compression reinforcement.

SECTION A.3:   **TIMBER DESIGN PROGRAMS**

A.3.1: HANK: Hankinson's formula.

A.3.2: BEAM: Beam slenderness factors.

A.3.3: COL: Column Slenderness factors.

SECTION A.4:   **STEEL DESIGN PROGRAMS**

A.4.1: FB: Allowable stress in slender beam.

A.4.2: CBX: Combined axial compression and bending.

A.4.3: MPCO: Plastic moment of member subject to combined axial compression and bending.

A.4.4: STPL: Column stiffener plates.

SECTION A.5:   **PROPERTIES OF SECTIONS**

A.5.1: IXX: Moment of inertia of composite sections.

A.5.2: RIG: Rigidity of walls.

A.1.1: TRD: DESIGN REINFORCED CONCRETE BEAM WITH TENSION REINFORCEMENT ONLY

PROGRAM LISTING:

KEYSTROKES	COMMENTARY
HOME ' CONC ' CRDIR ENTER	Create parent directory, CONC.
VAR CONC ' TRD ' CRDIR ENTER	Create subdirectory, TRD.
TRD	Open current directory, TRD.
≪ ' 12 ∗ M/(B ∗ D^2) ' EVAL ≫ ENTER ' K ' STO	Define the parameter, K. Store K.
≪ ' (1−√(1−2 ∗ K/(.9 ∗ .85 ∗ FC))) ∗ .85 ∗ FC/FY ' EVAL ≫ ENTER ' R ' STO	Define the reinforcement ratio, R. (ie. ρ of reference 1). Store R.
≪ IF FC 4 < THEN .85 ELSE IF FC 8 > THEN .65 ELSE ' .85−(FC−4)/20 ' EVAL END END ≫ ENTER ' B1 ' STO	Define the factor, B1 (i.e. β_1 of reference 2). Store B1.
≪ ' .75 ∗ .85 ∗ 87 ∗ B1 ∗ FC/(FY ∗ (87+FY)) ' EVAL ≫ ENTER ' RMAX ' STO	Define the maximum reinforcement ratio, RMAX (i.e. ρ_{max} of reference 3). Store RMAX.
≪ ' .2/FY ' EVAL ≫ ENTER ' RMIN ' STO	Define the minimum reinforcement ratio, RMIN (i.e. ρ_{min} of reference 4). Store RMIN.
≪ ' R ∗ B ∗ D ' EVAL ≫ ENTER ' AS ' STO	Define reinforcement area, AS. Store AS.
≪ ' .85 ∗ 2 ∗ B ∗ D ∗ √(1000 ∗ FC)/1000 ' EVAL ≫ ENTER ' PHVC ' STO	Define concrete design shear strength, PHVC (i.e. ϕV_c of reference 5). Store PHVC.
≪ ' 12 ∗ 50 ∗ B/(1000 ∗ FY) ' EVAL ≫ ENTER ' AVS ' STO	Define minimum shear reinforcement AVS (ie. A_v/s of reference 6). Store AVS.
≪{{M ≪' M ' STO ≫}{FY ≪' FY ' STO ≫}{FC ≪' FC ' STO ≫}{B ≪' B ' STO ≫ }{D ≪'D' STO ≫} K R B1 RMAX RMIN AS PHVC AVS} TMENU ≫ ENTER ' DAT ' STO	Set up an input menu for data entry Store the data menu

EXAMPLE

HOME VAR CONC TRD	Recall program TRD
DAT	Prepare data entry.
53.5 M	Required factored moment, kip ft.
60 FY	Reinforcement yield strength, ksi
3 FC	Concrete compressive strength, ksi.
12 B	Beam width, in.
14.4 D	Effective depth of reinforcement, in.
	Key output functions.
K	= 0.258 ksi
NEXT	
R	Reinforcement ratio, $\rho = 0.0051$
B1	Compression zone factor $\beta_1 = 0.85$
RMAX	Maximum reinforcement ratio, $\rho_{max} = 0.016$
RMIN	Minimum reinforcement ratio $\rho_{min} = 0.0033$
AS	Reinforcement area $A_s = 0.878$ in^2
PHVC	Concrete design shear strength, $\phi V_c = 16.09$ kip
NEXT	
AVS	Minimum required shear reinforcement $A_v/s = 0.12$ in^2/ft.

A.1.2: TRA: ANALYSIS OF REINFORCED CONCRETE BEAM WITH TENSION REINFORCEMENT ONLY

PROGRAM LISTING:

KEYSTROKES	COMMENTARY
HOME VAR CONC	Open parent directory, CONC.
' TRA ' CRDIR ENTER	Create subdirectory, TRA.
TRA	Open current directory, TRA.
≪ ' AS/(B * D) ' EVAL ≫ ENTER ' R ' STO	Define the reinforcement ratio, R. Store R.
≪ '.9 * AS * FY * D * (1−.59 * R * FY/ FC)/12 ' EVAL ≫ ENTER ' M ' STO	Define the design moment strength, M. (i.e ϕM_n of reference 1). Store M.

≪ IF FC 4 < THEN .85 ELSE IF FC 8 > THEN .65 ELSE ' .85-(FC-4)/20 ' EVAL END END ≫ ENTER

Define the factor, B1 (i.e. β_1 of reference 2).

' B1 ' STO

Store B1.

≪ ' .75 * .85 * 87 * B1 * FC/(FY * (87+ FY)) ' EVAL ≫ ENTER
' RMAX ' STO

Define the maximum reinforcement ratio, RMAX (i.e. ρ_{max} of reference 3). Store RMAX.

≪ ' .2/FY ' EVAL ≫ ENTER

Define the minimum reinforcement ratio, RMIN (i.e. ρ_{min} of reference 4).

' RMIN ' STO

Store RMIN

≪ ' .85 * 2 * B * D * $\sqrt{(1000 * FC)}$/1000 ' EVAL ≫ ENTER
' PHVC ' STO

Define concrete design shear strength, PHVC (i.e. ϕV_c of reference 5). Store PHVC

≪ ' 12 * 50 * B/(1000 * FY) ' EVAL ≫ ENTER
' AVS ' STO

Define minimum shear reinforcement, AVS (i.e. A_v/s of reference 6). Store AVS

≪{{AS ≪' AS ' STO ≫}{B ≪' B ' STO ≫}{D ≪' D ' STO ≫}{FY ≪' FY ' STO ≫}{FC≪' FC ' STO ≫} R M B1 RMAX RMIN PHVC AVS} TMENU ≫ ENTER

Set up an input menu for data entry.

' DAT ' STO

Store the data menu.

EXAMPLE

HOME VAR CONC TRA		Recall program TRA.
DAT		Prepare for data entry.
.878	AS	Reinforcement area, in².
12	B	Beam width, in.
14.4	D	Effective depth of reinforcement, in.
60	FY	Reinforcement yield strength, ksi.
3	FC	Concrete compressive strength, ksi.
		Key output functions.
R		Reinforcement ratio, $\rho = 0.0051$.
M		Design moment strength $\phi M_n = 53.5$ kip ft.
B1		Compression zone factor, $\beta_1 = 0.85$
RMAX		Maximum reinforcement ratio, $\rho_{max} = 0.016$

528

RMIN	Minimum reinforcement ratio, $\rho_{min} = 0.0033$
PHVC	Concrete design shear strength, $\phi V_c = 16.09$ kip.
AVS	Minimum required shear reinforcement, $A_v/s = 0.12$ in²/ft.

The actual reinforcement ratio lies between the maximum and minimum allowable values and is satisfactory.

A.1.3: CRD: DESIGN REINFORCED CONCRETE BEAM WITH COMPRESSION REINFORCEMENT

PROGRAM LISTING:

KEYSTROKES	**COMMENTARY**
HOME VAR CONC	Open parent directory, CONC.
' CRD ' CRDIR ENTER	Create subdirectory, CRD.
CRD	Open current directory, CRD.
≪ IF FC 4 < THEN .85 ELSE IF FC 8 > THEN .65 ELSE ' .85–(FC–4)/20 ' EVAL END END ≫ ENTER ' B1 ' STO	Define compression zone factor, B1 (i.e. β_1 of reference 2). Store B1.
≪ ' .75 * .85 * 87 * B1 * FC/(FY * (87+FY)) ' EVAL ≫ ENTER ' RMAX ' STO	Define the maximum reinforcement ratio for a singly reinforced section, RMAX Store RMAX.
≪ ' RMAX * B * D ' EVAL ≫ ENTER ' AMAX ' STO	Define the maximum reinforcement area for a singly reinforced section, AMAX. Store AMAX.
≪ ' .9 * AMAX * FY * D * (1–.59 * RMAX * FY/ FC)/12 ' EVAL ≫ ENTER ' MMAX ' STO	Define the maximum design moment strength for a singly reinforced section. Store MMAX.

≪ ' M–MMAX ' EVAL ≫ ENTER

' MR ' STO

Define the residual moment MR (i.e. factored applied moment – MMAX). Store MR.

≪ ' 12 ∗ MR/(.9 ∗ FY ∗ (D–D1)) ' EVAL ≫ ENTER
' ASTA ' STO

Define the additional area of tension reinforcement required, ASTA. Store ASTA.

≪ ' AMAX + ASTA ' EVAL ≫ ENTER

' AST ' STO

Define the total required area of tension reinforcement, AST. Store AST

≪ ' FY ∗ AMAX/(0.85 ∗ FC ∗ B) ' EVAL ≫ ENTER

' A ' STO

Define the stress block depth A by equating tensile force at balanced strain conditions to the total compressive force. Store A.

≪ ' A/B1 ' EVAL ≫ ENTER
' C ' STO

Define the neutral axis depth, C. Store C.

≪ IF ' 87 ∗ (1–D1/C) ' EVAL FY >
THEN FY ELSE ' 87 ∗ (1–D1/C) ' EVAL
END ≫ ENTER
' FSC ' STO

Define the stress in the compression reinforcement, FSC.

Store FSC.

≪ ' ASTA ∗ FY/FSC ' EVAL ≫ ENTER

' ASC ' STO

Define the area of compression reinforcement, ASC. Store ASC.

≪ ' .85 ∗ 2 ∗ B ∗ D ∗ √(1000 ∗ FC)/1000 '
EVAL ≫ ENTER
' PHVC ' STO

Define concrete design shear strength, PHVC (i.e. ϕV_c of reference 5). Store PHVC.

≪ ' 12 ∗ 50 ∗ B/(1000 ∗ FY) ' EVAL ≫
ENTER
' AVS ' STO

Define minimum shear reinforcement AVS (i.e. A_v/s of reference 6). Store AVS.

≪{{M ≪' M ' STO ≫}{FY ≪' FY ' STO
≫}{FC ≪' FC ' STO ≫}{B ≪' B ' STO
≫}{D ≪' D ' STO ≫}{D1 ≪' D1 ' STO
≫} B1 RMAX AMAX MMAX MR ASTA
AST A C FSC ASC PHVC AVS} TMENU
≫ ENTER
' DAT ' STO

Set up an input menu for data entry.

Store the data menu.

EXAMPLE

HOME VAR CONC CRD		Recall program CRD
DAT		Prepare for data entry.
190	M	Required factored moment, kip ft.
60	FY	Reinforcement yield strength, ksi.
5	FC	Concrete compressive strength, ksi.
12	B	Beam width, in.
12	D	Effective depth of reinforcement, in.
2.4	D1	Depth to compression reinforcement, in.
		Key output functions.
B1		Compression zone factor, $\beta_1 = 0.80$.
RMAX		Maximum reinforcement ratio, $\rho_{max} = 0.025$
AMAX		Maximum tension reinforcement area for singly reinforced beam $= 3.62$ in^2
MMAX		Maximum design moment strength for a singly reinforced section $= 160.76$ kip ft.
MR		Residual moment $= 29.24$ kip ft.
ASTA		Additional tension reinforcement area $= 0.68$ in^2
AST		Tension reinforcement area $= 4.30$ in^2
A		Stress block depth $= 4.26$ in
C		Neutral axis depth $= 5.33$ in
FSC		Stress in compression reinforcement $= 47.80$ ksi.
ASC		Compression reinforcement area $= 0.85$ in^2
PHVC		Concrete design shear strength, $\phi V_c = 17.31$ kip.
AVS		Minimum required shear reinforcement, $A_v/s = 0.12$ in^2/ft.

A.1.4: CRA: ANALYSIS OF REINFORCED CONCRETE BEAM WITH COMPRESSION REINFORCEMENT

PROGRAM LISTING:

KEYSTROKES	COMMENTARY
HOME VAR CONC	Open parent directory, CONC.
' CRA ' CRDIR ENTER	Create subdirectory, CRA.

CRA

Open current directory, CRA.

≪ IF FC 4 < THEN .85 ELSE IF FC 8 >
THEN .65 ELSE ' .85-(FC-4)/20 ' EVAL
END END ≫ ENTER
' B1 ' STO

Define compression zone factor, B1 (i.e. β_1 of reference 2)

Store B1

≪ IF ' 87 * (1-D1/C) ' EVAL FY >
THEN FY ELSE ' 87 * (1-D1/C) ' EVAL
END ≫ ENTER
' FSC ' STO

Define the stress in compression reinforcement, FSC (ie. f'_{sb} of reference 1).

Store FSC.

≪ ' 75 * .85 * 87 * B1 * FC/(FY * (87+
FY))+ASC * FSC/(B * D * FY) ' EVAL ≫
ENTER
' RMAX ' STO

Define the maximum reinforcement ratio, RMAX (i.e. ρ_{max} of reference 3).

Store RMAX

≪ ' AST/(B * D) ' EVAL ≫ ENTER
' R ' STO

Define the reinforcement ratio, R.
Store R.

≪ ' .85 * FC * B * C * B1 ' EVAL ≫
ENTER
' CC ' STO

Define the compressive force in the concrete compression zone, CC.
Store CC.

≪ ' ASC * FSC ' EVAL ≫ ENTER

' CS ' STO

Define the compressive force in the compression reinforcement, CS.
Store CS.

≪ ' AST * FY ' EVAL ≫ ENTER

' T ' STO

Define the tensile force, T, in the tension reinforcement.
Store T.

≪ ' .9 * (CC * (D-B1 * C/2)+CS *
(D-D1))/12 ' EVAL ≫ ENTER
' M ' STO

Define the design moment strength, M(i.e. ϕM_n of reference 1).
Store M.

≪ ' .85 * 2 * B * D * $\sqrt{(1000 * FC)}$/1000 '
EVAL ≫ ENTER
' PHVC ' STO

Define concrete design shear strength, PHVC (i.e. ϕV_c of reference 5)
Store PHVC

≪ ' 12 * 50 * B/(1000 * FY) ' EVAL ≫
ENTER
' AVS ' STO

Define minimum shear reinforcement, AVS (i.e. A_v/s of reference 6).
Store AVS.

≪{{AST ≪' AST ' STO ≫}{ASC ≪'
ASC ' STO ≫}{FY ≪' FY ' STO ≫}
{FC ≪' FC ' STO ≫}{B ≪' B ' STO ≫}
{D ≪' D ' STO ≫}{D1 ≪' D1 ' STO ≫}
{C ≪' C ' STO ≫} B1 FSC RMAX R CC
CS T M PHVC AVS} TMENU ≫ ENTER
' DAT ' STO

Set up an input menu for data entry.

Store the data menu.

EXAMPLE

HOME VAR CONC CRA		Recall program CRA
DAT		Prepare for data entry.
9.36	AST	Tension reinforcement area, in^2
6.24	ASC	Compression reinforcement area in^2
60	FY	Reinforcement yield strength, ksi.
4	FC	Concrete compressive strength, ksi.
24	B	Beam width, in.
27	D	Effective depth of reinforcement, in.
3	D1	Depth to compression reinforcement.
4.98	C	Initial estimate of depth to neutral axis, in.
		Key output functions.
B1		Compression zone factor, $\beta_1 = 0.85$.
FSC		Compression reinforcement stress $=34.59$ ksi
RMAX		Maximum reinforcement ratio, $\rho_{max} = 0.0269$
R		Reinforcement ratio, $\rho = 0.014 < \rho_{max}$ ... satisfactory
CC		Compressive force in the concrete compressive zone = 345.4 kip.
CS		Compressive force in the compression reinforcement = 215.8 kip.
T		Tensile force in the tension reinforcement = 561.6 kip $\approx$ CC + CS. Hence, the initial estimate for C is correct.
M		Design moment strength, $\phi M_n = 1033$ kip ft
PHVC		Concrete design shear strength, $\phi V_c = 69.67$ kip.
AVS		Minimum required shear reinforcement, $A_v/s = 0.24$ in^2/ft.

A.1.5: TBD: DESIGN OF REINFORCED CONCRETE T-BEAM

(Note: Assumes flange thickness is less than the depth of the equivalent rectangular stress block)

PROGRAM LISTING:

KEYSTROKES	COMMENTARY
HOME VAR CONC	Open parent directory, CONC.
' TBD ' CRDIR	Create subdirectory, TBD.
TBD	Open current directory, TBD.
≪ ' .85 ∗ FC ∗ (B–BW) ∗ HF ' EVAL ≫ ENTER ' CF ' STO	Define the compressive force in the flange, CF (i.e. C_f of reference 1). Store CF
≪ ' CF/FY ' EVAL ≫ ENTER ' ASF ' STO	Define the reinforcement area, ASF, required to balance the flange force (i.e. A_{sf} of reference 1). Store ASF.
≪ ' .9 ∗ ASF ∗ FY ∗ (D–.5 ∗ HF)/12 ' EVAL ≫ ENTER ' MF ' STO	Define the design moment strength, MF, due to ASF. Store MF.
≪ ' M–MF ' EVAL ≫ ENTER ' MW ' STO	Define the residual moment MW (i.e. the factored applied moment – MF) Store MW.
≪ ' 12 × MW/(BW ∗ D^2) ' EVAL ≫ ENTER ' KW ' STO	Define the parameter, KW. Store KW.
≪ ' (1–√(1–2 ∗ KW/(.9 ∗ .85 ∗ FC))) ∗ .85 ∗ FC/FY ' EVAL ≫ ENTER ' RW ' STO	Define the web reinforcement ratio, RW. Store RW.
≪ ' RW ∗ BW ∗ D ' EVAL ≫ ENTER ' ASW ' STO	Define the reinforcement area, ASW required to resist the residual moment. Store ASW.
≪ ' ASF + ASW ' EVAL ≫ ENTER ' AS ' STO	Define the total required area of tension reinforcement, AS. Store AS.

≪ IF FC 4 < THEN .85 ELSE IF FC 8 >
THEN .65 ELSE ' .85-(FC-4)/20 ' EVAL
END END ≫ ENTER
' B1 ' STO

Define the factor, B1 (i.e. β_1 of reference 2).

Store B1.

≪ ' .85 * 87 * B1 * FC/(FY * (87+FY)) '
EVAL ≫ ENTER
' RB ' STO

Define the web balanced reinforcement ratio, RB (i.e. $\bar{\rho}_b$ of reference 3).
Store RB.

≪ ' ASF/(BW * D) ' EVAL ≫ ENTER

' RF ' STO

Define the reinforcement ratio, RF, due to ASF (i.e. ρ_f of reference 1).
Store RF.

≪ ' .75 * BW * (RB+RF)/B ' EVAL ≫
ENTER
' RMAX ' STO

Define the maximum reinforcement ratio, RMAX (i.e. ρ_{max} of reference 3).
Store RMAX

≪ ' AS/(B * D) ' EVAL ≫ ENTER

' R ' STO

Define the reinforcement ratio, R(i.e. ρ of reference 1).
Store R.

≪ ' ASW * FY/(.85 * FC * BW) ' EVAL
≫ ENTER

' A ' STO

Define the depth of the equivalent rectangular stress block, A (i.e. a of reference 2).
Store A.

≪ ' .85 * 2 * BW * D * $\sqrt{(1000 * FC)}$/ 1000
' EVAL ≫ ENTER
' PHVC ' STO

Define concrete design shear strength, PHVC (i.e. ϕV_c of reference 5).
Store PHVC.

≪ ' 12 * 50 * BW/(1000 * FY) ' EVAL ≫
ENTER
' AVS ' STO

Define the minimum shear reinforcement, AVS (i.e. A_v/s of reference 6).
Store AVS.

≪{{M ≪' M ' STO ≫}{FY ≪' FY ' STO
≫}{FC ≪' FC ' STO ≫}{B ≪' B ' STO
≫}{BW ≪' BW ' STO ≫}{D ≪' D ' STO
≫}{HF ≪' HF ' STO ≫} CF ASF MF
MW KW RW ASW AS B1 RB RF RMAX
R A PHVC AVS} TMENU ≫ ENTER
' DAT ' STO

Set up an input menu for data entry.

Store the data menu.

535

EXAMPLE

HOME VAR CONC TBD		Recall program TBD.
DAT		Prepare for data entry.
400	M	Required factored moment, kip ft.
60	FY	Reinforcement yield strength, ksi.
4	FC	Concrete compressive strength, ksi.
30	B	Flange width, in.
10	BW	Web width, in.
19	D	Effective depth of reinforcement, in.
2.5	HF	Flange depth, in.
		Key output functions.
CF		Flange compressive force $C_f = 170$ kip.
ASF		Reinforcement area, $A_{sf} = 2.83$ in^2.
MF		Design moment due to $A_{sf} = 226$ kip ft.
MW		Residual moment = 174 kip ft.
KW		KW = 0.577
RW		Web reinforcement ratio = 0.012
ASW		Reinforcement area, ASW = 2.27 in^2
AS		Total reinforcement area = 5.1 in^2
B1		Compression zone factor, $\beta_1 = 0.85$
RB		Reinforcement ratio, $\bar{\rho}_b = 0.0285$
RF		Reinforcement ratio, $\rho_f = 0.0149$
RMAX		Reinforcement ratio, $\rho_{max} = 0.0109$
R		Actual reinforcement ratio, $\rho = 0.0090$ $< \rho_{max} \ldots$ satisfactory.
A		Depth of stress block, $a = 4.01" > h_f \ldots$ satisfactory.
PHVC		Concrete design shear strength $\phi V_c = 20.43$ kip.
AVS		Minimum required shear reinforcement $A_v / s = 0.0996$ in^2 / ft.

A.1.6: TBA: ANALYSIS OF REINFORCED CONCRETE T-BEAM

(Note: Assumes flange thickness is less than depth of the equivalent rectangular stress block)

PROGRAM LISTING:

KEYSTROKES	COMMENTARY
HOME VAR CONC	Open parent directory, CONC.
' TBA ' CRDIR ENTER	Create subdirectory, TBA.
TBA	Open current directory, TBA
≪ ' .85 * FC * (B−BW) * HF ' EVAL≫ ENTER ' CF ' STO	Define the compressive force in the flange, CF(i.e. C_f of reference 1). Store CF.
≪ ' CF/FY ' EVAL ≫ ENTER ' ASF ' STO	Define the reinforcement area, ASF, required to balance the flange force (i.e. A_{st} of reference 1). Store ASF.
≪ ' .9 * ASF * FY * (D−.5 * HF)/12 ' EVAL ≫ ENTER ' MF ' STO	Define the design moment strength, MF, due to ASF. Store MF.
≪ ' AS − ASF ' EVAL ≫ ENTER ' ASW ' STO	Define the reinforcement area, ASW, required to resist the residual moment. Store ASW.
≪ ' ASW/(BW * D) ' EVAL ≫ ENTER ' RW ' STO	Define the web reinforcement ratio, RW, required to resist the residual moment. Store RW.
≪ ' .9 * ASW * FY * D * (1−.59 * RW * FY/FC)/12 ' EVAL ≫ ENTER ' MW ' STO	Define the residual moment, MW. Store MW.
≪ ' MF + MW ' EVAL ≫ ENTER ' M ' STO	Define the design moment strength, M. (i.e. ϕM_n of reference 1). Store M

« IF FC 4 < THEN .85 ELSE IF FC 8 > THEN .65 ELSE ' .85–(FC–4)/20 ' EVAL END END » ENTER ' B1 ' STO	Define the factor, B1 (i.e. β_1 of reference 2). Store B1.
« ' .85 * 87 * B1 * FC/(FY * (87+FY)) ' EVAL » ENTER ' RB ' STO	Define the web balanced reinforcement ratio, RB(i.e. $\bar{\rho}_b$ of reference 3). Store RB.
« ' ASF/(BW * D) ' EVAL » ENTER ' RF ' STO	Define the reinforcement ratio, RF, due to ASF (i.e. ρ_f of reference 1). Store RF.
« ' .75 * BW * (RB+RF)/B ' EVAL » ENTER ' RMAX ' STO	Define the maximum reinforcement ratio, RMAX (i.e. ρ_{max} of reference 3). Store RMAX.
« ' AS/(B * D) ' EVAL » ENTER ' R ' STO	Define the reinforcement ratio, R(i.e. ρ of reference 1). Store R.
« ' RW * FY * D/(.85 * FC) ' EVAL » ENTER ' A ' STO	Define the depth of the equivalent rectangular stress block, A (i.e. a of reference 2). Store A
« ' .85 * 2 * BW * D * $\sqrt{(1000 * FC)}$ ' EVAL » ENTER ' PHVC ' STO	Define concrete design shear strength, PHVC (i.e. A_v/s of reference 6). Store PHVC
« ' 12 * 50 * BW/(1000 * FC) ' EVAL » ENTER ' AVS ' STO	Define the minimum shear reinforcement, AVS (i.e. A_v/s of reference 6). Store AVS.
«{{AS «' AS ' STO »}{FY«' FY ' STO »}{ «FC ' FC ' STO »}{B «' B ' STO »}{BW «' BW ' STO »}{D «' D ' STO »}{HF«' HF ' STO »} CF ASF MF ASW RW MW M B1 RB RF RMAX R A PHVC AVS} TMENU » ENTER ' DAT ' STO	Set up an input menu for the data entry. Store the data menu

EXAMPLE

HOME VAR CONC TBA		Recall program TBA.
DAT		Prepare for data entry.
5	AS	Reinforcement area, in².
60	FY	Reinforcement yield strength, ksi.
4	FC	Concrete compressive strength, ksi.
30	B	Flange width, in.
10	BW	Web width, in.
19	D	Effective depth of reinforcement, in.
2.5	HF	Flange depth, in.
		Key output functions.
CF		Flange compressive force, $C_f = 170$ kip.
ASF		Reinforcement area, $A_{sf} = 2.83$ in².
MF		Design moment due to $A_{sf} = 226$ kip ft.
ASW		Reinforcement area, ASW = 2.17 in².
RW		Web reinforcement ratio = 0.011.
MW		Residual moment = 167 kip ft.
M		Design moment strength, $\phi M_n = 393$ kip ft.
B1		Compression zone factor, $\beta_1 = 0.85$.
RB		Reinforcement ratio, $\rho_b = 0.0285$.
RF		Reinforcement ratio, $\rho_f = 0.0149$.
RMAX		Reinforcement ratio, $\rho_{max} = 0.0109$
R		Actual reinforcement ratio, $\rho = 0.00877$ $< \rho_{max}$ ... satisfactory.
A		Depth of stress block, a = 3.82 in > h_f ... satisfactory.
PHVC		Concrete design shear strength, $\phi V_c = 20.43$ kip.
AVS		Minimum required shear reinforcement, $A_v/s = 0.0996$ in²/ft

A.2.1: EAS: ANALYSIS OF MASONRY BEAM TO DETERMINE STRESSES

PROGRAM LISTING:

KEYSTROKES	COMMENTARY
HOME ' MAS ' CRDIR ENTER	Create parent directory, MAS.
VAR MAS ' EAS ' CRDIR ENTER	Create subdirectory, EAS.
EAS	Open current directory, EAS.
≪ ' ES/EM ' EVAL ≫ ENTER	Define the modular ratio, N (i.e n of reference 7).
' N ' STO	Store N.
≪ ' AS/(B * D) ' EVAL ≫ ENTER	Define the reinforcement ratio, P (i.e. p of reference 7).
' P ' STO	Store P.
≪ ' $\sqrt{(2*N*P + P^2*N^2)}-N*P$ ' EVAL ≫ ENTER	Define the neutral axis depth ratio, K (i.e. k of equation (9–5), reference 8).
' K ' STO	Store K.
≪ ' 1–K/3 ' EVAL ≫ ENTER	Define lever arm ratio, J (i.e. j of equation (9–7), reference 8).
' J ' STO	Store J.
≪ ' $24*M/(J*K*B*D^2)$ ' EVAL ≫ ENTER	Define the compressive stress in the masonry, FM (i.e. f_b of equation (9–3), reference 8).
' FM ' STO	Store FM.
≪ ' $12*M/(AS*J*D)$ ' EVAL ≫ ENTER	Define the tensile stress in the reinforcement, FS (i.e. f_s of equation (9–4), reference 8).
' FS ' STO	Store FS.
≪{{M ≪' M ' STO ≫}{AS ≪' AS ' STO ≫}{ES ≪' ES ' STO ≫}{EM ≪' EM ' STO ≫}{B ≪' B ' STO ≫}{D ≪' D ' STO ≫} TMENU ≫ ENTER	Set up an input menu for data entry.
' DAT ' STO	Store the data menu.

EXAMPLE

HOME VAR MAS EAS		Recall program EAS.
DAT		Prepare data entry.
30	M	Applied moment, kip ft.
.88	AS	Reinforcement area, in².
29000	ES	Modulus of elasticity of reinforcement, ksi
1100	EM	Modulus of elasticity of the masonry, ksi.
9.63	B	Beam width in.
34	D	Effective depth of reinforcement, in.
		Key output functions.
N		Modular ratio, n = 26.36
P		Reinforcement ratio, p = 0.0027.
K		Neutral axis depth ratio, k = 0.312.
J		Lever arm ratio, j = 0.896.
FM		Masonry compressive stress f_b = 0.23 ksi.
FS		Reinforcement tensile stress f_s = 13.43 ksi

A.2.2: EAM: ANALYSIS OF MASONRY BEAM TO DETERMINE ALLOWABLE MOMENT

PROGRAM LISTING:

KEYSTROKES	COMMENTARY
HOME VAR MAS	Open parent directory, MAS.
' EAM ' CRDIR ENTER	Create subdirectory, EAM.
EAM	Open current directory, EAM.
≪ ' ES/EM ' EVAL ≫ ENTER	Define the modular ratio, N (i.e. n of reference 7).
' N ' STO	Store N.
≪ ' AS/(B * D) ' EVAL ≫ ENTER	Define the reinforcement ratio, P (i.e. p of reference 7).
' P ' STO	Store P.

≪ ' √(2 * N * P + P^2 * N^2)−N * P '
EVAL ≫ ENTER
' K ' STO

≪ ' 1−K/3 ' EVAL ≫ ENTER

' J ' STO

≪ ' FM * J * K * B * D^2/24 ' EVAL ≫
ENTER
' MM ' STO

≪ ' FS * AS * J * D/12 ' EVAL ≫ ENTER

' MS ' STO

≪{{AS ≪' AS ' STO ≫}{FS ≪' FS '
STO ≫}{FM ≪' FM ' STO ≫}{ES ≪' ES
' STO ≫}{EM ≪' EM ' STO ≫}{B ≪' B
' STO ≫}{D ≪' D ' STO ≫} N P K J
MM MS} TMENU ≫ ENTER
' DAT ' STO

Define the neutral axis depth ratio, K (i.e. k of equation (9-5), reference 8). Store K.	
Define the lever arm ratio, J (i.e. j of equation (9-7), reference 8). Store J.	
Define the moment capacity of the masonry, MM (i.e. M_m of reference 7). Store MM.	
Define the moment capacity of the reinforcement, MS(i.e. M_s of reference 7). Store MS.	
Set up an input menu for data entry.	
Store the data menu.	

EXAMPLE:

HOME VAR MAS EAM   Recall program EAM.
DAT   Prepare data entry.

.33	AS	Reinforcement area, in².
24	FS	Allowable reinforcement stress, ksi.
.3	FM	Allowable masonry stress, ksi.
29000	ES	Modulus of elasticity of reinforcement, ksi
1350	EM	Modulus of elasticity of the masonry, ksi.
12	B	Beam width, in.
5	D	Effective depth of reinforcement, in.
		Key output functions
	N	Modular ratio, n = 21.48.
	P	Reinforcement ratio, p = 0.0055.
	K	Neutral axis depth ratio, k = 0.382.
	J	Lever arm ratio, j = 0.873.
	MM	Moment capacity of the masonry, M_m = 1.250 kip ft....governs.
	MS	Moment capacity of the reinforcement, M_s = 2.880 kip ft.

542

A.2.3: EDC: DESIGN OF MASONRY BEAM WITH COMPRESSION REINFORCEMENT

PROGRAM LISTING:

KEYSTROKES	COMMENTARY
HOME VAR MAS	Open parent directory, MAS.
' EDC ' CRDIR ENTER	Create subdirectory, EDC.
EDC	Open current directory, EDC.
≪ ' ES/EM ' EVAL ≫ ENTER ' N ' STO	Define the modular ratio, N (i.e. n of reference 7). Store N.
≪ ' N ∗ FM/(2 ∗ FS ∗ (N+FS/FM)) ' EVAL ≫ ENTER ' PB ' STO	Define the reinforcement ratio, PB, for balanced stress conditions (i.e. p_b of equation (6–7), reference 9). Store PB.
≪ ' $\sqrt{(2 \ast N \ast PB + PB\hat{\,}2 \ast N\hat{\,}2)}$−N ∗ PB ' EVAL ≫ ENTER ' KB ' STO	Define the neutral axis depth ratio, KB, for balanced stress conditions. Store KB.
≪ ' 1−KB/3 ' EVAL ≫ ENTER ' JB ' STO	Define the lever arm ratio, JB. Store JB.
≪ ' FM ∗ JB ∗ KB ∗ B ∗ D^2/24 ' EVAL ≫ ENTER ' MB ' STO	Define the moment capacity of the masonry, MB, for balanced stress conditions. Store MB.
≪ ' 12 ∗ MB/(FS ∗ JB ∗ D) ' EVAL ≫ ENTER ' AB ' STO	Define the tensile reinforcement area, AB, required for balanced stress conditions. Store AB.
≪ ' M−MB ' EVAL ≫ ENTER ' MR ' STO	Define the residual moment MR (i.e. the applied moment − MB). Store MR.

≪ IF ' FS * (KB–D1/D)/(1–KB) ' EVAL
FS > THEN FS ELSE '
FS * (KB–D1/D)/(1–KB) ' EVAL END ≫
ENTER
' FSC ' STO

Define the stress in the compression reinforcement, FSC, at balanced stress condition.

Store FSC.

≪ ' 12 * MR/(FSC * (D–D1) * (N–1)/N) '
EVAL ≫ ENTER
' ASC ' STO

Define the required area of compression reinforcement, ASC.
Store ASC

≪ ' AB + 12 * MR/(FS * (D–D1)) ' EVAL
≫ ENTER
' AST ' STO

Define the required area of tension reinforcement, AST.
Store AST.

≪{{M ≪' M ' STO ≫}{FS ≪' FS ' STO
≫}{FM ≪' FM ' STO ≫}{ES ≪' ES '
STO ≫}{EM ≪' EM ' STO ≫}{B ≪' B '
STO ≫}{D ≪' D ' STO ≫}{D1 ≪' D1 '
STO ≫} N PB KB JB MB AB MR FSC
ASC AST} TMENU ENTER
' DAT ' STO

Set up an input menu for data entry.

Store the data menu.

EXAMPLE

HOME VAR MAS EDC
DAT

Recall program EDC.
Prepare data entry.

33	M	Applied moment, kip ft.
24	FS	Allowable reinforcement stress, ksi.
.5	FM	Allowable masonry stress, ksi.
29000	ES	Modulus of elasticity of the reinforcement, ksi.
1125	EM	Modulus of elasticity of the masonry, ksi.
7.63	B	Beam width, in.
24.5	D	Effective depth of reinforcement, in.
4.5	D1	Depth to the compression reinforcement, in.

	Key output functions.
N	Modular ratio, n = 25.78
PB	Balanced reinforcement ratio, p_b = 0.00364
KB	Neutral axis depth ratio = 0.349
JB	Lever arm ratio = 0.884
MB	Moment capacity of the masonry = 29.45 kip ft.
AB	Tension reinforcement area for balanced stress conditions = 0.68 in^2
MR	Residual moment = 3.55
FSC	Stress in compression reinforcement = 6.113 ksi.
ASC	Compression reinforcement area = 0.36 in^2
AST	Tension reinforcement area = 0.77 in^2

A.3.1: HANK: HANKINSON'S FORMULA

PROGRAM LISTING:

KEYSTROKES	COMMENTARY
HOME ' WOOD ' CRDIR ENTER	Create parent directory, WOOD.
VAR WOOD ' HANK ' CRDIR ENTER	Create subdirectory, HANK.
HANK	Open current directory, HANK.
≪ ' P ∗ Q/(P ∗ (SIN(θ))^2 + Q ∗ (COS(θ))^2) ' →NUM ≫ ENTER	Define the allowable bolt load, N, at an angle $\theta°$ with the direction of the grain (i.e. $F_{e\theta}$ of reference 10).
' N ' STO	Store N.
≪{{P ≪' P ' STO ≫}{Q ≪' Q ' STO ≫}{θ ≪' θ ' STO ≫} N} TMENU ≫ ENTER	Set up an input menu for data entry.
' DAT ' STO	Store the data menu.

EXAMPLE

HOME VAR WOOD HANK		Recall program HANK.
DAT		Prepare data entry.
1400	P	Allowable bolt load parallel to the grain, lb.
630	Q	Allowable bolt load perpendicular to the grain, lb.
19	θ	Angle of inclination between the applied load and the direction of the grain, $\theta°$.
		Key output functions.
N		Allowable load, N = 1239 lb.

A.3.2: BEAM: ALLOWABLE BENDING DESIGN STRESS[10,11]

PROGRAM LISTING:

KEYSTROKES	COMMENTARY
HOME VAR WOOD	Open parent directory, WOOD.
' BEAM ' CRDIR ENTER	Create subdirectory, BEAM.
BEAM	Open current directory, BEAM.
≪ ' √(12 * LE * D/B^2) ' EVAL ≫ ENTER ' RB ' STO	Define the slenderness ratio. Store R_b
≪ ' FB * CD * CM * CT * CS ' EVAL ≫ ENTER ' FAB ' STO	Define the adjusted tabulated bending value. Store F_b^*
≪ ' 1000000 * E * CM * CT ' EVAL ≫ ENTER ' E1 ' STO	Define the allowable modulus of elasticity. Store E'
≪ ' KBE * E1/RB^2 ' EVAL ≫ ENTER ' FBE ' STO	Define the critical buckling design value. Store F_{bE}
≪ ' FBE/FAB ' EVAL ≫ ENTER ' F ' STO	Define F_{bE}/F_b^* Store this value.
≪ ' (1 + F)/1.9 – √(((1 + F)/1.9)^2 – F/0.95) ' EVAL ≫ ENTER ' CL ' STO	Define the beam stability factor. Store C_L
≪ ' KL * (1291.5/(B * D * L))^(1/X) ' EVAL ≫ ENTER ' CV ' STO	Define the volume factor. Store C_V
≪ IF CL CV > THEN ' FAB * CV ' EVAL ELSE ' FAB * CL ' EVAL END ≫ ENTER ' F1B ' STO	Define the allowable bending design stress. Store F_b'

≪{{FB ≪' FB ' STO ≫}{E ≪' E ' STO
≫}{KBE ≪' KBE ' STO ≫}{B ≪' B '
STO ≫}{D ≪' D ' STO ≫}{LE ≪' LE '
STO ≫}{X ≪' X ' STO ≫}{KL ≪' KL '
STO ≫}{L ≪' L ' STO ≫}{CD ≪' CD '
STO ≫}{CM ≪' CM ' STO ≫}{CT ≪'
CT ' STO ≫}{CS ≪' CS ' STO ≫} RB
FAB E1 FBE CL CV F1B} TMENU ≫
ENTER

Set up an input menu for data entry.

' DAT ' STO

Store the data menu

EXAMPLE

HOME VAR WOOD BEAM
DAT

Recall program BEAM.
Prepare data entry.

2400	FB	

Tabulated bending stress F_b for a 24F- V5 visually graded DF/HF (western species) glued–laminated beam.

1.5	E
0.69	KBE

Modulus of elasticity E_{xx} x 10^6, psi.
Euler buckling coefficient K_{bE} for glued–laminated beams.

5.125	B
36	D
8 * 1.63 + 3 * 3	LE

Beam width, in.
Beam depth, in.
Effective length ℓ_e for a uniformly loaded, 32 ft span simply supported beam, braced at 8 ft on centers.

10	X
1.0	KL
32	L
1.25	CD
1.0	CM
1.0	CT
1.0	CS

x = 10 for western species glulam beams.
Load condition coefficient for distributed load
Length between points of zero moment.
Load duration factor C_D for roof live load.
Wet service factor C_M for interior use.
Temperature factor C_t for normal temperature
Size factor C_F (not applicable to glulam beams)
Key output functions.

RB
FAB
E1
FBE
CL
CV
F1B

Slenderness factor R_B = 19.04
Adjusted bending stress F_b^* = 3000 psi.
Allowable modulus of elasticity E' = 1.5 $\times 10^6$ psi.
Critical buckling design value F_{bE} = 2520 psi
Beam stability factor C_L = 0.737
Volume factor C_V = 0.859
Allowable bending stress F_b' = 2210 psi.

A.3.3: COL: ALLOWABLE COMPRESSION DESIGN STRESS[10,11]

PROGRAM LISTING:

KEYSTROKES	COMMENTARY
HOME VAR WOOD	Open parent directory, WOOD.
' COL ' CRDIR ENTER	Create subdirectory, COL.
COL	Open current directory, COL.
≪ ' FC * CD * CM * CT * CS ' EVAL ≫ ENTER ' FAC ' STO	Define the adjusted tabulated compression value. Store F_c^*
≪ ' 1000000 * E * CM * CT' EVAL ≫ ENTER ' E1 ' STO	Define the allowable modulus of elasticity. Store E'
≪ ' KCE * E1/SRA^2 ' EVAL ≫ ENTER ' FCE ' STO	Define the critical buckling design value. Store F_{cE}
≪ ' FCE/FAC ' EVAL ≫ ENTER ' F ' STO	Define F_{cE}/F_c^* Store this value.
≪ ' (1 + F)/(2 * C) – $\sqrt{(((1 + F)/(2 * C))^2 - F/C)}$ ' EVAL ≫ ENTER ' CP ' STO	Define the column stability factor. Store C_P
≪ ' FAC * CP ' EVAL ≫ ENTER ' F1C ' STO	Define the allowable compressive design stress. Store F_c'
≪ IF 11 SRF > THEN 0 ELSE IF K SRF > THEN ' (SRF–11)/(K–11) ' EVAL ELSE 1 END END ≫ ENTER ' J ' STO	Define the column stability factor for flexural effects, J(i.e. J of reference 14). Store J.
≪{{FC ≪' FC ' STO ≫}{E ≪' E ' STO ≫}{SRA ≪' SRA ' STO ≫}{KCE ≪' KCE ' STO ≫}{C ≪' C ' STO ≫}{CD ≪' CD ' STO ≫}{CM ≪' CM ' STO ≫}{CT ≪' CT ' STO ≫}{CS ≪' CS ' STO ≫}{CTS ≪' CTS ' STO ≫} FAC E1 FCE CP F1} TMENU ≫ ENTER ' DAT ' STO	Set up an input menu for data entry. Store the data menu.

EXAMPLE

HOME VAR WOOD COL		Recall program COL.
DAT		Prepare data entry.
1300	FC	Tabulated compression value F_c Douglas fir-larch No. 2 grade, psi.
1.6	E	Tabulated modulus of elasticity $\times 10^6$, psi.
1 * 12 * 12/7.25	SRA	Slenderness ratio $K_e l/d$ for a 4×8 pin ended column, 12 feet long, braced about the weak axis.
0.3	KCE	Euler buckling coefficient K_{cE} for visually graded lumber.
0.8	C	Column parameter c for sawn lumber.
1.0	CD	Load duration factor C_D for normal live load duration.
1.0	CM	Wet service factor C_M for interior use.
1.0	CT	Temperature factor C_t for normal temperatures.
1.05	CS	Size factor C_F
		Key output functions.
FAC		Adjusted tabulated compression stress $F_c^* = 1365$ psi.
E1		Allowable modulus of elasticity $E' = 1.6 \times 10^6$ psi.
FCE		Critical buckling design value $F_{cE} = 1217$ psi.
CP		Column stability factor $C_P = 0.650$
F1C		Allowable compression stress $F_c' = 887$ psi.

A.4.1: FB: ALLOWABLE STRESS IN SLENDER BEAM

PROGRAM LISTING:

KEYSTROKES	COMMENTARY
HOME ' STEEL ' CRDIR ENTER	Create parent directory, STEEL.
VAR STEEL ' FB ' CRDIR ENTER	Create subdirectory, FB.
FB	Open current directory, FB
≪ ' CB * 12000/(L * DAF) ' EVAL ≫ ENTER ' FB2 ' STO	Define the allowable bending stress, FB2 (i.e. F_b of equation (F1-8), reference 12). Store FB2
≪ ' L/RT ' EVAL ≫ ENTER ' LRT ' STO	Define the factor, LRT (i.e. ℓ/r_t of reference 12). Store LRT.
≪ ' √(102000 * CB/FY) ' EVAL ≫ ENTER ' B0 ' STO	Define the factor, B0 Store B0
≪ ' B0 * √5 ' EVAL ≫ ENTER ' B1 ' STO	Define the factor, B1. Store B1.
≪ IF LRT B0 <THEN ' .6 * FY ' EVAL ELSE IF LRT B1 > THEN ' 170000 * CB/LRT^2 ' EVAL ELSE ' FY * (.667-FY * LRT^2/(1530000 * CB)) ' EVAL END END ≫ ENTER ' FB1 STO	Define the allowable bending stress, FB1 (i.e. F_b of equations (F1-6) and (F1-7), reference 12). Store FB1.
≪{{FY ≪' FY ' STO ≫}{L ≪' L ' STO ≫}{CB ≪' CB ' STO ≫}{DAF ≪' DAF ' STO ≫}{RT ≪' RT ' STO ≫} FB2 LRT B0 B1 FB1} TMENU ≫ ENTER ' DAT ' STO	Set up an input menu for data entry. Store the data menu.

EXAMPLE

HOME VAR STEEL FB		Recall program FB.
DAT		Prepare data entry.
36	FY	Specified yield stress, F_y ksi.
20x12	L	Unbraced length, ℓ, ins.
1	CB	Bending coefficient, C_b.
4.23	DAF	Ratio, depth of beam/area compression flange, d/A_f, in^{-1}.
2.57	RT	Radius of gyration of compression area about the web, r_t.
		Key output functions.
FB2		Allowable bending stress from equation (F1-8), $F_b = 11.82$ ksi.
LRT		$\ell/r_t = 93.39$.
B0		$B0 = 53.2 < 93.39$.
B1		$B1 = 119.0 > 93.39 \ldots$ eq. (F1-6) applies.
FB1		Allowable bending stress from equation (F1-7), $F_b = 16.62$ ksi ... governs.

A.4.2: CBX: COMBINED AXIAL COMPRESSION AND BENDING

PROGRAM LISTING:

KEYSTROKES	**COMMENTARY**
HOME VAR STEEL	Open parent directory, STEEL.
' CBX ' CRDIR ENTER	Create subdirectory, CBX.
CBX	Open current directory, CBX.
≪ ' FA/FPA +CM * FB/(FPB * (1–FA/FE)) ' EVAL ≫ ENTER ' E31A ' STO	Define the value of E31A (i.e. left hand side of equation (H1-1), reference 13). Store E31A.
≪ ' FA/(.6 * FY) + FB/FPB ' EVAL ≫ ENTER ' E31B ' STO	Define the value of E31B (i.e. left hand side of equation (H1-2), reference 13). Store E31b.
≪ ' FA/FPA + FB/FPB ' EVAL ENTER ' E32 STO	Define the value of E32 (i.e. the left hand side of equation (H1-3), reference 13). Store E32

≪{{FA ≪' FA ' STO ≫}}{FPA ≪' FPA ' STO ≫}}{CM ≪' CM ' STO ≫}}{FB ≪' FB ' STO ≫}}{FPB ≪' FPB ' STO ≫}}{FE ≪' FE ' STO ≫}}{FY ≪' FY ' STO ≫} E31A E31B} TMENU ≫ ENTER	Set up an input menu for data entry.
' DAT ' STO	Store the data menu.

EXAMPLE

HOME VAR STEEL CBX		Recall program CBX.
DAT		Prepare data entry.
4.82	FA	Computed axial stress, f_a, ksi.
15.1	FPA	Axial compressive stress permitted in the absence of bending moment, F_a, ksi...f_a/F_a > 0.15; equation (H1-3) is not applicable.
.85	CM	Coefficient applied to bending term, C_m.
17.68	FB	Computed bending stress, f_b ksi.
22	FPB	Bending stress permitted in the absence of axial force, F_b, ksi.
22.2	FE	Euler stress divided by factor of safety, F'_e, ksi.
36	FY	Specified yield stress, F_y, ksi.
		Key output functions.
E31A		Left hand side of equation (H1-1) = 1.19.
E31B		Left hand side of equation (H1-2) = 1.03.

A.4.3: MPCO: PLASTIC MOMENT OF MEMBER SUBJECT TO COMBINED AXIAL COMPRESSION AND BENDING.

PROGRAM LISTING:

KEYSTROKES	**COMMENTARY**
HOME VAR STEEL	Open parent directory, STEEL.
' MPCO ' CRDIR ENTER	Create subdirectory, MPCO.

MPCO	Open current directory, MPCO.
≪ ' (1–P/PCR) * (1–P/PE)/CM ' EVAL ≫ ENTER	Define the value of M2, the ratio of the maximum factored moment to the critical moment which can be resisted by the member in the absence of axial load (i.e. M/M_m of equation (N4–2), reference 14).
' M2 ' STO	Store M2.
≪ ' (1–P/PY) * 1.18 ' EVAL ≫ ENTER	Define the value of M3, the ratio of the maximum factored moment to the fully plastic moment (i.e, M/M_p of equation (N4–3), reference 14).
' M3 ' STO	Store M3.
≪{{P ≪' P ' STO ≫}{PCR ≪' PCR ' STO ≫}{PE ≪' PE ' STO ≫}{PY ≪' PY ' STO ≫}{CM ≪' CM ' STO ≫} M2 M3} TMENU ≫ ENTER	Set up an input menu for data entry.
' DAT ' STO	Store the data menu.

EXAMPLE

HOME VAR STEEL MPCO	Recall program MPCO.
DAT	Prepare data entry.
300 P	Applied factored axial load, kips.
1230 PCR	Maximum strength of an axially loaded compression member, kips (i.e. P_{cr} of equation (N4–2), reference 14).
8240 PE	Euler buckling load, P_e, kips.
1340 PY	Plastic axial load, P_y, kips.
.85 CM	Coefficient applied to bending term, C_m.
	Key output functions.
M2	$M = 0.857 M_m$, kip ft.
M3	$M = 0.916 M_p$, kip ft.

A.4.4: STPL: COLUMN STIFFENER PLATES

PROGRAM LISTING:

KEYSTROKES	COMMENTARY
HOME VAR STEEL	Open parent directory, STEEL.
' STPL ' CRDIR ENTER	Create subdirectory, STPL.
STPL	Open current directory, STPL.
≪ ' (PBF–FYC * T * (TBF+5 * K))/ FYST ' EVAL ≫ ENTER	Define the required area of column–web stiffeners (i.e. A_{st} of equation (K1-9) reference 15).
' AST ' STO	Store AST
≪ ' 4100 * TCW^3 * √FYC/PBF ' EVAL ≫ ENTER	Define the column–web clear depth, DC, which, when exceeded, requires stiffeners opposite the compression flange (i.e. d_c of equation (K1-8) reference 15).
' DC ' STO	Store DC.
≪ ' .4 * √(1.8 * PBF/FYC) ' EVAL ≫ ENTER	Define the column flange thickness, TCF, below which stiffeners are required opposite the tension flange (i.e. t_f of equation (K1-1) reference 15, as modified by reference 16 for special moment resisting space frames in seismic zones 3 and 4).
' TCF ' STO	Store TCF.
≪{{PBF ≪' PBF ' STO ≫}{FYC ≪' FYC ' STO ≫}{T ≪' T ' STO ≫}{TBF ≪' TBF ' STO ≫}{K ≪' K ' STO ≫}{FYST ≪' FYST ' STO ≫}{TCW ≪' TCW ' STO ≫} AST DC TCF} TMENU ≫ ENTER	Set up an input menu for data entry.
' DAT ' STO	Store the data menu.

EXAMPLE

HOME VAR STEEL STPL		Recall program STPL.
DAT		Prepare data entry.

189	PBF	Force in beam flange, P_{bf}, kips.
36	FYC	Column yield stress, F_{yc}, ksi.
.91	T	Column web thickness + doubler plate thickness, in.
.585	TBF	Beam flange thickness, t_b, in.
1.69	K	Column flange thickness plus depth of fillet, k, in.
36	FYST	Stiffener yield stress.
.66	TCW	Column web thickness, t, in.
		Key output functions.
AST		$A_{st} = -2.98$ sq in ... stiffeners are not required.
DC		$d_c = 37.42$ in ... stiffeners are required opposite the compression flange if the column web clear depth exceeds 37.42 in.
TCF		$t_f = 1.23$ in ... stiffeners are required opposite the tension flange if the column flange thickness is less than 1.23 in.

A.5.1: IXX: MOMENT OF INERTIA OF COMPOSITE SECTIONS

PROGRAM LISTING:

KEYSTROKES	**COMMENTARY**
HOME ' PROPS ' CRDIR ENTER	Create parent directory, PROPS.
VAR PROPS ' IXX ' CRDIR ENTER	Create subdirectory, IXX.
IXX	Open current directory, IXX.
≪ 0 ' ΣI ' STO 0 ' ΣA ' STO 0 ' ΣAY ' STO 0 ' ΣAYY ' STO SUBR ≫ ENTER	Set to zero the values of ΣI, ΣA, Σ(A * Y) and Σ(A * Y^2) and recall the subroutine, SUBR.
' STRT ' STO	Store the above routine in memory under the label, STRT.
≪{{A ≪' A ' STO ≫}{Y ≪' Y ' STO ≫}{I ≪' I ' STO ≫} ΣA ΣAY ΣAYY ΣI YBAR IXX SBOT RGYR SUBR} TMENU HALT I ' ΣI ' STO+ A DUP ' ΣA ' STO+ Y * DUP ' ΣAY ' STO+ Y * ' ΣAYY ' STO+ SUBR ≫ ENTER	Set up on input menu for data entry and, for each element of the cross section in turn, input the relevant data whilst the program operation is suspended. Then, key CONT to increment the values of ΣI, ΣA, Σ(A * Y) and Σ(A * Y^2).
' SUBR ' STO	Store the subroutine SUBR.
≪ ' ΣAY/ΣA ' EVAL ≫ ENTER	Define the height of the centroid, YBAR.
' YBAR ' STO	Store YBAR.
≪ ' ΣI + ΣAYY – ΣA * YBAR^2 ' EVAL ≫ ENTER	Define the moment of inertia, IXX.
' IXX ' STO	Store IXX.
≪ ' IXX/YBAR ' EVAL ≫ ENTER	Define the bottom fibre section modulus, SBOT.
' SBOT ' STO	Store SBOT
≪ ' √(IXX/ΣA) ' EVAL ≫ ENTER	Define the radius of gyration, RGYR.
' RGYR ' STO	Store RGYR.

EXAMPLE

HOME VAR PROPS IXX	Recall program IXX.
STRT	Initialize all storage registers to zero and prepare for data entry for element no.1.
34.2 A	Area of element no.1, sq.in.
25.56 Y	Height of centroid of element no.1, in.
26 I	Moment of inertia of element no.1, in^4.
CONT	Increment the values of ΣA, ΣI, $\Sigma(A*Y)$ and $\Sigma(A*Y^2)$ and suspend program operation to allow data entry for element no.2.
16.7 A	Area of element no.2, sq.in.
10.58 Y	Height of centroid of element no.2, in.
1170 I	Moment of inertia of element no.2, in^4.
CONT	Increment the values of ΣA, ΣI, $\Sigma(A*Y)$ and $\Sigma(A*Y^2)$.
USER	Display output functions
ΣA	Total area = 50.9 sq. in.
ΣAY	$\Sigma(A*Y)$ = 1,051 in^3.
ΣAYY	$\Sigma(A*Y^2)$ = 24,212 in^4.
NEXT	
ΣI	ΣI =1,196 in^4
YBAR	Height of centroid of composite section = 20.65 in.
IXX	Momement of inertia of composite section = 3,714 in^4.
SBOT	Bottom fibre section modulus = 180 in^3.
RGYR	Radius of gyration = 8.54 in.

A.5.2: RIG: RIGIDITY OF WALLS

PROGRAM LISTING:

KEYSTROKES	**COMMENTARY**
HOME VAR PROPS	Open parent directory, PROPS.
' RIG ' CRDIR ENTER	Create subdirectory, RIG.

RIG	Open current directory, RIG.
≪ ' 3 * H/L ' EVAL ≫ ENTER	Define the deflection due to shear deformation in a wall with an applied unit load, SHR.
' SHR ' STO	Store SHR.
≪ IF WTYP C SAME THEN ' 4 * (H/L)^3 ' EVAL ELSE ' (H/L)^3 ' EVAL END ≫ ENTER	Define the deflection due to flexural deformation, flex, in either a cantilever wall or a fixed ended wall with an applied unit load.
' FLEX ' STO	Store FLEX.
≪ ' SHR+FLEX ' EVAL ≫ ENTER ' DEFL ' STO	Define the wall flexibility, DEFL. Store DEFL
≪ ' 1/DEFL ' EVAL ≫ ENTER ' RIG ' STO	Define the wall rigidity, RIG. Store RIG
≪{{H ≪' H ' STO ≫}{L ≪' L ' STO ≫}{WTYP ≪' WTYP ' STO ≫} SHR FLEX DEFL RIG} TMENU ≫ ENTER	Set up an input menu for data entry.
' DAT ' STO	Store the data menu.

EXAMPLE

HOME VAR PROPS RIG		Recall program RIG.
DAT		Prepare for data entry.
20	H	Height of wall, ft.
35	L	Length of wall, ft.
F	WTYP	The wall is fixed ended.
		Key output functions.
SHR		Shear deflection = 1.7143/Et in/kip where, E = Youngs modulus of wall, ksi. t = thickness of wall, ins.
FLEX		Flexural deflection = 0.1866/Et in/kip.
DEFL		Wall flexibility = 1.9009/Et in/kip
RIG		Wall rigidity = 0.5261 * Et kip/in.

REFERENCES

1. American Concrete Institute. *Commentary on Building Code Requirements for Reinforced Concrete (ACI 318-83)*. Section 10.3. Detroit, MI, 1985.

2. International Conference of Building Officials. *Uniform Building Code – 1994*. Section 1910.2. Whittier, CA 1994.

3. American Concrete Institute. *Commentary on Building Code Requirements for Reinforced Concrete (ACI 318-83)*. Table 10.3.2. Detroit, MI, 1985.

4. International Conference of Building Officials. *Uniform Building Code – 1994*. Section 1910.5. Whittier, CA, 1994.

5. International Conference of Building Officials. *Uniform Building Code – 1994*. Equations (11–1), (11–2), (11–3). Whittier, CA 1994.

6. International Conference of Building Officials. *Uniform Building Code – 1994*. Equation (11–14). Whittier, CA, 1994.

7. International Conference of Building Officials. *Uniform Building Code – 1994*. Section 2101.4. Whittier, CA, 1994.

8. International Conference of Building Officials. *Uniform Building Code – 1994*. Section 2107.2. Whittier, CA, 1994.

9. Schneider R.R. & Dickey W.L. *Reinforced Masonry Design*. p158. Prentice–Hall, Inc., Englewood, Cliffs, NJ, 1987.

10. American Forest and Paper Association. *National Design Specification for Wood Construction, Eleventh Edition (ANSI/NFoPA NDS-1991)*. Washington, 1991.

11. American Forest and Paper Association. *Commentary on the National Design Specification for Wood Construction, Eleventh Edition (ANSI/NFoPA NDS-1991)*. Washington, 1991.

12. International Conference of Building Officials. *Uniform Building Code – 1994.* Section 2251 F1.3. Whittier, CA, 1994.

13. International Conference of Building Officials. *Uniform Building Code – 1994.* Section 2251 H1. Whittier, CA, 1994.

14. International Conference of Building Officials. *Uniform Building Code – 1994.* Section 2251 N4. Whittier, CA, 1994.

15. International Conference of Building Officials. *Uniform Building Code – 1994.* Section 2251 K1. Whittier, CA 1994.

16. International Conference of Building Officials. *Uniform Building Code – 1994.* Section 2211.7.4. Whittier, CA, 1994.

This page left blank intentionally

562

INDEX

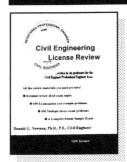

Civil Engineering License Review

Donald G. Newnan, Ph.D., P.E., Civil Engineer, Editor

Written by Six Professors, each with a Ph.D. in Civil Engineering.

A **complete license review book** written for the National Civil Engineering/Professional engineer exam.

☆ The 12 sections cover all the essential materials on the exam.

☆ A detailed discussion of the exam and how to prepare for it!

☆ 335 exam essay and multiple-choice problems with a total of 650 individual questions

☆ A complete 24 problem Sample Exam

☆ Printed in two colors for quicker understanding of the content

This is a new license review book written specifically for the latest version of the Civil Engineering/Professional Engineer (principles and practice) exam. The 12 sections cover the topics on the exam: Buildings, Bridges, Foundations, Retaining Structures, Seismic Design, Hydraulics, Engineering Hydrology, Water Treatment, Wastewater Treatment, Geotechnical/Soils, and Transportation Engineering. In addition, an appendix discusses Engineering Economy, which may appear as a component in some problems. A complete 24-problem sample exam (12 essay and 12 multiple-choice problems) is the final chapter in the book. Since some states do not allow books containing solutions to be taken into the CE/PE exam, the end-of-chapter problems do not have the solutions in this book. Instead, the problems are restated together with detailed solutions in the companion *book: Civil Engineering License Problems and Solutions* listed below. 80% Text 20% Problems. 2 Color **ISBN 0-910554-90-0**
Hardbound   Yellow/Blue 8-1/2 x 11

Order No. 900   728 Pgs 1995 12th Edition $50.50

Civil Engineering
License Problems and Solutions

Donald G. Newnan, Ph.D., P.E., Civil Engineer, Editor

Written by Six Professors, each with a Ph.D. in Civil Engineering

☆ A detailed description of the examination and suggestions on preparing for it

☆ 195 exam essay and multiple choice problems with a total of 510 individual questions

☆ A complete 24 problem Sample Exam

☆ A detailed Step-by-Step solution for every problem in the book

☆ Printed in two colors

This book may be used alone or in conduction with *Civil Engineering License Review,* 12th edition (Order No. 900). Either way, this book is a complete separate book. The chapter topics match those of *the License Review* book. Here all the problems have been reproduced for each chapter, this time followed by detailed Step-by-Step solutions. Similarly, the 24-Problem Sample Exam (12 essay and 12 multiple-choice problems) is given, followed by Step-by-Step solutions to the exam. Engineers wanting a CE/PE review with problems and solutions, will buy both books. If they only want an elaborate set of exam problems and a sample exam, plus detailed solutions to every problem, then they will purchase this book. 100% Problems and Solutions

ISBN 0-910554-91-9 Hardbound   Yellow/Blue   2 color   8-1/2 x 11

Order No. 919   344 Pgs   1995   12th Edition $29.50

Civil Engineering Problem Solving Flowcharts

Jorge L. Rodriguez, P.E., Civil Engineer   Revised Second Edition

89 Flowcharts To Solve Typical Problems On the Civil Engr./Prof Engr. Examination

The flowcharts describe the procedure to follow when solving problems that frequently appear on the CE/PE exam. They combine theory and formulas to show the logical steps in the solution of the problem. Flowcharts are provided for problems in Fluid Mechanics; Hydraulic Machines; Open Channel Flow; Hydrology; Water Supply and Waste Water Engineering; Soils; Foundations; Steel and Reinforced Concrete Design; and other topics. With each flowchart there is an example problem to illustrate the flowcharts Step-by-Step solution procedure. Second Edition. Hardbound   Yellow/Blue Cover Full Merchandising Cover 8-1/2 x 11. ISBN 1-57645-012-0

Order No. 120   160 pgs   1997 $27.50

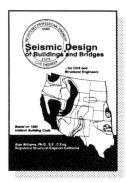

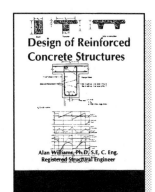

For the latest information...
Check out our Web site at
Http://www.engrpress.com

Order Toll Free   1(800) 800-1651
M-F 9-5 EST   Use your Visa/MC

Order Form   Catalog No. Struct1   FAX ORDER 1(800) 700-1651

Quantity	Order No.	Title		Price
If items total $250.00 or more (excluding Rentals, Videos and shipping), Deduct 11 % Here				
Regular Shipping (10-14 working days)				$4.75
If Air Shipping Desired, Add $4.75/Book				
HOT ORDER PROCESSING add $4.00				
Texas Residents Add Sales Tax				
Order Total				

Name

Address

City   State   Zip

Day Phone
Evening Phone

I authorize Engineering Press to charge current Pricing and shipping charges to my credit card.
Visa Or MasterCard Information:
Card Number   Expires

Signature

To Order By Mail:
Enclose a check, money order, or Visa or MasterCard number and expiration
date, with the order form and mail to:

Engineering Press Bookstore
P.O. Box 200129
Austin, TX 78720-0129

Order Toll Free   1(800) 800-1651
M-F 9-5 EST   Use your Visa/MC

Engineering Press Bookstore
P.O. Box 200129
Austin, TX 78720-0129

Engineering Press Bookstore
P.O. Box 200129
Austin, TX 78720-0129

Engineering Press Bookstore
P.O. Box 200129
Austin, TX 78720-0129